Dodge Daytona & Chrysler Laser Automotive Repair Manual

by Larry Warren and John H Haynes

Member of the Guild of Motoring Writers

Models covered:

Dodge Daytona and Chrysler Laser
2.2 and 2.5 liter fuel injection and turbocharged engines with 5-speed manual and 3-speed automatic transmission
1984 through 1989

(8X8 - 1140)

Haynes Publishing Group
Sparkford Nr Yeovil
Somerset BA22 7JJ England

Haynes North America, Inc
861 Lawrence Drive
Newbury Park
California 91320 USA

Acknowledgements

We are grateful for the help and cooperation of the Chrysler Corporation for their assistance with technical information, certain illustrations and vehicle photos, and the Champion Spark Plug Company who supplied the illustrations of various spark plug conditions.

© **Haynes North America, Inc.   1986, 1988, 1990**
With permission from J. H. Haynes & Co. Ltd.

A book in the **Haynes Automotive Repair Manual Series**

Printed in the USA

All rights reserved. No part of this book may be reproduced or transmitted in any form or by any means, electronic or mechanical, including photocopying, recording or by any information storage or retrieval system, without permission in writing from the copyright holder.

ISBN 1 85010 707 6

Library of Congress Catalog Card Number 90-82723

While every attempt is made to ensure that the information in this manual is correct, no liability can be accepted by the authors or publishers for loss, damage or injury caused by any errors in, or omissions from, the information given.

Contents

Introductory pages
    About this manual   5
    Introduction to the Dodge Daytona and Chrysler Laser   5
    Vehicle identification numbers   7
    Buying parts   9
    Maintenance techniques, tools and working facilities   9
    Booster battery (jump) starting   16
    Jacking and towing   16
    Safety first!   18
    Conversion factors   19
    Automotive chemicals and lubricants   20
    Troubleshooting   21

Chapter 1
Tune-up and routine maintenance   28   **1**

Chapter 2 Part A
2.2L and 2.5L engines   54   **2A**

Chapter 2 Part B
General engine overhaul procedures   72   **2B**

Chapter 3
Cooling, heating and air conditioning systems   91   **3**

Chapter 4 Part A
Fuel and exhaust systems   102   **4A**

Chapter 4 Part B
Turbocharger   117   **4B**

Chapter 5
Engine electrical systems   121   **5**

Chapter 6
Emissions control systems   131   **6**

Chapter 7 Part A
Manual transaxle   137   **7A**

Chapter 7 Part B
Automatic transaxle   142   **7B**

Chapter 8
Clutch and driveaxles   150   **8**

Chapter 9
Brakes   167   **9**

Chapter 10
Steering and suspension systems   181   **10**

Chapter 11
Body   201   **11**

Chapter 12
Chassis electrical system   214   **12**

Chapter 13 Supplement: Revisions and information on 1987 and later models   229   **13**

Wiring diagrams   236

Index   259

1985 Dodge Daytona Turbo Z

About this manual

Its purpose

The purpose of this manual is to help you get the best value from your vehicle. It can do so in several ways. It can help you decide what work must be done, even if you choose to have it done by a dealer service department or a repair shop; it provides information and procedures for routine maintenance and servicing; and it offers diagnostic and repair procedures to follow when trouble occurs.

It is hoped that you will use the manual to tackle the work yourself. For many simpler jobs, doing it yourself may be quicker than arranging an appointment to get the vehicle into a shop and making the trips to leave it and pick it up. More importantly, a lot of money can be saved by avoiding the expense the shop must pass on to you to cover its labor and overhead costs. An added benefit is the sense of satisfaction and accomplishment that you feel after having done the job yourself.

Using the manual

The manual is divided into Chapters. Each Chapter is divided into numbered Sections, which are headed in bold type between horizontal lines. Each Section consists of consecutively numbered paragraphs.

At the beginning of each numbered section you will be referred to any illustrations which apply to the procedures in that section. The reference numbers used in illustration captions pinpoint the pertinent Section and the Step within that section. That is, illustration 3.2 means the illustration refers to Section 3 and Step (or paragraph) 2 within that Section.

Procedures, once described in the text, are not normally repeated. When it is necessary to refer to another Chapter, the reference will be given as Chapter and Section number i.e. Chapter 1/16). Cross references given without use of the word "Chapter" apply to Sections and/or paragraphs in the same Chapter. For example, "see Section 8" means in the same Chapter.

Reference to the left or right side of the vehicle is based on the assumption that one is sitting in the driver's seat, facing forward.

Even though extreme care has been taken during the preparation of this manual, neither the publisher nor the author can accept responsibility for any errors in, or omissions from, the information given.

NOTE

A Note provides information necessary to properly complete a procedure or information which will make the steps to be followed easier to understand.

CAUTION

A Caution indicates a special procedure or special steps which must be taken in the course of completing the procedure in which the **Caution** is found which are necessary to avoid damage to the assembly being worked on.

WARNING

A Warning indicates a special procedure or special steps which must be taken in the course of completing the procedure in which the **Warning** is found which are necessary to avoid injury to the person performing the procedure.

Introduction to the Dodge Daytona and Chrysler Laser

The Dodge Daytona and Chrysler Laser are based on the K-car chassis and drivetrain and share many components with Chrysler Corporation's popular front wheel drive cars.

The transverse mounted, four-cylinder overhead cam engine is available in two displacements — 2.2 liter and 2.5 liter. The 2.2 liter engine is available with single-point fuel injection or turbocharged with multi-point fuel injection. The single-point fuel injected 2.5L features counter-rotating balance shafts mounted below the crankshaft for smoother operation. The engine drives the front wheels through a choice of either a manual 5-speed or automatic 3-speed transaxle via equal (turbocharged models) or unequal length driveaxles. The rack and pinion steering gear is mounted behind the engine.

The brakes are disc at the front and drum-type at the rear, with vacuum assist as standard equipment.

1984 Chrysler Laser

Vehicle identification numbers

The vehicle identification number (arrow) is visible through the driver's side of the windshield

The body code plate (arrow) can be found in front of the radiator

Modifications are a continuing and unpublicized process in vehicle manufacturing. Since parts manuals and lists are compiled on a numerical basis, the individual vehicle numbers are essential to correctly identify the component required.

Vehicle identification number (VIN)

This very important identification number is located on a plate attached to the top left corner of the dashboard of the vehicle. The VIN also appears on the Vehicle Certificate of Title and Registration. It contains valuable information such as where and when the vehicle was manufactured, the model year and the body style.

Body code plate

This metal plate is located on the top side of the radiator support. Like the VIN, it contains valuable information concerning the production of the vehicle as well as information about the way in which the vehicle is equipped. This plate is especially useful for matching the color and type of paint during repair work.

Engine identification numbers

The engine identification number (EIN) on the 2.2 liter engine is stamped into the rear of the block, just above the bellhousing. On the 2.5 liter engine it is on the radiator side of the block, between the core plug and the rear of the block.

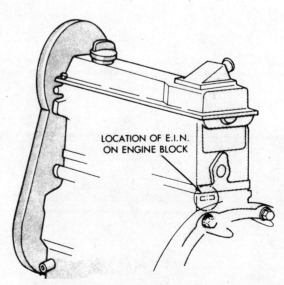

Engine identification number location (2.2L engine)

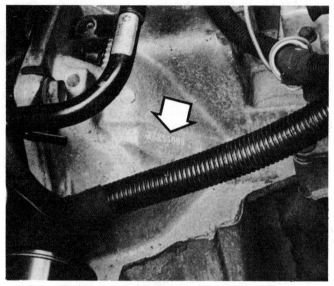

Manual transaxle identification number location

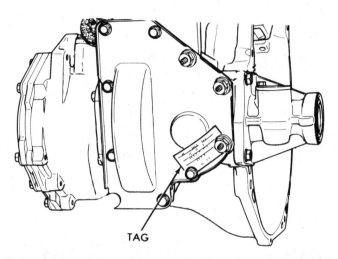

The manual transaxle assembly part number is located on a metal tag

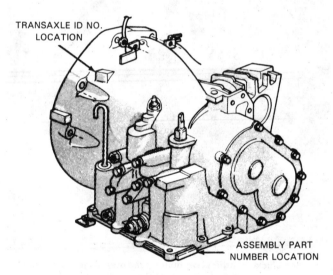

Automatic transaxle number locations

Engine serial numbers

In addition to the EIN, a serial number, which is required when buying replacement parts, is also used. On the 2.2 liter engine it is located just below the EIN on the block. On the 2.5 liter engine it is located on the rear side of the engine block, adjacent to the exhaust manifold stud (firewall side in vehicle).

Transaxle identification number

The transaxle identification number is stamped into the boss on the upper surface of the housing.

Transaxle serial numbers

The transaxle serial number, also called the assembly part number, is required when buying parts. On manual transaxles it is located on a metal tag attached to the front side of the transaxle. On automatics it is located on a pad just above the oil pan at the rear of the transaxle.

Vehicle Emissions Control Information label

The Emissions Control Information label is attached to the front edge of the hood, on the underside (see Chapter 6 for an illustration of the label and its location).

Buying parts

Replacement parts are available from many sources, which generally fall into one of two categories – authorized dealer parts departments and independent retail auto parts stores. Our advice concerning these parts is as follows:

Retail auto parts stores: Good auto parts stores will stock frequently needed components which wear out relatively fast, such as clutch components, exhaust systems, brake parts, tune-up parts, etc. These stores often supply new or reconditioned parts on an exchange basis, which can save a considerable amount of money. Discount auto parts stores are often very good places to buy materials and parts needed for general vehicle maintenance such as oil, grease, filters, spark plugs, belts, touch-up paint, bulbs, etc. They also usually sell tools and general accessories, have convenient hours, charge lower prices and can often be found not far from home.

Authorized dealer parts department: This is the best source for parts which are unique to the vehicle and not generally available elsewhere (such as major engine parts, transmission parts, trim pieces, etc.).

Warranty information: If the vehicle is still covered under warranty, be sure that any replacement parts purchased – regardless of the source – do not invalidate the warranty!

To be sure of obtaining the correct parts, have engine and chassis numbers available and, if possible, take the old parts along for positive identification.

Maintenance techniques, tools and working facilities

Maintenance techniques

There are a number of techniques involved in maintenance and repair that will be referred to throughout this manual. Application of these techniques will enable the home mechanic to be more efficient, better organized and capable of performing the various tasks properly, which will ensure that the repair job is thorough and complete.

Fasteners

Fasteners are nuts, bolts, studs and screws used to hold two or more parts together. There are a few things to keep in mind when working with fasteners. Almost all of them use a locking device of some type, either a lockwasher, locknut, locking tab or thread adhesive. All threaded fasteners should be clean and straight, with undamaged threads and undamaged corners on the hex head where the wrench fits. Develop the habit of replacing all damaged nuts and bolts with new ones. Special locknuts with nylon or fiber inserts can only be used once. If they are removed, they lose their locking ability and must be replaced with new ones.

Rusted nuts and bolts should be treated with a penetrating fluid to ease removal and prevent breakage. Some mechanics use turpentine in a spout-type oil can, which works quite well. After applying the rust penetrant, let it work for a few minutes before trying to loosen the nut or bolt. Badly rusted fasteners may have to be chiseled or sawed off or removed with a special nut breaker, available at tool stores.

If a bolt or stud breaks off in an assembly, it can be drilled and removed with a special tool commonly available for this purpose. Most automotive machine shops can perform this task, as well as other repair procedures, such as the repair of threaded holes that have been stripped out.

Flat washers and lockwashers, when removed from an assembly, should always be replaced exactly as removed. Replace any damaged washers with new ones. Never use a lockwasher on any soft metal surface (such as aluminum), thin sheet metal or plastic.

Maintenance techniques, tools and working facilities

Fastener sizes

For a number of reasons, automobile manufacturers are making wider and wider use of metric fasteners. Therefore, it is important to be able to tell the difference between standard (sometimes called U.S. or SAE) and metric hardware, since they cannot be interchanged.

All bolts, whether standard or metric, are sized according to diameter, thread pitch and length. For example, a standard 1/2 — 13 x 1 bolt is 1/2 inch in diameter, has 13 threads per inch and is 1 inch long. An M12 — 1.75 x 25 metric bolt is 12 mm in diameter, has a thread pitch of 1.75 mm (the distance between threads) and is 25 mm long. The two bolts are nearly identical, and easily confused, but they are not interchangeable.

In addition to the differences in diameter, thread pitch and length, metric and standard bolts can also be distinguished by examining the bolt heads. To begin with, the distance across the flats on a standard bolt head is measured in inches, while the same dimension on a metric bolt is sized in millimeters (the same is true for nuts). As a result, a standard wrench should not be used on a metric bolt and a metric wrench should not be used on a standard bolt. Also, most standard bolts have slashes radiating out from the center of the head to denote the grade or strength of the bolt, which is an indication of the amount of torque that can be applied to it. The greater the number of slashes, the greater the strength of the bolt. Grades 0 through 5 are commonly used on automobiles. Metric bolts have a property class (grade) number, rather than a slash, molded into their heads to indicate bolt strength. In this case, the higher the number, the stronger the bolt. Property class numbers 8.8, 9.8 and 10.9 are commonly used on automobiles.

Strength markings can also be used to distinguish standard hex nuts from metric hex nuts. Many standard nuts have dots stamped into one side, while metric nuts are marked with a number. The greater the number of dots, or the higher the number, the greater the strength of the nut.

Metric studs are also marked on their ends according to property class (grade). Larger studs are numbered (the same as metric bolts),

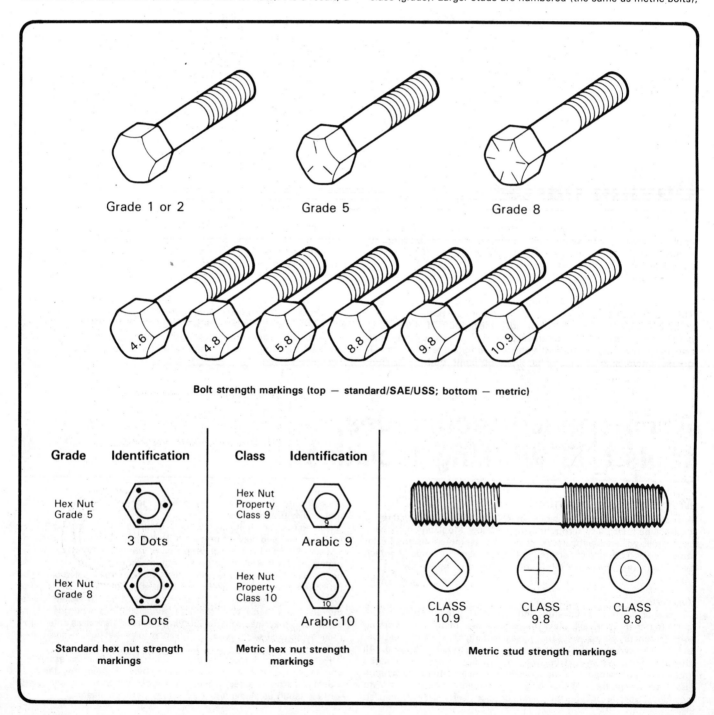

Maintenance techniques, tools and working facilities

while smaller studs carry a geometric code to denote grade.

It should be noted that many fasteners, especially Grades 0 through 2, have no distinguishing marks on them. When such is the case, the only way to determine whether it is standard or metric is to measure the thread pitch or compare it to a known fastener of the same size.

Standard fasteners are often referred to as SAE, as opposed to metric. However, it should be noted that SAE technically refers to a non-metric *fine thread* fastener only. Coarse thread non-metric fasteners are referred to as USS sizes.

Since fasteners of the same size (both standard and metric) may have different strength ratings, be sure to reinstall any bolts, studs or nuts removed from your vehicle in their original locations. Also, when replacing a fastener with a new one, make sure that the new one has a strength rating equal to or greater than the original.

Tightening sequences and procedures

Most threaded fasteners should be tightened to a specific torque value (torque is the twisting force applied to a threaded component such as a nut or bolt). Overtightening the fastener can weaken it and cause it to break, while undertightening can cause it to eventually come loose. Bolts, screws and studs, depending on the material they are made of and their thread diameters, have specific torque values, many of which are noted in the Specifications at the beginning of each Chapter. Be sure to follow the torque recommendations closely. For fasteners not assigned a specific torque, a general torque value chart is presented here as a guide. These torque values are for dry (unlubricated) fasteners threaded into steel or cast iron (not aluminum). As was previously mentioned, the size and grade of a fastener determine the amount of torque that can safely be applied to it. The figures listed here are approximate

	Ft-lb	Nm/m
Metric thread sizes		
M-6	6 to 9	9 to 12
M-8	14 to 21	19 to 28
M-10	28 to 40	38 to 54
M-12	50 to 71	68 to 96
M-14	80 to 140	109 to 154
Pipe thread sizes		
1/8	5 to 8	7 to 10
1/4	12 to 18	17 to 24
3/8	22 to 33	30 to 44
1/2	25 to 35	34 to 47
U.S. thread sizes		
1/4 – 20	6 to 9	9 to 12
5/16 – 18	12 to 18	17 to 24
5/16 – 24	14 to 20	19 to 27
3/8 – 16	22 to 32	30 to 43
3/8 – 24	27 to 38	37 to 51
7/16 – 14	40 to 55	55 to 74
7/16 – 20	40 to 60	55 to 81
1/2 – 13	55 to 80	75 to 108

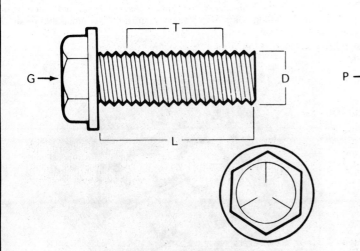

Standard (SAE and USS) bolt dimensions/grade marks

- G Grade marks (bolt strength)
- L Length (in inches)
- T Thread pitch (number of threads per inch)
- D Nominal diameter (in inches)

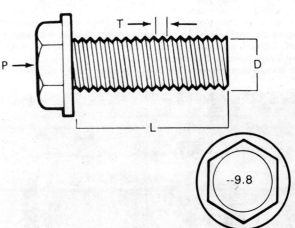

Metric bolt dimensions/grade marks

- P Property class (bolt strength)
- L Length (in millimeters)
- T Thread pitch (distance between threads in millimeters)
- D Diameter

for Grade 2 and Grade 3 fasteners. Higher grades can tolerate higher torque values.

Fasteners laid out in a pattern, such as cylinder head bolts, oil pan bolts, differential cover bolts, etc., must be loosened or tightened in sequence to avoid warping the component. This sequence will normally be shown in the appropriate Chapter. If a specific pattern is not given, the following procedures can be used to prevent warping.

Initially, the bolts or nuts should be assembled finger-tight only. Next, they should be tightened one full turn each, in a criss-cross or diagonal pattern. After each one has been tightened one full turn, return to the first one and tighten them all one-half turn, following the same pattern. Finally, tighten each of them one-quarter turn at a time until each fastener has been tightened to the proper torque. To loosen and remove the fasteners, the procedure would be reversed.

Component disassembly

Component disassembly should be done with care and purpose to help ensure that the parts go back together properly. Always keep track of the sequence in which parts are removed. Make note of special characteristics or marks on parts that can be installed more than one way, such as a grooved thrust washer on a shaft. It is a good idea to lay the disassembled parts out on a clean surface in the order that they were removed. It may also be helpful to make sketches or take instant photos of components before removal.

When removing fasteners from a component, keep track of their locations. Sometimes threading a bolt back in a part, or putting the washers and nut back on a stud, can prevent mix-ups later. If nuts and bolts cannot be returned to their original locations, they should be kept in a compartmented box or a series of small boxes. A cupcake or muffin tin is ideal for this purpose, since each cavity can hold the bolts and nuts from a particular area (i.e. oil pan bolts, valve cover bolts, engine mount bolts, etc.). A pan of this type is especially helpful when working on assemblies with very small parts, such as the carburetor, alternator, valve train or interior dash and trim pieces. The cavities can be marked with paint or tape to identify the contents.

Whenever wiring looms, harnesses or connectors are separated, it is a good idea to identify the two halves with numbered pieces of masking tape so they can be easily reconnected.

Gasket sealing surfaces

Throughout any vehicle, gaskets are used to seal the mating surfaces between two parts and keep lubricants, fluids, vacuum or pressure contained in an assembly.

Many times these gaskets are coated with a liquid or paste-type gasket sealing compound before assembly. Age, heat and pressure can sometimes cause the two parts to stick together so tightly that they are very difficult to separate. Often, the assembly can be loosened by striking it with a soft-face hammer near the mating surfaces. A regular hammer can be used if a block of wood is placed between the hammer and the part. Do not hammer on cast parts or parts that could be easily damaged. With any particularly stubborn part, always recheck to make sure that every fastener has been removed.

Avoid using a screwdriver or bar to pry apart an assembly, as they can easily mar the gasket sealing surfaces of the parts, which must remain smooth. If prying is absolutely necessary, use an old broom handle, but keep in mind that extra clean up will be necessary if the wood splinters.

After the parts are separated, the old gasket must be carefully scraped off and the gasket surfaces cleaned. Stubborn gasket material can be soaked with rust penetrant or treated with a special chemical to soften it so it can be easily scraped off. A scraper can be fashioned from a piece of copper tubing by flattening and sharpening one end. Copper is recommended because it is usually softer than the surfaces to be scraped, which reduces the chance of gouging the part. Some gaskets can be removed with a wire brush, but regardless of the method used, the mating surfaces must be left clean and smooth. If for some reason the gasket surface is gouged, then a gasket sealer thick enough to fill scratches will have to be used during reassembly of the components. For most applications, a non-drying (or semi-drying) gasket sealer should be used.

Hose removal tips

Warning: *If the vehicle is equipped with air conditioning, do not disconnect any of the A/C hoses without first having the system depressurized by a dealer service department or an air conditioning specialist.*

Hose removal precautions closely parallel gasket removal precautions. Avoid scratching or gouging the surface that the hose mates against or the connection may leak. This is especially true for radiator hoses. Because of various chemical reactions, the rubber in hoses can bond itself to the metal spigot that the hose fits over. To remove a hose, first loosen the hose clamps that secure it to the spigot. Then, with slip-joint pliers, grab the hose at the clamp and rotate it around the spigot. Work it back and forth until it is completely free, then pull it off. Silicone or other lubricants will ease removal if they can be applied between the hose and the outside of the spigot. Apply the same lubricant to the inside of the hose and the outside of the spigot to simplify installation.

As a last resort (and if the hose is to be replaced with a new one anyway), the rubber can be slit with a knife and the hose peeled from the spigot. If this must be done, be careful that the metal connection is not damaged.

If a hose clamp is broken or damaged, do not reuse it. Wire-type clamps usually weaken with age, so it is a good idea to replace them with screw-type clamps whenever a hose is removed.

Tools

A selection of good tools is a basic requirement for anyone who plans to maintain and repair his or her own vehicle. For the owner who has few tools, the initial investment might seem high, but when compared to the spiraling costs of professional auto maintenance and repair, it is a wise one.

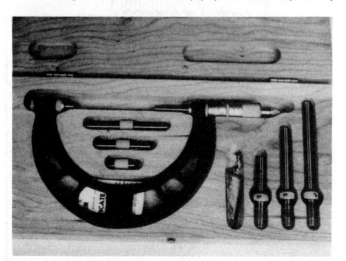

Micrometer set

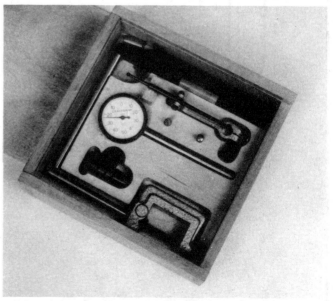

Dial indicator set

Maintenance techniques, tools and working facilities

Dial caliper

Hand-operated vacuum pump

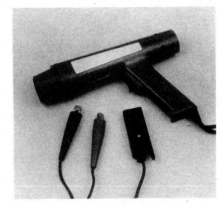
Timing light

Compression gauge with spark plug hole adapter

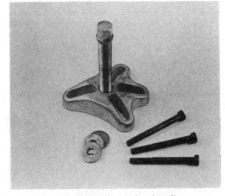
Damper/steering wheel puller

General purpose puller

Hydraulic lifter removal tool

Valve spring compressor

Valve spring compressor

Ridge reamer

Piston ring groove cleaning tool

Ring removal/installation tool

Ring compressor

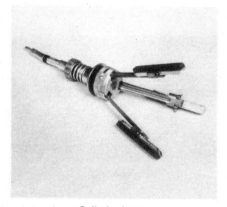
Cylinder hone

Brake hold-down spring tool

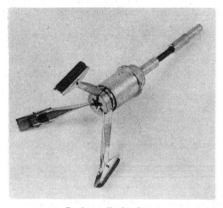
Brake cylinder hone

Clutch plate alignment tool

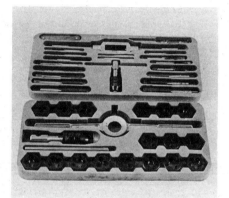
Tap and die set

To help the owner decide which tools are needed to perform the tasks detailed in this manual, the following tool lists are offered: *Maintenance and minor repair, Repair/overhaul* and *Special*.

The newcomer to practical mechanics should start off with the maintenance and minor repair tool kit, which is adequate for the simpler jobs performed on a vehicle. Then, as confidence and experience grow, the owner can tackle more difficult tasks, buying additional tools as they are needed. Eventually the basic kit will be expanded into the repair and overhaul tool set. Over a period of time, the experienced do-it-yourselfer will assemble a tool set complete enough for most repair and overhaul procedures and will add tools from the special category when it is felt that the expense is justified by the frequency of use.

Maintenance and minor repair tool kit

The tools in this list should be considered the minimum required for performance of routine maintenance, servicing and minor repair work. We recommend the purchase of combination wrenches (box-end and open-end combined in one wrench). While more expensive than open end wrenches, they offer the advantages of both types of wrench.

Combination wrench set (1/4-inch to 1 inch or 6 mm to 19 mm)
Adjustable wrench, 8 inch
Spark plug wrench with rubber insert
Spark plug gap adjusting tool
Feeler gauge set
Brake bleeder wrench
Standard screwdriver (5/16-inch x 6 inch)
Phillips screwdriver (No. 2 x 6 inch)
Combination pliers — 6 inch
Hacksaw and assortment of blades
Tire pressure gauge
Grease gun
Oil can
Fine emery cloth
Wire brush
Battery post and cable cleaning tool
Oil filter wrench
Funnel (medium size)
Safety goggles
Jackstands (2)
Drain pan

Note: *If basic tune-ups are going to be part of routine maintenance, it will be necessary to purchase a good quality stroboscopic timing light and combination tachometer/dwell meter. Although they are included in the list of special tools, it is mentioned here because they are absolutely necessary for tuning most vehicles properly.*

Repair and overhaul tool set

These tools are essential for anyone who plans to perform major repairs and are in addition to those in the maintenance and minor repair tool kit. Included is a comprehensive set of sockets which, though expensive, are invaluable because of their versatility, especially when various extensions and drives are available. We recommend the 1/2-inch drive over the 3/8-inch drive. Although the larger drive is bulky and more expensive, it has the capacity of accepting a very wide range of large sockets. Ideally, however, the mechanic should have a 3/8-inch drive set and a 1/2-inch drive set.

Socket set(s)
Reversible ratchet
Extension — 10 inch
Universal joint
Torque wrench (same size drive as sockets)
Ball peen hammer — 8 ounce
Soft-face hammer (plastic/rubber)
Standard screwdriver (1/4-inch x 6 inch)
Standard screwdriver (stubby — 5/16-inch)
Phillips screwdriver (No. 3 x 8 inch)
Phillips screwdriver (stubby — No. 2)

Maintenance techniques, tools and working facilities

Pliers — vise grip
Pliers — lineman's
Pliers — needle nose
Pliers — snap-ring (internal and external)
Cold chisel — 1/2-inch
Scribe
Scraper (made from flattened copper tubing)
Centerpunch
Pin punches (1/16, 1/8, 3/16-inch)
Steel rule/straightedge — 12 inch
Allen wrench set (1/8 to 3/8-inch or 4 mm to 10 mm)
A selection of files
Wire brush (large)
Jackstands (second set)
Jack (scissor or hydraulic type)

Note: *Another tool which is often useful is an electric drill motor with a chuck capacity of 3/8-inch and a set of good quality drill bits.*

Special tools

The tools in this list include those which are not used regularly, are expensive to buy, or which need to be used in accordance with their manufacturer's instructions. Unless these tools will be used frequently, it is not very economical to purchase many of them. A consideration would be to split the cost and use between yourself and a friend or friends. In addition, most of these tools can be obtained from a tool rental shop on a temporary basis.

This list primarily contains only those tools and instruments widely available to the public, and not those special tools produced by the vehicle manufacturer for distribution to dealer service departments. Occasionally, references to the manufacturer's special tools are inluded in the text of this manual. Generally, an alternative method of doing the job without the special tool is offered. However, sometimes there is no alternative to their use. Where this is the case, and the tool cannot be purchased or borrowed, the work should be turned over to the dealer service department or an automotive repair shop.

Valve spring compressor
Piston ring groove cleaning tool
Piston ring compressor
Piston ring installation tool
Cylinder compression gauge
Cylinder ridge reamer
Cylinder surfacing hone
Cylinder bore gauge
Micrometers and/or dial calipers
Hydraulic lifter removal tool
Balljoint separator
Universal-type puller
Impact screwdriver
Dial indicator set
Stroboscopic timing light (inductive pick-up)
Hand operated vacuum/pressure pump
Tachometer/dwell meter
Universal electrical multimeter
Cable hoist
Brake spring removal and installation tools
Floor jack

Buying tools

For the do-it-yourselfer who is just starting to get involved in vehicle maintenance and repair, there are a number of options available when purchasing tools. If maintenance and minor repair is the extent of the work to be done, the purchase of individual tools is satisfactory. If, on the other hand, extensive work is planned, it would be a good idea to purchase a modest tool set from one of the large retail chain stores. A set can usually be bought at a substantial savings over the individual tool prices, and they often come with a tool box. As additional tools are needed, add-on sets, individual tools and a larger tool box can be purchased to expand the tool selection. Building a tool set gradually allows the cost of the tools to be spread over a longer period of time and gives the mechanic the freedom to choose only those tools that will actually be used.

Tool stores will often be the only source of some of the special tools that are needed, but regardless of where tools are bought, try to avoid cheap ones, especially when buying screwdrivers and sockets, because they won't last very long. The expense involved in replacing cheap tools will eventually be greater than the initial cost of quality tools.

Care and maintenance of tools

Good tools are expensive, so it makes sense to treat them with respect. Keep them clean and in usable condition and store them properly when not in use. Always wipe off any dirt, grease or metal chips before putting them away. Never leave tools lying around in the work area. Upon completion of a job, always check closely under the hood for tools that may have been left there so they won't get lost during a test drive.

Some tools, such as screwdrivers, pliers, wrenches and sockets, can be hung on a panel mounted on the garage or workshop wall, while others should be kept in a tool box or tray. Measuring instruments, gauges, meters, etc. must be carefully stored where they cannot be damaged by weather or impact from other tools.

When tools are used with care and stored properly, they will last a very long time. Even with the best of care, though, tools will wear out if used frequently. When a tool is damaged or worn out, replace it. Subsequent jobs will be safer and more enjoyable if you do.

Working facilities

Not to be overlooked when discussing tools is the workshop. If anything more than routine maintenance is to be carried out, some sort of suitable work area is essential.

It is understood, and appreciated, that many home mechanics do not have a good workshop or garage available, and end up removing an engine or doing major repairs outside. It is recommended, however, that the overhaul or repair be completed under the cover of a roof.

A clean, flat workbench or table of comfortable working height is an absolute necessity. The workbench should be equipped with a vise that has a jaw opening of at least four inches.

As mentioned previously, some clean, dry storage space is also required for tools, as well as the lubricants, fluids, cleaning solvents, etc. which will soon become necessary.

Sometimes waste oil and fluids, drained from the engine or cooling system during normal maintenance or repairs, present a disposal problem. To avoid pouring them on the ground or into a sewage system, pour the used fluids into large containers, seal them with caps and take them to an authorized disposal site or recycling center. Plastic jugs, such as old antifreeze containers, are ideal for this purpose.

Always keep a supply of old newspapers and clean rags available. Old towels are excellent for mopping up spills. Many mechanics use rolls of paper towels for most work because they are readily available and disposable. To help keep the area under the vehicle clean, a large cardboard box can be cut open and flattened to protect the garage or shop floor.

Whenever working over a painted surface, such as when leaning over a fender to service something under the hood, always cover it with an old blanket or bedspread to protect the finish. Vinyl covered pads, made especially for this purpose, are available at auto parts stores.

Booster battery (jump) starting

Certain precautions must be observed when using a booster battery to jump start a vehicle.
  a) Before connecting the booster battery, make sure that the ignition switch is in the Off position.
  b) Turn off the lights, heater and other electrical loads.
  c) The eyes should be shielded. Safety goggles are a good idea.
  d) Make sure the booster battery is the same voltage as the dead one in the vehicle.
  e) The two vehicles must not touch each other.
  f) Make sure the transaxle is in Neutral (manual transaxle) or Park (automatic transaxle).
  g) If the booster battery is not a maintenance-free type, remove the vent caps and lay a cloth over the vent holes.

Connect the red jumper cable to the *positive* (+) terminals of each battery.

Connect one end of the black jumper cable to the *negative* (–) terminal of the booster battery. The other end of this cable should be connected to a good ground on the vehicle to be started, such as a bolt or bracket on the engine block. Use caution to insure that the cable will not come into contact with the fan, drivebelts or other moving parts of the engine.

Start the engine using the booster battery, then, with the engine running at idle speed, disconnect the jumper cables in the reverse order of connection.

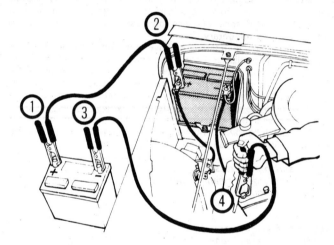

Make the booster battery cable connections in the numerical order shown (note that the negative cable of the booster *is not* attached to the negative terminal of the dead battery)

Jacking and towing

Jacking

The jack supplied with the vehicle should only be used for raising the vehicle when changing a tire or placing jackstands under the frame.
Warning: *Never work under the vehicle or start the engine while this jack is being used as the only means of support.*

The vehicle should be on level ground with the wheels blocked and the transaxle in Park (automatic) or Reverse (manual). If the tire is to be changed, pry off the hub cap (if equipped) using the tapered end of the lug wrench. If the wheel is being replaced, loosen the wheel nuts one-half turn and leave them in place until the wheel is raised off the ground. Refer to Chapter 10 for information related to removing and installing the tire.

Place the jack under the side of the vehicle in the indicated position and raise it until the jack head hole fits over the rocker flange jack locator pin. Operate the jack with a slow, smooth motion until the wheel is raised off the ground.

Lower the vehicle, remove the jack and tighten the nuts (if loosened or removed) in a criss-cross sequence by turning the wrench clockwise. Replace the hub cap (if equipped) by placing it in position and using the heel of your hand or a rubber mallet to seat it.

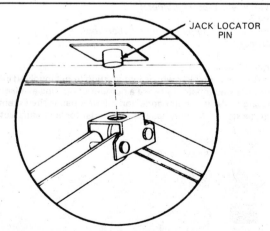

The jack supplied with the vehicle fits over a pin on the underside of the vehicle, which secures it in place

Jacking and Towing

Towing

Turbo Z models

Caution: *The 1984 Turbo Z cannot be towed from either the front or rear without damage to the front or rear of the body. The 1985 and 1986 Turbo Z models cannot be towed from the front but can be towed from the rear with the front end placed on a proper towing dolly. Care must be taken not to damage the front fascia and license plate bracket when lifting on and off the dolly.*

All other models

Vehicles with an automatic transaxle can be towed with all four wheels on the ground, provided that speeds do not exceed 25 mph and the distance is not over 15 miles, otherwise transaxle damage can result. Vehicles with a manual transaxle can be towed at legal highway speeds for any distance. If the vehicle has a damaged transaxle, tow it only with the front wheels off the ground.

Towing equipment specifically designed for this purpose should be used and should be attached to the main structural members of the vehicle and not the bumper or brackets.

Safety is a major consideration when towing and all applicable state and local laws must be obeyed. A safety chain system must be used for all towing.

While towing, the parking brake should be released and the transmission must be in Neutral. The steering must be unlocked (ignition switch in the Off position). Remember that power steering and power brakes will not work with the engine off.

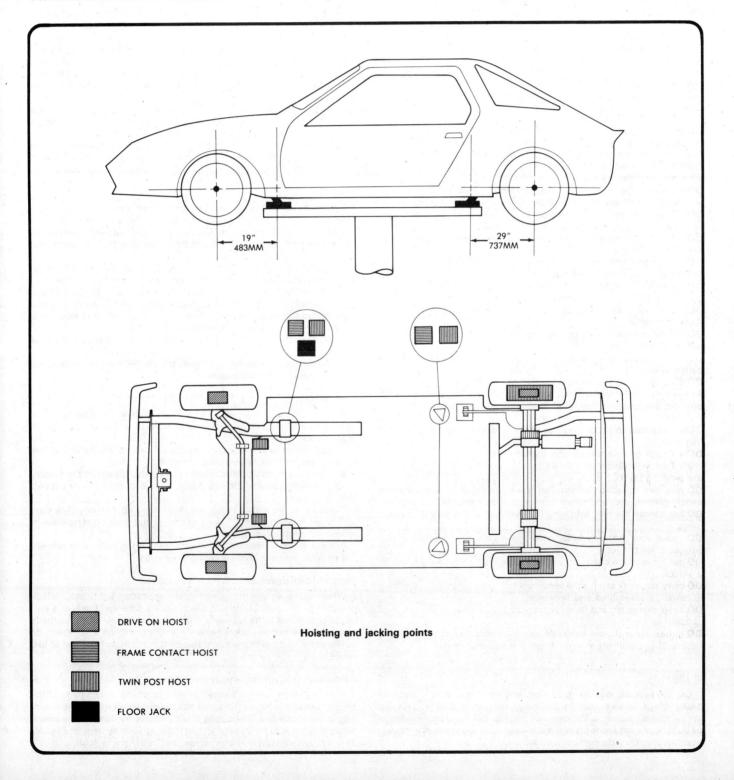

Hoisting and jacking points

Safety first!

Regardless of how enthusiastic you may be about getting on with the job at hand, take the time to ensure that your safety is not jeopardized. A moment's lack of attention can result in an accident, as can failure to observe certain simple safety precautions. The possibility of an accident will always exist, and the following points should not be considered a comprehensive list of all dangers. Rather, they are intended to make you aware of the risks and to encourage a safety conscious approach to all work you carry out on your vehicle.

Essential DOs and DON'Ts

DON'T rely on a jack when working under the vehicle. Always use approved jackstands to support the weight of the vehicle and place them under the recommended lift or support points.
DON'T attempt to loosen extremely tight fasteners (i.e. wheel lug nuts) while the vehicle is on a jack — it may fall.
DON'T start the engine without first making sure that the transmission is in Neutral (or Park where applicable) and the parking brake is set.
DON'T remove the radiator cap from a hot cooling system — let it cool or cover it with a cloth and release the pressure gradually.
DON'T attempt to drain the engine oil until you are sure it has cooled to the point that it will not burn you.
DON'T touch any part of the engine or exhaust system until it has cooled sufficiently to avoid burns.
DON'T siphon toxic liquids such as gasoline, antifreeze and brake fluid by mouth, or allow them to remain on your skin.
DON'T inhale brake lining dust — it is potentially hazardous (see *Asbestos* below)
DON'T allow spilled oil or grease to remain on the floor — wipe it up before someone slips on it.
DON'T use loose fitting wrenches or other tools which may slip and cause injury.
DON'T push on wrenches when loosening or tightening nuts or bolts. Always try to pull the wrench toward you. If the situation calls for pushing the wrench away, push with an open hand to avoid scraped knuckles if the wrench should slip.
DON'T attempt to lift a heavy component alone — get someone to help you.
DON'T rush or take unsafe shortcuts to finish a job.
DON'T allow children or animals in or around the vehicle while you are working on it.
DO wear eye protection when using power tools such as a drill, sander, bench grinder, etc. and when working under a vehicle.
DO keep loose clothing and long hair well out of the way of moving parts.
DO make sure that any hoist used has a safe working load rating adequate for the job.
DO get someone to check on you periodically when working alone on a vehicle.
DO carry out work in a logical sequence and make sure that everything is correctly assembled and tightened.
DO keep chemicals and fluids tightly capped and out of the reach of children and pets.
DO remember that your vehicle's safety affects that of yourself and others. If in doubt on any point, get professional advice.

Asbestos

Certain friction, insulating, sealing, and other products — such as brake linings, brake bands, clutch linings, torque converters, gaskets, etc. — contain asbestos. *Extreme care must be taken to avoid inhalation of dust from such products since it is hazardous to health.* If in doubt, assume that they *do* contain asbestos.

Fire

Remember at all times that gasoline is highly flammable. Never smoke or have any kind of open flame around when working on a vehicle. But the risk does not end there. A spark caused by an electrical short circuit, by two metal surfaces contacting each other, or even by static electricity built up in your body under certain conditions, **can ignite gasoline vapors, which in a confined space are highly explosive.** Do not, under any circumstances, use gasoline for cleaning parts. Use an approved safety solvent.

Always disconnect the battery ground (–) cable *at the battery* before working on any part of the fuel system or electrical system. **Never risk spilling fuel on a hot engine or exhaust component.**

It is strongly recommended that a fire extinguisher suitable for use on fuel and electrical fires be kept handy in the garage or workshop at all times. Never try to extinguish a fuel or electrical fire with water.

Fumes

Certain fumes are highly toxic and can quickly cause unconsciousness and even death if inhaled to any extent. Gasoline vapor falls into this category, as do the vapors from some cleaning solvents. Any draining or pouring of such volatile fluids should be done in a well ventilated area.

When using cleaning fluids and solvents, read the instructions on the container carefully. Never use materials from unmarked containers.

Never run the engine in an enclosed space, such as a garage. Exhaust fumes contain carbon monoxide, which is extremely poisonous. If you need to run the engine, always do so in the open air, or at least have the rear of the vehicle outside the work area.

If you are fortunate enough to have the use of an inspection pit, never drain or pour gasoline and never run the engine while the vehicle is over the pit. The fumes, being heavier than air, will concentrate in the pit with possibly lethal results.

The battery

Never create a spark or allow a bare light bulb near the battery. The battery normally gives off a certain amount of hydrogen gas, which is highly explosive.

Always disconnect the battery ground (–) cable *at the battery* before working on the fuel or electrical systems.

If possible, loosen the filler caps or cover when charging the battery from an external source. Do not charge at an excessive rate or the battery may burst.

Take care when adding water and when carrying a battery. The electrolyte, even when diluted, is very corrosive and should not be allowed to contact clothing or skin.

Always wear eye protection when cleaning the battery to prevent the caustic deposits from entering your eyes.

Household current

When using an electric power tool, inspection light, etc., which operates on household current, always make sure that the tool is correctly connected to its plug and that, where necessary, it is properly grounded. Do not use such items in damp conditions and, again, do not create a spark or apply excessive heat in the vicinity of fuel or fuel vapor.

Secondary ignition system voltage

A severe electric shock can result from touching certain parts of the ignition system (such as the spark plug wires) when the engine is running or being cranked, particularly if components are damp or the insulation is defective. In the case of an electronic ignition system, the secondary system voltage is much higher and could prove fatal.

Conversion factors

Length (distance)
Inches (in)	X	25.4	= Millimetres (mm)	X 0.0394	= Inches (in)
Feet (ft)	X	0.305	= Metres (m)	X 3.281	= Feet (ft)
Miles	X	1.609	= Kilometres (km)	X 0.621	= Miles

Volume (capacity)
Cubic inches (cu in; in^3)	X	16.387	= Cubic centimetres (cc; cm^3)	X 0.061	= Cubic inches (cu in; in^3)
Imperial pints (Imp pt)	X	0.568	= Litres (l)	X 1.76	= Imperial pints (Imp pt)
Imperial quarts (Imp qt)	X	1.137	= Litres (l)	X 0.88	= Imperial quarts (Imp qt)
Imperial quarts (Imp qt)	X	1.201	= US quarts (US qt)	X 0.833	= Imperial quarts (Imp qt)
US quarts (US qt)	X	0.946	= Litres (l)	X 1.057	= US quarts (US qt)
Imperial gallons (Imp gal)	X	4.546	= Litres (l)	X 0.22	= Imperial gallons (Imp gal)
Imperial gallons (Imp gal)	X	1.201	= US gallons (US gal)	X 0.833	= Imperial gallons (Imp gal)
US gallons (US gal)	X	3.785	= Litres (l)	X 0.264	= US gallons (US gal)

Mass (weight)
Ounces (oz)	X	28.35	= Grams (g)	X 0.035	= Ounces (oz)
Pounds (lb)	X	0.454	= Kilograms (kg)	X 2.205	= Pounds (lb)

Force
Ounces-force (ozf; oz)	X	0.278	= Newtons (N)	X 3.6	= Ounces-force (ozf; oz)
Pounds-force (lbf; lb)	X	4.448	= Newtons (N)	X 0.225	= Pounds-force (lbf; lb)
Newtons (N)	X	0.1	= Kilograms-force (kgf; kg)	X 9.81	= Newtons (N)

Pressure
Pounds-force per square inch (psi; lbf/in^2; lb/in^2)	X	0.070	= Kilograms-force per square centimetre (kgf/cm^2; kg/cm^2)	X 14.223	= Pounds-force per square inch (psi; lbf/in^2; lb/in^2)
Pounds-force per square inch (psi; lbf/in^2; lb/in^2)	X	0.068	= Atmospheres (atm)	X 14.696	= Pounds-force per square inch (psi; lbf/in^2; lb/in^2)
Pounds-force per square inch (psi; lbf/in^2; lb/in^2)	X	0.069	= Bars	X 14.5	= Pounds-force per square inch (psi; lbf/in^2; lb/in^2)
Pounds-force per square inch (psi; lbf/in^2; lb/in^2)	X	6.895	= Kilopascals (kPa)	X 0.145	= Pounds-force per square inch (psi; lbf/in^2; lb/in^2)
Kilopascals (kPa)	X	0.01	= Kilograms-force per square centimetre (kgf/cm^2; kg/cm^2)	X 98.1	= Kilopascals (kPa)

Torque (moment of force)
Pounds-force inches (lbf in; lb in)	X	1.152	= Kilograms-force centimetre (kgf cm; kg cm)	X 0.868	= Pounds-force inches (lbf in; lb in)
Pounds-force inches (lbf in; lb in)	X	0.113	= Newton metres (Nm)	X 8.85	= Pounds-force inches (lbf in; lb in)
Pounds-force inches (lbf in; lb in)	X	0.083	= Pounds-force feet (lbf ft; lb ft)	X 12	= Pounds-force inches (lbf in; lb in)
Pounds-force feet (lbf ft; lb ft)	X	0.138	= Kilograms-force metres (kgf m; kg m)	X 7.233	= Pounds-force feet (lbf ft; lb ft)
Pounds-force feet (lbf ft; lb ft)	X	1.356	= Newton metres (Nm)	X 0.738	= Pounds-force feet (lbf ft; lb ft)
Newton metres (Nm)	X	0.102	= Kilograms-force metres (kgf m; kg m)	X 9.804	= Newton metres (Nm)

Power
Horsepower (hp)	X	745.7	= Watts (W)	X 0.0013	= Horsepower (hp)

Velocity (speed)
Miles per hour (miles/hr; mph)	X	1.609	= Kilometres per hour (km/hr; kph)	X 0.621	= Miles per hour (miles/hr; mph)

*Fuel consumption**
Miles per gallon, Imperial (mpg)	X	0.354	= Kilometres per litre (km/l)	X 2.825	= Miles per gallon, Imperial (mpg)
Miles per gallon, US (mpg)	X	0.425	= Kilometres per litre (km/l)	X 2.352	= Miles per gallon, US (mpg)

Temperature
Degrees Fahrenheit = (°C x 1.8) + 32   Degrees Celsius (Degrees Centigrade; °C) = (°F - 32) x 0.56

*It is common practice to convert from miles per gallon (mpg) to litres/100 kilometres (l/100km), where mpg (Imperial) x l/100 km = 282 and mpg (US) x l/100 km = 235

Automotive chemicals and lubricants

A number of automotive chemicals and lubricants are available for use during vehicle maintenance and repair. They include a wide variety of products ranging from cleaning solvents and degreasers to lubricants and protective sprays for rubber, plastic and vinyl.

Cleaners

Carburetor cleaner and choke cleaner is a strong solvent for gum, varnish and carbon. Most carburetor cleaners leave a dry-type lubricant film which will not harden or gum up. Because of this film it is not recommended for use on electrical components.

Brake system cleaner is used to remove grease and brake fluid from the brake system where clean surfaces are absolutely necessary. It leaves no residue and often eliminates brake squeal caused by contaminants.

Electrical cleaner removes oxidation, corrosion and carbon deposits from electrical contacts, restoring full current flow. It can also be used to clean spark plugs, carburetor jets, voltage regulators and other parts where an oil-free surface is desired.

Demoisturants remove water and moisture from electrical components such as alternators, voltage regulators, electrical connectors and fuse blocks. It is non-conductive, non-corrosive and non-flammable.

Degreasers are heavy-duty solvents used to remove grease from the outside of the engine and from chassis components. They can be sprayed or brushed on, and, depending on the type, are rinsed off either with water or solvent.

Lubricants

Motor oil is the lubricant formulated for use in engines. It normally contains a wide variety of additives to prevent corrosion and reduce foaming and wear. Motor oil comes in various weights (viscosity ratings) from 5 to 80. The recommended weight of the oil depends on the season, temperature and the demands on the engine. Light oil is used in cold climates and under light load conditions. Heavy oil is used in hot climates and where high loads are encountered. Multi-viscosity oils are designed to have characteristics of both light and heavy oils and are available in a number of weights from 5W-20 to 20W-50.

Gear oil is designed to be used in differentials, manual transaxles and other areas where high-temperature lubrication is required.

Chassis and wheel bearing grease is a heavy grease used where increased loads and friction are encountered, such as for wheel bearings, balljoints, tie rod ends and universal joints.

High temperature wheel bearing grease is designed to withstand the extreme temperatures encountered by wheel bearings in disc brake equipped vehicles. It usually contains molybdenun disulfide (moly), which is a dry-type lubricant.

White grease is a heavy grease for metal to metal applications where water is a problem. White grease stays soft under both low and high temperatures (usually from −100°F to +190°F), and will not wash off or dilute in the presence of water.

Assembly lube is a special extreme pressure lubricant, usually containing moly, used to lubricate high-load parts such as main and rod bearings and cam lobes for initial start-up of a new engine. The assembly lube lubricates the parts without being squeezed out or washed away until the engine oiling system begins to function.

Silicone lubricants are used to protect rubber, plastic, vinyl and nylon parts.

Graphite lubricants are used where oils cannot be used due to contamination problems, such as in locks. The dry graphite will lubricate metal parts while remaining uncontaminated by dirt, water, oil or acids. It is electrically conductive and will not foul electrical contacts in locks such as the ignition switch.

Moly penetrants loosen and lubricate frozen, rusted and corroded fasteners and prevent future rusting or freezing.

Heat-sink grease is a special electrically non-conductive grease that is used for mounting HEI ignition modules where it is essential that heat be transferred away from the module.

Sealants

RTV sealant is one of the most widely used gasket compounds. Made from silicone, RTV is air curing, it seals, bonds, waterproofs, fills surface irregularities, remains flexible, doesn't shrink, is relatively easy to remove, and is used as a supplementary sealer with almost all low and medium temperature gaskets.

Anaerobic sealant is much like RTV in that it can be used either to seal gaskets or to form gaskets by itself. It remains flexible, is solvent resistant and fills surface imperfections. The difference between an anaerobic sealant and an RTV-type sealant is in the curing. RTV cures when exposed to air, while an anaerobic sealant cures only in the absence of air. This means that an anaerobic sealant cures only after the assembly of parts, sealing them together.

Thread and pipe sealant is used for sealing hydraulic and pneumatic fittings and vacuum lines. It is usually made from a teflon compound, and comes in a spray, a paint-on liquid and as a wrap-around tape.

Chemicals

Anti-seize compound prevents seizing, galling, cold welding, rust and corrosion in fasteners. High temperature anti-seize, usually made with copper and graphite lubricants, is used for exhaust system and manifold bolts.

Anaerobic locking compounds are used to keep fasteners from vibrating or working loose, and cure only after installation, in the absence of air. Medium strength locking compound is used for small nuts, bolts and screws that you expect to be removing later. High strength locking compound is for large nuts, bolts and studs which you don't intend to be removing on a regular basis.

Oil additives range from viscosity index improvers to chemical treatments that claim to reduce internal engine friction. It should be noted that most oil manufacturers caution against using additives with their oils.

Gas additives perform several functions, depending on their chemical makeup. They usually contain solvents that help dissolve gum and varnish that build up on carburetor and intake parts. They also serve to break down carbon deposits that form on the inside surfaces of the combustion chambers. Some additives contain upper cylinder lubricants for valves and piston rings, and others chemicals to remove condensation from the gas tank.

Other

Brake fluid is specially formulated hydraulic fluid that can withstand the heat and pressure encountered in brake systems. Care must be taken that this fluid does not come in contact with painted surfaces or plastics. An opened container should always be resealed to prevent contamination by water or dirt.

Weatherstrip adhesive is used to bond weatherstripping around doors, windows and trunk lids. It is sometimes used to attach trim pieces.

Undercoating is a petroleum-based tar-like substance that is designed to protect metal surfaces on the underside of the vehicle from corrosion. It also acts as a sound-deadening agent by insulating the bottom of the vehicle.

Waxes and polishes are used to help protect painted and plated surfaces from the weather. Different types of paint may require the use of different types of wax and polish. Some polishes utilize a chemical or abrasive cleaner to help remove the top layer of oxidized (dull) paint on older vehicles. In recent years many non-wax polishes that contain a wide variety of chemicals such as polymers and silicones have been introduced. These non-wax polishes are usually easier to apply and last longer than conventional waxes and polishes.

Troubleshooting

Refer to Chapter 13 for information on 1987 and later models

Contents

Symptom	Section
Engine	
Engine backfires	14
Engine diesels (continues to run) after switching off	16
Engine hard to start when cold	4
Engine hard to start at temperatures below 32°F at altitudes above 5000 feet (1985 models only)	5
Engine hard to start when hot	6
Engine lacks power	13
Engine lopes while idling or idles erratically	9
Engine misses at idle speed	10
Engine misses throughout driving speed range	11
Engine rotates but will not start	2
Engine stalls	12
Engine starts but stops immediately	8
Engine will not rotate when attempting to start	1
Pinging or knocking engine sounds during acceleration or uphill	15
Starter motor noisy or excessively rough in engagement	7
Starter motor operates without rotating engine	3
Engine electrical system	
Battery will not hold a charge	17
Ignition light fails to come on when key is turned on	19
"Power Loss/Power Limited" light comes on	20
Ignition light fails to go out	18
Fuel system	
Excessive fuel consumption	21
Fuel leakage and/or fuel odor	22
Cooling system	
Coolant loss	27
External coolant leakage	25
Internal coolant leakage	26
Overcooling	24
Overheating	23
Poor coolant circulation	28
Clutch	
Clutch slips (engine speed increases with no increase in vehicle speed)	30
Clutch pedal stays on the floor when disengaged	34
Fails to release (pedal pressed to the floor — shift lever does not move freely in and out of Reverse)	29
Grabbing (chattering) as clutch is engaged	31
Squeal or rumble with the clutch fully disengaged (pedal depressed)	33

Symptom	Section
Squeal or rumble with the clutch fully engaged (pedal released)	32
Manual transaxle	
Difficulty in engaging gears	39
Noisy in all gears	36
Noisy in Neutral with the engine running	35
Noisy in one particular gear	37
Oil leakage	40
Slips out of high gear	38
Automatic transaxle	
Fluid leakage	44
General shift mechanism problems	41
Transaxle slips, shifts rough, is noisy or has no drive in forward or reverse gears	43
Transaxle will not downshift with the accelerator pedal pressed to the floor	42
Driveaxles	
Clicking noise in turns	45
Knock or clunk when accelerating after coasting	46
Shudder or vibration during acceleration	47
Rear axle	
Noise	48
Brakes	
Brake pedal feels spongy when depressed	52
Brake pedal pulsates during brake application	55
Excessive brake pedal travel	51
Excessive effort required to stop the vehicle	53
Noise (high-pitched squeal with the brakes applied)	50
Pedal travels to the floor with little resistance	54
Vehicle pulls to one side during braking	49
Suspension and steering systems	
Excessive pitching and/or rolling around corners or during braking	58
Excessive play in the steering	60
Excessive tire wear (not specific to one area)	62
Excessive tire wear on inside edge	64
Excessive tire wear on outside edge	63
Excessively stiff steering	59
Lack of power assistance	61
Shimmy, shake or vibration	57
Tire tread worn in one place	65
Vehicle pulls to one side	56

This section provides an easy reference guide to the more common problems which may occur during the operation of your vehicle. These problems and possible causes are grouped under various components or systems; i.e. Engine, Cooling System, etc., and also refer to the Chapter and/or Section which deals with the problem.

Remember that successful troubleshooting is not a mysterious *black art* practiced only by professional mechanics. It's simply the result of a bit of knowledge combined with an intelligent, systematic approach to the problem. Always work by a process of elimination, starting with the simplest solution and working through to the most complex — and never overlook the obvious. Anyone can forget to fill the gas tank or leave the lights on overnight, so don't assume that you are above such oversights.

Finally, always get clear in your mind why a problem has occurred and take steps to ensure that it doesn't happen again. If the electrical system fails because of a poor connection, check all other connections in the system to make sure that they don't fail as well. If a particular fuse continues to blow, find out why — don't just go on replacing fuses. Remember, failure of a small component can often be indicative of potential failure or incorrect functioning of a more important component or system.

Engine

1 Engine will not rotate when attempting to start

1 Battery terminal connections loose or corroded. Check the cable terminals at the battery. Tighten the cable or remove corrosion as necessary.
2 Battery discharged or faulty. If the cable connections are clean and tight on the battery posts, turn the key to the On position and switch on the headlights and/or windshield wipers. If they fail to function, the battery is discharged.
3 Automatic transaxle not completely engaged in Park or clutch not completely depressed.
4 Broken, loose or disconnected wiring in the starting circuit. Inspect all wiring and connectors at the battery, starter solenoid and ignition switch.
5 Starter motor pinion jammed in flywheel ring gear. If manual transaxle, place transaxle in gear and rock the vehicle to manually turn the engine. Remove starter and inspect pinion and flywheel at earliest convenience.
6 Starter solenoid faulty (Chapter 5).
7 Starter motor faulty (Chapter 5).
8 Ignition switch faulty (Chapter 12).

2 Engine rotates but will not start

1 Fuel tank empty.
2 Fault in the EFI system (Chapters 4 and 5).
3 Battery discharged (engine rotates slowly). Check the operation of electrical components as described in the previous Section.
4 Battery terminal connections loose or corroded. See the previous Section.
5 Fuel injector or fuel pump faulty (Chapter 4).
6 Excessive moisture on, or damage to, ignition components (Chapter 5).
7 Worn, faulty or incorrectly gapped spark plugs (Chapter 1).
8 Broken, loose or disconnected wiring in the starting circuit (see the previous Section).
9 Distributor loose, causing ignition timing to change. Turn the distributor as necessary to start the engine, then set the ignition timing as soon as possible (Chapter 1).
10 Broken, loose or disconnected wires at the ignition coil or faulty coil (Chapter 5).

3 Starter motor operates without rotating engine

1 Starter pinion sticking. Remove the starter (Chapter 5) and inspect.
2 Starter pinion or flywheel teeth worn or broken. Remove the cover at the rear of the engine and inspect.

4 Engine hard to start when cold

1 Battery discharged or low. Check as described in Section 1.
2 Fault in the EFI system (Chapters 4 and 5).
3 Fuel injection system in need of overhaul (Chapter 4).
4 Distributor rotor carbon tracked and/or damaged.

5 Engine hard to start at temperatures below 32°F at altitudes above 5000 feet (1985 models only)

A Chrysler Corporation Technical Service Bulletin concerning this problem has been issued. Take your vehicle to your dealer and inform him of the problem.

6 Engine hard to start when hot

1 Air filter clogged (Chapter 1).
2 Fault in the EFI system (Chapters 4 and 5).
3 Fuel not reaching the fuel injector (see Section 2).

7 Starter motor noisy or excessively rough in engagement

1 Pinion or flywheel gear teeth worn or broken. Remove the cover at the rear of the engine (if so equipped) and inspect.
2 Starter motor mounting bolts loose or missing.

8 Engine starts but stops immediately

1 Loose or faulty electrical connections at distributor, coil or alternator.
2 Fault in the EFI system (Chapters 4 and 5).
3 Insufficient fuel reaching the fuel injector. Have the fuel injection pressure checked by your dealer or a properly equipped shop.
4 Vacuum leak at the gasket surfaces of the intake manifold and fuel injection unit on non-turbocharged models. Make sure that all mounting bolts/nuts are tightened securely and that all vacuum hoses connected to the fuel injection unit and manifold are positioned properly and in good condition.

9 Engine lopes while idling or idles erratically

1 Vacuum leakage. Check the mounting bolts/nuts at the fuel injection unit and intake manifold for tightness. Make sure that all vacuum hoses are connected and in good condition. Use a stethoscope or a length of fuel hose held against your ear to listen for vacuum leaks while the engine is running. A hissing sound will be heard. A soapy water solution will also detect leaks. Check the fuel injector and intake manifold gasket surfaces.
2 Fault in the EFI system (Chapters 4 and 5).
3 Leaking EGR valve or plugged PCV valve (see Chapters 1 and 6).
4 Air filter clogged (Chapter 1).
5 Fuel pump not delivering sufficient fuel to the fuel injector (see Chapter 4).
6 Fuel injection system out of adjustment (Chapter 4).
7 Leaking head gasket. If this is suspected, check the compression (Chapter 1).
8 Timing belt or sprockets worn (Chapter 2).
9 Camshaft lobes worn (Chapter 2).

10 Engine misses at idle speed

1 Spark plugs worn or not gapped properly (Chapter 1).
2 Fault in the EFI system (Chapters 4 and 5).
3 Faulty spark plug wires (Chapter 1).

Troubleshooting

11 Engine misses throughout driving speed range

1 Fuel filter clogged and/or impurities in the fuel system (Chapter 1).
2 Faulty or incorrectly gapped spark plugs (Chapter 1).
3 Fault in the EFI system (Chapters 4 and 5).
4 Incorrect ignition timing (Chapter 1).
5 Check for cracked distributor cap, disconnected distributor wires and damaged distributor components (Chapter 1).
6 Leaking spark plug wires (Chapter 1).
7 Faulty emissions system components (Chapter 6).
8 Low or uneven cylinder compression pressures. Remove the spark plugs and test the compression with a gauge (Chapter 1).
9 Weak or faulty ignition system (Chapter 5).
10 Vacuum leaks at the fuel injection unit, intake manifold or vacuum hoses (see Section 9).

12 Engine stalls

1 Idle speed incorrect (Chapter 1).
2 Fuel filter clogged and/or water and impurities in the fuel system (Chapter 1).
3 Distributor components damp or damaged (Chapter 5).
4 Fault in the EFI system or sensors (Chapters 4 and 5).
5 Faulty emissions system components (Chapter 6).
6 Faulty or incorrectly gapped spark plugs (Chapter 1). Also check the spark plug wires (Chapter 1).
7 Vacuum leak at the fuel injection unit, intake manifold or vacuum hoses. Check as described in Section 9.

13 Engine lacks power

1 Incorrect ignition timing (Chapter 1).
2 Fault in the EFI system (Chapters 4 and 5).
3 Excessive play in the distributor shaft. At the same time, check for a damaged rotor, faulty distributor cap, wires, etc. (Chapters 1 and 5).
4 Faulty or incorrectly gapped spark plugs (Chapter 1).
5 Fuel injection unit not adjusted properly or excessively worn (Chapter 4).
6 Faulty coil (Chapter 5).
7 Brakes binding (Chapter 1).
8 Automatic transaxle fluid level incorrect (Chapter 1).
9 Clutch slipping (Chapter 8).
10 Fuel filter clogged and/or impurities in the fuel system (Chapter 1).
11 Emissions control system not functioning properly (Chapter 6).
12 Use of substandard fuel. Fill the tank with the proper octane fuel.
13 Low or uneven cylinder compression pressures. Test with a compression tester, which will detect leaking valves and/or a blown head gasket (Chapter 1).

14 Engine backfires

1 Emissions system not functioning properly (Chapter 6).
2 Fault in the EFI system (Chapters 4 and 5).
3 Ignition timing incorrect (Chapter 1).
4 Faulty secondary ignition system (cracked spark plug insulator, faulty plug wires, distributor cap and/or rotor) (Chapters 1 and 5).
5 Fuel injection unit in need of adjustment or worn excessively (Chapter 4).
6 Vacuum leak at the fuel injection unit, intake manifold or vacuum hoses. Check as described in Section 9.
7 Valves sticking (Chapter 2).

15 Pinging or knocking engine sounds during acceleration or uphill

1 Incorrect grade of fuel. Fill the tank with fuel of the proper octane rating.
2 Fault in the EFI system (Chapters 4 and 5).
3 Ignition timing incorrect (Chapter 1).
4 Fuel injection unit in need of adjustment (Chapter 4).
5 Improper spark plugs. Check the plug type against the Emissions Control Information label located in the engine compartment. Also check the plugs and wires for damage (Chapter 1).
6 Worn or damaged distributor components (Chapter 5).
7 Faulty emissions system (Chapter 6).
8 Vacuum leak. Check as described in Section 9.

16 Engine diesels (continues to run) after switching off

1 Idle speed too high (Chapter 1).
2 Fault in the EFI system (Chapters 4 and 5).
3 Ignition timing incorrectly adjusted (Chapter 1).
4 Thermo-controlled air cleaner heat valve (non-turbocharged models) not operating properly (Chapter 6).
5 Excessive engine operating temperature. Probable causes of this are a malfunctioning thermostat, clogged radiator, faulty water pump (Chapter 3).

Engine electrical system

17 Battery will not hold a charge

1 Alternator drivebelt defective or not adjusted properly (Chapter 1).
2 Electrolyte level low or battery discharged (Chapter 1).
3 Battery terminals loose or corroded (Chapter 1).
4 Alternator not charging properly (Chapter 5).
5 Loose, broken or faulty wiring in the charging circuit (Chapter 5).
6 Short in the vehicle wiring causing a continual drain on battery.
7 Battery defective internally.

18 Ignition light fails to go out

1 Fault in the alternator or charging circuit (Chapter 5).
2 Alternator drivebelt defective or not properly adjusted (Chapter 1).

19 Ignition light fails to come on when key is turned on

1 Warning light bulb defective (Chapter 12).
2 Alternator faulty (Chapter 5).
3 Fault in the printed circuit, dash wiring or bulb holder (Chapter 12).

20 "Power Loss/Power Limited" light comes on

See Chapter 5.

Fuel system

21 Excessive fuel consumption

1 Dirty or clogged air filter element (Chapter 1).
2 Incorrectly set ignition timing (Chapter 1).
3 Choke sticking or improperly adjusted (Chapter 1).
4 Emissions system not functioning properly (not all vehicles, see Chapter 6).
5 Fault in the EFI system (Chapters 4 and 5).
6 Fuel injection internal parts excessively worn or damaged (Chapter 4).
7 Low tire pressure or incorrect tire size (Chapter 1).

22 Fuel leakage and/or fuel odor

1 Leak in a fuel feed or vent line (Chapter 4).
2 Tank overfilled. Fill only to automatic shut-off.
3 Emissions system filter clogged (Chapter 1).
4 Vapor leaks from system lines (Chapter 4).
5 Fuel injection internal parts excessively worn or out of adjustment (Chapter 4).

Cooling system

23 Overheating

1 Insufficient coolant in the system (Chapter 1).
2 Water pump drivebelt defective or not adjusted properly (Chapter 1).
3 Radiator core blocked or radiator grille dirty and restricted (Chapter 3).
4 Thermostat faulty (Chapter 3).
5 Fan blades broken or cracked (Chapter 3).
6 Radiator cap not maintaining proper pressure. Have the cap pressure tested by gas station or repair shop.
7 Ignition timing incorrect (Chapter 1).

24 Overcooling

1 Thermostat faulty (Chapter 3).
2 Inaccurate temperature gauge (Chapter 12).

25 External coolant leakage

1 Deteriorated or damaged hoses or loose clamps. Replace hoses and/or tighten the clamps at the hose connections (Chapter 1).
2 Water pump seals defective. If this is the case, water will drip from the weep hole in the water pump body (Chapter 1).
3 Leakage from radiator core or header tank. This will require the radiator to be professionally repaired (see Chapter 3 for removal procedures).
4 Engine drain plugs or water jacket core plugs leaking (see Chapter 2).

26 Internal coolant leakage

Note: *Internal coolant leaks can usually be detected by examining the oil. Check the dipstick and inside of the rocker arm cover for water deposits and an oil consistency like that of a milkshake.*

1 Leaking cylinder head gasket. Have the cooling system pressure tested.
2 Cracked cylinder bore or cylinder head. Dismantle the engine and inspect (Chapter 2).

27 Coolant loss

1 Too much coolant in the system (Chapter 1).
2 Coolant boiling away due to overheating (see Section 23).
3 Internal or external leakage (see Sections 25 and 26).
4 Faulty radiator cap. Have the cap pressure tested.

28 Poor coolant circulation

1 Inoperative water pump. A quick test is to pinch the top radiator hose closed with your hand while the engine is idling, then let it loose. You should feel the surge of coolant if the pump is working properly (Chapter 1).
2 Restriction in the cooling system. Drain, flush and refill the system (Chapter 1). If necessary, remove the radiator (Chapter 3) and have it reverse flushed.
3 Water pump drivebelt defective or not adjusted properly (Chapter 1).
4 Thermostat sticking (Chapter 3).

Clutch

29 Fails to release (pedal pressed to the floor — shift lever does not move freely in and out of Reverse)

1 Improper linkage free play adjustment (Chapter 8).
2 Damaged clutch arm or cable. Look under the vehicle, on the left side of transaxle.
3 Clutch plate warped or damaged (Chapter 8).

30 Clutch slips (engine speed increases with no increase in vehicle speed)

1 Linkage out of adjustment (Chapter 8).
2 Clutch plate oil soaked or lining worn. Remove clutch (Chapter 8) and inspect.
3 Clutch plate not seated. It may take 30 or 40 normal starts for a new one to seat.

31 Grabbing (chattering) as clutch is engaged

1 Oil on clutch plate lining. Remove and inspect (Chapter 8). Correct any leakage source.
2 Worn or loose engine or transaxle mounts. These units move slightly when the clutch is released. Inspect the mounts and bolts.
3 Worn splines on clutch plate hub. Remove the clutch components (Chapter 8) and inspect.
4 Warped pressure plate or flywheel. Remove the clutch components and inspect.

32 Squeal or rumble with the clutch fully engaged (pedal released)

1 Improper adjustment; no free play (Chapter 1).
2 Release bearing binding on transaxle bearing retainer. Remove the clutch components (Chapter 8) and check the bearing. Remove any burrs or nicks, clean and relubricate before reinstallation.
3 Weak linkage return spring. Replace the spring.

33 Squeal or rumble with the clutch fully disengaged (pedal depressed)

1 Worn, defective or broken release bearing (Chapter 8).
2 Worn or broken pressure plate springs (or diaphragm fingers) (Chapter 8).

34 Clutch pedal stays on the floor when disengaged

1 Bind in linkage or release bearing. Inspect the linkage or remove the clutch components as necessary.
2 Linkage springs being over-extended. Check the adjuster for damage or wear (Chapter 8).

Troubleshooting

Manual transaxle

35 Noisy in Neutral with the engine running

1. Input shaft bearing worn.
2. Damaged main drive gear bearing.
3. Worn countershaft bearings.
4. Worn or damaged countershaft end play shims.

36 Noisy in all gears

1. Any of the above causes, and/or:
2. Insufficient lubricant (see the checking procedures in Chapter 1).

37 Noisy in one particular gear

1. Worn, damaged or chipped gear teeth for that particular gear.
2. Worn or damaged synchronizer for that particular gear.

38 Slips out of high gear

1. Transaxle loose on clutch housing (Chapter 7).
2. Damaged mainshaft pilot bearing.
3. Dirt between the transaxle case and engine or misalignment of the transaxle (Chapter 7).
4. Worn or improperly adjusted cable (Chapter 7).

39 Difficulty in engaging gears

1. Clutch not releasing completely (see Chapter 8).
2. Loose, damaged or out-of-adjustment shift cable. Make a thorough inspection, replacing parts as necessary (Chapter 7).

40 Oil leakage

1. Excessive amount of lubricant in the transaxle (see Chapter 1 for correct checking procedure). Drain lubricant as required.
2. Cover loose or gasket damaged.
3. Rear oil seal or speedometer oil seal in need of replacement (Chapter 7).

Automatic transaxle

Note: *Due to the complexity of the automatic transaxle, it is difficult for the home mechanic to properly diagnose and service this component. For problems other than the following, the vehicle should be taken to a dealer or reputable mechanic.*

41 General shift mechanism problems

1. Chapter 7 deals with checking and adjusting the shift cable on automatic transaxles. Common problems which may be attributed to poorly adjusted linkage are:
   Engine starting in gears other than Park or Neutral.
   Indicator on shifter pointing to a gear other than the one actually being used.
   Vehicle moves when in Park.
2. Refer to Chapter 7 to adjust the cable.

42 Transaxle will not downshift with the accelerator pedal pressed to the floor

Chapter 7 deals with adjusting the throttle cable to enable the transaxle to downshift properly.

43 Transaxle slips, shifts rough, is noisy or has no drive in forward or reverse gears

1. There are many probable causes for the above problems, but the home mechanic should be concerned with only one possibility — fluid level.
2. Before taking the vehicle to a repair shop, check the level and condition of the fluid as described in Chapter 1. Correct fluid level as necessary or change the fluid and filter if needed. If the problem persists, have a professional diagnose the probable cause.

44 Fluid leakage

1. Automatic transaxle fluid is a deep red color. Fluid leaks should not be confused with engine oil, which can easily be blown by air flow to the transaxle.
2. To pinpoint a leak, first remove all built-up dirt and grime from around the transaxle. Degreasing agents and/or steam cleaning will achieve this. With the underside clean, drive the vehicle at low speeds so air flow will not blow the leak far from its source. Raise the vehicle and determine where the leak is coming from. Common areas of leakage are:
   a) Pan: Tighten the mounting bolts and/or replace the pan gasket as necessary (see Chapters 1 and 7).
   b) Filler pipe: Replace the rubber seal where the pipe enters the transaxle case.
   c) Transaxle oil lines: Tighten the connectors where the lines enter the transaxle case and/or replace the lines.
   d) Vent pipe: Transaxle overfilled and/or water in fluid (see checking procedures, Chapter 1).
   e) Speedometer connector: Replace the O-ring where the speedometer cable enters the transaxle case (Chapter 7).

Driveaxles

45 Clicking noise in turns

Worn or damaged outboard joint. Check for cut or damaged seals. Repair as necessary (Chapter 8).

46 Knock or clunk when accelerating after coasting

Worn or damaged inboard joint. Check for cut or damaged seals. Repair as necessary (Chapter 8).

47 Shudder or vibration during acceleration

1. Excessive joint angle. Check and correct as necessary (Chapter 8).
2. Worn or damaged inboard or outboard joints. Repair or replace as necessary (Chapter 8).
3. Sticking inboard joint assembly. Correct or replace as necessary (Chapter 8).

Rear axle

48 Noise

1 Road noise. No corrective procedures available.
2 Tire noise. Inspect tires and check tire pressures (Chapter 1).
3 Rear wheel bearings loose, worn or damaged (Chapter 10).

Brakes

Note: *Before assuming that a brake problem exists, make sure that the tires are in good condition and inflated properly (see Chapter 1), that the front end alignment is correct and that the vehicle is not loaded with weight in an unequal manner.*

49 Vehicle pulls to one side during braking

1 Defective, damaged or oil contaminated disc brake pads on one side. Inspect as described in Chapter 9.
2 Excessive wear of brake pad material or disc on one side. Inspect and correct as necessary.
3 Loose or disconnected front suspension components. Inspect and tighten all bolts to the specified torque (Chapter 10).
4 Defective caliper assembly. Remove the caliper and inspect for a stuck piston or other damage (Chapter 9).

50 Noise (high-pitched squeal with the brakes applied)

Disc brake pads worn out. The noise comes from the wear sensor rubbing against the disc (does not apply to all vehicles) or the actual pad backing plate itself if the material is completely worn away. Replace the pads with new ones immediately (Chapter 9). If the pad material has worn completely away, the brake rotors should be inspected for damage as described in Chapter 9.

51 Excessive brake pedal travel

1 Partial brake system failure. Inspect the entire system (Chapters 1 and 9) and correct as required.
2 Insufficient fluid in the master cylinder. Check (Chapter 1), add fluid and bleed the system if necessary (Chapter 9).
3 Rear brakes not adjusting properly. Make a series of starts and stops while the transaxle is in Reverse. If this does not correct the situation, remove the drums and inspect the self-adjusters (Chapter 9).

52 Brake pedal feels spongy when depressed

1 Air in the hydraulic lines. Bleed the brake system (Chapter 9).
2 Faulty flexible hoses. Inspect all system hoses and lines. Replace parts as necessary.
3 Master cylinder mounting bolts/nuts loose.
4 Master cylinder defective (Chapter 9).

53 Excessive effort required to stop the vehicle

1 Power brake booster not operating properly (Chapter 9).
2 Excessively worn linings or pads. Inspect and replace if necessary (Chapter 9).
3 One or more caliper pistons or wheel cylinders seized or sticking. Inspect and rebuild as required (Chapter 9).
4 Brake linings or pads contaminated with oil or grease. Inspect and replace as required (Chapter 9).
5 New pads or shoes installed and not yet seated. It will take a while for the new material to seat against the drum (or rotor).

54 Pedal travels to the floor with little resistance

Little or no fluid in the master cylinder reservoir caused by leaking wheel cylinder(s), leaking caliper piston(s), loose, damaged or disconnected brake lines. Inspect the entire system and correct as necessary.

55 Brake pedal pulsates during brake application

1 Wheel bearings not adjusted properly or in need of replacement (Chapter 1).
2 Caliper not sliding properly due to improper installation or obstructions. Remove and inspect (Chapter 9).
3 Rotor defective. Remove the rotor (Chapter 9) and check for excessive lateral runout and parallelism. Have the rotor resurfaced or replace it with a new one.

Suspension and steering systems

56 Vehicle pulls to one side

1 Tire pressures uneven (Chapter 1).
2 Defective tire (Chapter 1).
3 Excessive wear in suspension or steering components (Chapter 10).
4 Front end in need of alignment.
5 Front brakes dragging. Inspect the brakes as described in Chapter 9.

57 Shimmy, shake or vibration

1 Tire or wheel out-of-balance or out-of-round. Have professionally balanced.
2 Loose, worn or out-of-adjustment wheel bearings (Chapters 1 and 8).
3 Shock absorbers and/or suspension components worn or damaged (Chapter 10).

58 Excessive pitching and/or rolling around corners or during braking

1 Defective shock absorbers. Replace as a set (Chapter 10).
2 Broken or weak springs and/or suspension components. Inspect as described in Chapter 10.

59 Excessively stiff steering

1 Lack of fluid in power steering fluid reservoir (Chapter 1).
2 Incorrect tire pressures (Chapter 1).
3 Lack of lubrication at the steering joints (Chapter 1).
4 Front end out of alignment.
5 See also Section titled *Lack of power assistance.*

60 Excessive play in the steering

1 Loose front wheel bearings (Chapter 1).
2 Excessive wear in suspension or steering components (Chapter 10).
3 Steering gearbox damaged (Chapter 10).

Troubleshooting

61 Lack of power assistance

1 Steering pump drivebelt faulty or not adjusted properly (Chapter 1).
2 Fluid level low (Chapter 1).
3 Hoses or lines restricted. Inspect and replace parts as necessary.
4 Air in power steering system. Bleed the system (Chapter 10).

62 Excessive tire wear (not specific to one area)

1 Incorrect tire pressures (Chapter 1).
2 Tires out of balance. Have professionally balanced.
3 Wheels damaged. Inspect and replace as necessary.
4 Suspension or steering components excessively worn (Chapter 10).

63 Excessive tire wear on outside edge

1 Inflation pressures incorrect (Chapter 1).
2 Excessive speed in turns.
3 Front end alignment incorrect (excessive toe-in). Have professionally aligned.
4 Suspension arm bent or twisted (Chapter 10).

64 Excessive tire wear on inside edge

1 Inflation pressures incorrect (Chapter 1).
2 Front end alignment incorrect (toe-out). Have professionally aligned.
3 Loose or damaged steering components (Chapter 10).

65 Tire tread worn in one place

1 Tires out of balance.
2 Damaged or buckled wheel. Inspect and replace if necessary.
3 Defective tire (Chapter 1).

Chapter 1 Tune-up and routine maintenance

Refer to Chapter 13 for information on 1987 and later models

Contents

Air filter element and PCV valve maintenance	15
Battery check and maintenance	8
Brake checks	27
Combustion chamber conditioner application (Canadian models)	18
Chassis lubrication	11
Clutch pedal free play check	23
Compression check	30
Cooling system check	21
Cooling system servicing (draining, flushing and refilling)	28
Drivebelt check, adjustment and replacement	14
Exhaust system check	22
Fuel filter replacement	10
Fluid level checks	4
Fuel system check	20
General information	1
Heated inlet air system general check	19
Ignition timing check and adjustment	29
Introduction to routine maintenance	2
Manual transmission fluid change	31
Oil and oil filter change	9
Power loss or power limited lamp general information	See Chapter 5
Routine maintenance schedule	3
Spark plug replacement	12
Spark plug wire, distributor cap and rotor check and replacement	13
Steering shaft seal lubrication	25
Suspension and steering check	24
Throttle body mounting nut torque check	17
Tires and tire pressure checks	5
Tire rotation	6
Underhood hose check and replacement	16
Wheel bearing check, adjustment and repacking	26
Wiper blade element removal and installation	7

Specifications

Note: *Additional Specifications and torque requirements can be found in each individual Chapter.*

Recommended lubricants and fluids

Engine oil	Consult your owner's manual or local dealer for recommendations on the particular service grade and viscosity oil recommended for your area, special driving conditions and climatic parameters
Manual and automatic transaxle fluid	Dexron II ATF
Transaxle shift linkage	NLGI No. 2 chassis grease
Clutch linkage	NLGI No. 2 chassis grease
Power steering reservoir	Mopar 4-253 power steering fluid or equivalent
Brake system	DOT 3 brake fluid
Combustion chamber conditioner	Autopar combustion chamber conditioner No. VU788 or equivalent
Engine coolant	50/50 mixture of ethylene glycol-based antifreeze and water
Parking brake mechanism	White lithium-based grease NLGI No. 2
Chassis lubrication	NLGI No. 2 EP chassis grease
Steering shaft seal	NLGI No. 2 EP grease
Rear wheel bearings	NLGI No. 2 EP grease
Steering gear	API GL-4 SAE 90 oil
Hood and door hinges/liftgate hinges	Engine oil
Door hinge half and check spring	NLGI No. 2 multi-purpose grease
Key lock cylinders	Graphite spray
Hood latch assembly	Mopar Lubriplate or equivalent
Door latch striker	Mopar Door Ease No. 3744859 or equivalent

Chapter 1 Tune-up and routine maintenance

Quick reference capacities

Engine oil (including filter)
- non-turbocharged engine .... 4.0 qts (3.8 liters)
- turbocharged engine .... 5.0 qts (4.7 liters)

Fuel tank .... 14 gal (53.0 liters)

Automatic transaxle
- from dry, including torque converter .... 8.9 qts (8.4 liters)
- transaxle drain and refill .... 3.8 qts (3.6 liters)

Manual transaxle .... 2.3 qts (2.2 liters)
Power steering system .... 2.5 pts (1.2 liters)
Cooling system .... 9.5 qts (9.0 liters)

Ignition system

Spark plug type
- 1984 .... Champion RN12Y or RN65PR
- 1985 and 1986 .... Champion RN12Y

Spark plug gap .... 0.035 in (0.9 mm)
Spark plug wire resistance .... 3000 ohms per foot minimum/7200 ohms per foot maximum
Ignition timing .... See *Emission Control Information* label in engine compartment
Firing order 1-3-4-2

Drivebelt deflection

Alternator
- new .... 1/8 in (3 mm)
- used .... 1/4 in (6 mm)

Power steering pump
- new .... 1/4 in (6 mm)
- used .... 7/16 in (11 mm)

Water Pump
- new .... 1/8 in (3 mm)
- used .... 1/4 in (6 mm)

Air conditioning compressor
- new .... 5/16 in (8 mm)
- used .... 3/8 in (9 mm)

General

Idle speed .... See *Emission Control Information* label in engine compartment
Compression pressure .... 130 to 150 psi (100 psi minimum)
Maximum variation between cylinders
- 1984 and 1985 .... 20 psi
- 1986 .... 25 psi

Torque specifications

	Ft-lbs	Nm
Throttle body nuts	17	23
Manual transaxle filler plug	24	33
Oil pan drain plug	20	27
Spark plugs	26	35
Wheel lug nuts	95	129

Cylinder location and distributor rotation

1 General information

Warning: *The electric fan on some models can start at any time, even when the engine is turned off. Consequently, the negative battery cable should be disconnected whenever you are working in the vicinity of the fan.*

Front wheel drive vehicles incorporate several features which require special maintenance techniques. Among these are the driveaxles, transaxle and cooling system. Consult this Chapter and the *General information* Sections of each Chapter to determine which components are unique to these vehicles.

2 Introduction to routine maintenance

Refer to illustrations 2.2a, 2.2b and 2.2c

This Chapter was designed to help the home mechanic maintain his (or her) vehicle for peak performance, economy, safety and longevity.

On the following pages you will find a maintenance schedule, along with Sections which deal specifically with each item on the schedule. Included are visual checks, adjustments and item replacements. Refer to the accompanying underhood and underside illustrations to locate the various items mentioned in each procedure.

Servicing your vehicle using the time/mileage maintenance schedule and the sequenced Sections will give you a planned program of maintenance. Keep in mind that it is a full plan, and maintaining only a few items at the specified intervals will not give you the same results.

You will find as you service your vehicle that many of the procedures can, and should, be grouped together, due to the nature of the job at hand. Examples of this are as follows:

If the vehicle is fully raised for a chassis inspection, for example, this is the ideal time for the following checks: manual transaxle fluid, exhaust system, suspension, steering and fuel system.

If the tires and wheels are removed, as during a routine tire rotation, check the brakes and wheel bearings at the same time.

If you must borrow or rent a torque wrench, it is a good idea to service the spark plugs and repack (or replace) the rear wheel bearings all in the same day to save time and money.

The first step of this, or any, maintenance plan is to prepare yourself before the actual work begins. Read through the appropriate Sections for all work that is to be performed before you begin. Gather together all the necessary parts and tools. If it appears you could have a problem during a particular job, don't hesitate to ask advice from your local parts man or dealer service department.

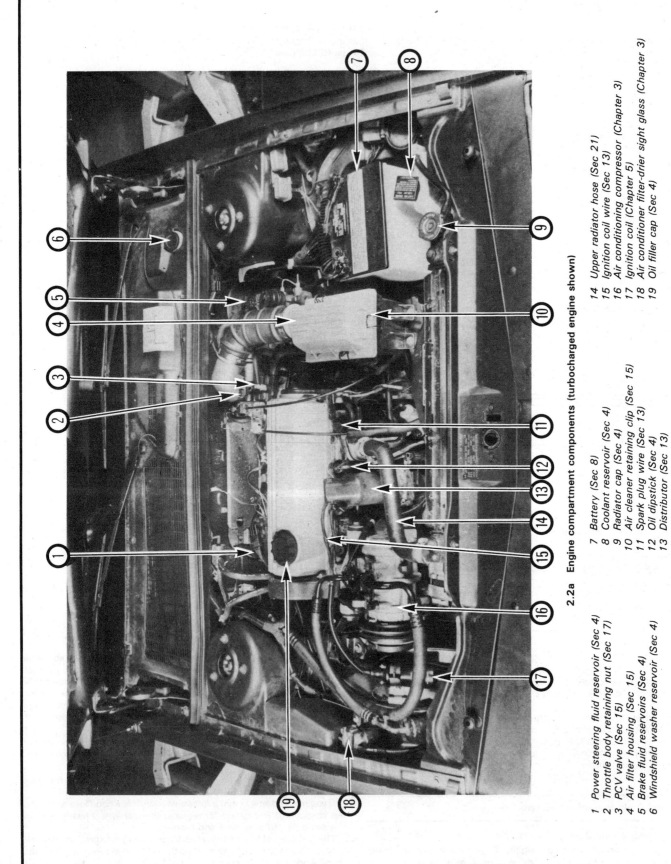

2.2a Engine compartment components (turbocharged engine shown)

1 Power steering fluid reservoir (Sec 4)
2 Throttle body retaining nut (Sec 17)
3 PCV valve (Sec 15)
4 Air filter housing (Sec 15)
5 Brake fluid reservoirs (Sec 4)
6 Windshield washer reservoir (Sec 4)
7 Battery (Sec 8)
8 Coolant reservoir (Sec 4)
9 Radiator cap (Sec 4)
10 Air cleaner retaining clip (Sec 15)
11 Spark plug wire (Sec 13)
12 Oil dipstick (Sec 4)
13 Distributor (Sec 13)
14 Upper radiator hose (Sec 21)
15 Ignition coil wire (Sec 13)
16 Air conditioning compressor (Chapter 3)
17 Ignition coil (Chapter 5)
18 Air conditioner filter-drier sight glass (Chapter 3)
19 Oil filler cap (Sec 4)

2.2b Engine compartment underside components (turbocharged engine shown)

1 Water pump (Chapter 3)
2 Alternator (Chapter 5)
3 Lower radiator hose (Sec 21)
4 Manual transaxle filler plug (Sec 4)
5 Brake hose (Sec 27)
6 Driveaxle CV joint boot (Sec 4)
7 Oil drain plug (Sec 9)
8 Intermediate shaft (Chapter 8)
9 Exhaust pipe-to-manifold flange bolt (Sec 22)
10 Brake caliper (Sec 27)
11 EVAP system canister (Chapter 6)

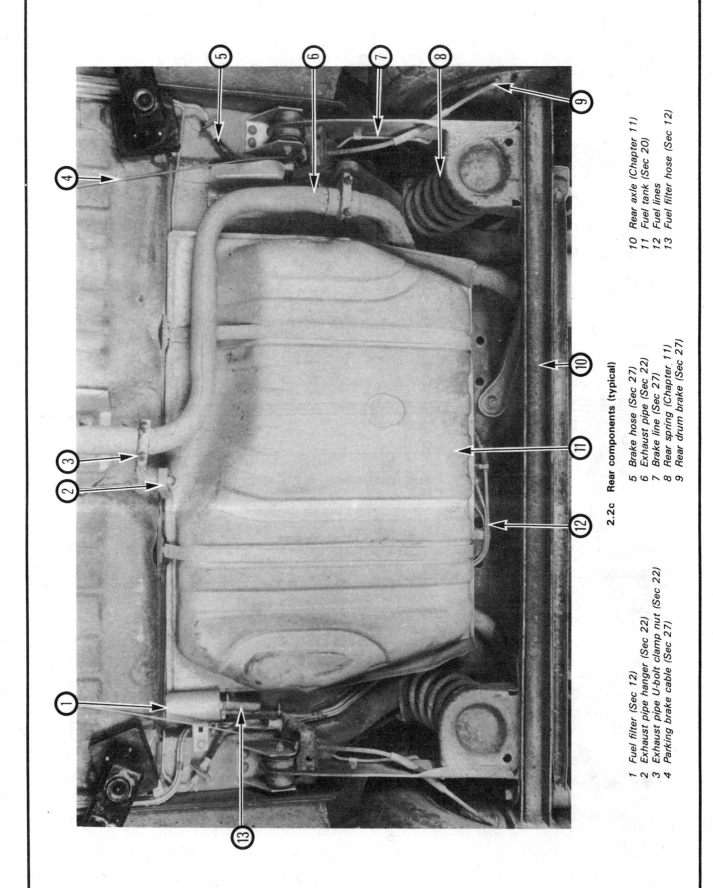

2.2c Rear components (typical)

1 Fuel filter (Sec 12)
2 Exhaust pipe hanger (Sec 22)
3 Exhaust pipe U-bolt clamp nut (Sec 22)
4 Parking brake cable (Sec 27)
5 Brake hose (Sec 27)
6 Exhaust pipe (Sec 22)
7 Brake line (Sec 27)
8 Rear spring (Chapter 11)
9 Rear drum brake (Sec 27)
10 Rear axle (Chapter 11)
11 Fuel tank (Sec 20)
12 Fuel lines
13 Fuel filter hose (Sec 12)

Chapter 1 Tune-up and routine maintenance

3 Routine maintenance schedule

The following recommendations are given with the assumption that the vehicle owner will be doing the maintenance or service work, as opposed to having a dealer service department do the work. The majority of the following intervals are factory recommendations. However, subject to the preference of the individual owner interested in keeping his or her vehicle in peak condition at all times and with the vehicle's ultimate resale in mind, many of the operations may be performed more often. We encourage such owner initiative.

When the vehicle is new it should be serviced initially by a factory authorized dealer service department to protect the factory warranty. In some cases the initial maintenance check is done at no cost to the owner.

Weekly or every 250 miles (400 km)

Check the engine oil level and add oil as necessary (Sec 4)
Check the wiper blade condition (Sec 7)
Check the engine coolant level and add coolant as necessary (Sec 4)
Check the tires and tire pressures (Sec 5)
Check the automatic transaxle fluid level (Sec 4)
Check the power steering fluid level (Sec 4)
Check the brake fluid level (Sec 4)
Check the battery condition (Sec 8)
Check the windshield washer fluid level (Sec 4)
Check the operation of all lights
Check horn operation

Every 2000 miles (3200 km) or 2 months, whichever comes first

Change the engine oil and filter (all models) (Sec 9)
Check the suspension balljoint and steering linkage boots for damage and lubricant leakage (Sec 24)
Check the steering gear boots for cracks and lubricant leakage (Sec 24)
Inspect the driveaxle CV joints and boots for damage, wear and lubricant leakage (Sec 4)

Every 7500 miles (12,000 km) or 6 months, whichever comes first

Check the manual transaxle fluid level (Sec 4)
Check the deflection of all drivebelts (Sec 14)
Check the fuel hoses, lines and connections for leaks and damage (Sec 20)
Check the brake hoses and lines for leaks and damage (Sec 27)
Clean the combustion chamber with combustion conditioner (Canadian models only) (Sec 18)

Every 15,000 miles (24,000 km) or 12 months, whichever comes first

Check the cooling system hoses and connections for leaks and damage (Sec 21)
Check for free play in the steering linkage and balljoints (Sec 24)
Check the hoses and connections in the fuel evaporative emission system (Chapter 6)
Check the exhaust pipes and hangers (Sec 22)
Check the EGR system components for proper operation (Chapter 6)
Check the condition of the vacuum hoses and connections (Sec 16)
Check the condition of the wiring harness connections (Chapter 12)
Check and clean the battery (Sec 8)
Check the condition of the ignition wiring and spark plug wires (Sec 13)
Check the distributor cap and rotor for cracks, wear and damage (Sec 13)
Rotate the tires (Sec 6)
Replace the spark plugs (vehicles *without* catalytic converter) (Sec 12)
Check/replace the wiper blade elements (Sec 7)

Every 22,500 miles (36,000 km) or 18 months, whichever comes first

Check the front disc brake pads (Sec 28)
Lubricate the front suspension and steering balljoints (Sec 11)
Check the air conditioning hoses, belts and sight glass (Chapter 3)
Check the rear brake linings and drums for wear and damage (Sec 27)
Inspect and adjust the rear wheel bearings (Chapter 10)

Every 30,000 miles (48,000 km) or 24 months, whichever comes first

Drain and replace the engine coolant (Sec 28)
Check the throttle body mounting nut torque (Sec 17)
Check the cylinder compression (Sec 30)
Change the automatic transaxle fluid (Chapter 7B)
Adjust the automatic transaxle bands (Chapter 7B)
Check the parking brake operation (Chapter 9)
Check the steering shaft seal and lubricate as necessary (Sec 25)
Replace the spark plugs (vehicles *with* catalytic converter) (Sec 12)

Every 52,500 miles (84,000 km) or 60 months, whichever comes first

Inspect/replace the gas tank cap (Sec 20)
Replace the fuel filter (Sec 10)
Replace the air filter element (Sec 15)
Replace the spark plug wires, distributor cap and rotor (Sec 13)
Replace the oxygen sensor (Chapter 6)
Replace the PCV valve (Sec 15)

Severe operating conditions

Severe operating conditions are defined as:
Stop-and-go driving
Driving in dusty conditions
Extensive idling
Frequent short trips
Sustained high speed driving during hot weather (over 90°F/32°C)

Severe operating conditions maintenance intervals:
Change the automatic transaxle fluid and filter and adjust the bands every 15,000 miles (24,000 km)
Change the engine oil and filter every 2000 miles (3200 km) (turbocharged engines every 1000 miles/1600 km)
Inspect the disc brake linings every 9000 miles (14,000 km)
Inspect the rear brakes and drums every 9000 miles (14,000 km)
Lubricate and adjust the rear wheel bearings every 9000 miles (14,000 km)
Inspect the driveaxle CV joint boots and front suspension boots every 2000 miles (3200 km)
Lubricate the tie-rod ends every 15,000 miles (24,000 km)
Replace the air filter element every 15,000 miles (24,000 km)
Change the manual transaxle fluid and clean the pan magnet every 15,000 miles (24,000 km)

4 Fluid level checks

Refer to illustrations 4.4, 4.5, 4.6, 4.8, 4.12, 4.15, 4.16, 4.24, 4.30, 4.31, 4.32, 4.36, 4.37, 4.42, 4.49a, 4.49b, 4.49c and 4.50

1 There are a number of components on a vehicle which rely on the use of fluids to perform their job. Through the normal operation of the vehicle, these fluids are used up and must be replenished before damage occurs. See *Recommended lubricants and fluids* for the specific fluid

4.4 The engine oil level dipstick is located on the front side of the engine

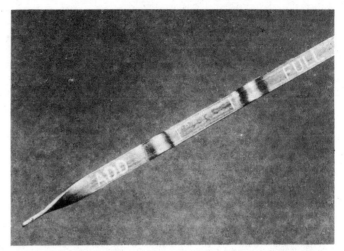

4.5 The oil level must be between the Add and Full marks

4.8 The coolant level (which should be kept between the MAX and MIN marks) is checked at the reservoir — do not remove the radiator cap when the engine is hot!

4.6 Oil is added to the engine through the filler cap on the cylinder head cover

to be used when adding is required. When checking fluid levels it is important that the vehicle be on a level surface.

Engine oil

2 The engine oil level is checked with a dipstick which is located on the front side of the engine block. This dipstick travels through a tube and into the oil pan at the bottom of the engine.

3 The oil level should preferably be checked before the vehicle has been driven or about 15 minutes after the engine has been shut off. If the oil is checked immediately after driving the vehicle, some of the oil will remain in the upper engine components, giving an inaccurate reading on the dipstick.

4 Pull the dipstick from its tube (see illustration) and wipe all the oil from the end with a clean rag. Insert the clean dipstick all the way back into the tube and pull it out again. Observe the oil at the end of the dipstick. At its highest point, the level should be within the Full range mark.

5 It takes approximately 1 quart of oil to raise the level from the Add mark to the Full range mark on the dipstick (see illustration). Do not allow the level to drop below the Add mark as this may cause engine damage due to oil starvation. On the other hand, do not overfill the engine by adding oil above the Full range mark, as this may result in oil fouled spark plugs, oil leaks or oil seal failures.

6 Oil is added to the engine after removing a twist off cap located on the cylinder head cover (see illustration). The cap should be marked *Engine oil* or *oil*. An oil can spout or funnel will reduce spills as the oil is added.

7 Checking the oil level can also be a step towards preventive maintenance. If you notice the oil level dropping consistently it is an indication of oil leakage or internal engine wear which should be corrected. If there are water droplets in the oil, or if it is milky looking, it indicates component failure and the engine should be checked immediately. The condition of the oil should also be checked along with the level. With the dipstick removed from the engine, wipe your thumb and index finger up the dipstick, looking for small dirt or metal particles clinging to the dipstick. This is an indication that the oil should be drained and fresh oil added (Section 9).

Engine coolant

8 All vehicles are equipped with a pressurized coolant recovery system which makes coolant level checks easy. A coolant reservoir attached to the inner fender panel is connected by a hose to the radiator filler neck (see illustration). As the engine heats during operation, coolant is forced from the radiator, through the connecting tube and into the reservoir. As the engine cools this coolant is automatically drawn back into the radiator to maintain the correct level.

9 The coolant level should be checked when the engine is idling and

Chapter 1 Tune-up and routine maintenance

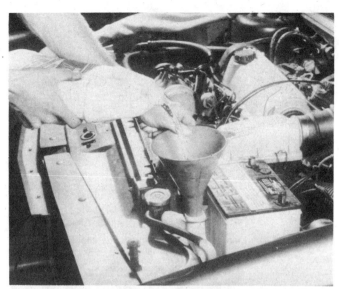

4.12 A funnel can be used to prevent spills when adding coolant to the reservoir

4.15 The windshield washer fluid reservoir (arrow) is located under the hood

4.16 Use a funnel when adding fluid to the rear washer reservoir to avoid spilling

at normal operating temperature. Observe the level of fluid in the reservoir, which should be at or near the *Max* mark on the side of the reservoir.

10 The coolant level can also be checked by removing the radiator cap. **Warning:** *The cap should not, under any circumstances, be removed while the system is hot, as escaping steam could cause serious injury. Wait until the engine has cooled, then wrap a thick cloth around the cap and turn it to its first stop. If any steam escapes from the cap, allow the engine to cool further, then remove the cap and check the level in the radiator.*

11 If only a small amount of coolant is required to bring the system up to the proper level, plain water can be used. However, to maintain the proper antifreeze/water mixture in the system, both should be mixed together to replenish a low level. Antifreeze offering protection to –20°F should be mixed with water in the proportion specified on the container. These vehicles have aluminum cylinder heads, so use of the proper coolant is critical to avoid corrosion. Do not allow antifreeze to come into contact with your skin or painted surfaces of the vehicle. Flush contacted areas immediately with plenty of water.

12 The coolant should be added to the reservoir with a funnel to reduce the risk of spills (see illustration).

13 As the coolant level is checked, note the condition of the coolant. It should be relatively clear. If the fluid is brown or rust colored, it is an indication that the system should be drained, flushed and refilled (Section 28).

14 If the cooling system requires repeated additions to maintain the proper level, have the radiator cap checked for proper sealing. Also check for leaks in the system such as cracked hoses, loose hose connections, leaking gaskets, etc.

Windshield and rear window washer fluid

15 The fluid for the windshield and rear window washer systems is located in plastic reservoirs. The level inside the reservoirs should be maintained about one inch below the filler cap. The reservoirs are accessible after opening the hood and rear hatch (see illustration).

16 An approved windshield washer solvent should be added to the reservoir whenever replenishing is required (see illustration). Do not use plain water in this system, especially in cold climates where the water could freeze.

Battery electrolyte

Warning: *Certain precautions must be followed when checking or servicing the battery. Hydrogen gas, which is highly flammable, is produced in the cells, so keep lighted tobacco, open flames, bare light bulbs and sparks away from the battery. The electrolyte inside the battery is dilute sulfuric acid, which can burn your skin and cause serious injury if splashed in the eyes (wear safety glasses). It will also ruin clothes and painted surfaces. Remove all metal jewelry which could contact the positive terminal and a ground, causing a short circuit.*

17 Vehicles equipped with maintenance-free batteries require no level checking or water addition, as the battery case is sealed.

18 If a maintenance-type battery is installed, the caps on the top of the battery should be removed periodically to check for a low electrolyte level.

19 Remove each of the caps and add distilled water to bring the level in each cell to the split ring in the filler opening.

20 At the same time the battery water level is checked, the overall condition of the battery and its related components should be noted. If corrosion is found on the cable ends or battery terminals, remove the cables and remove the corrosion with a baking soda/water solution and a wire brush cleaning tool designed for this purpose. See Section 8 for complete battery care and servicing procedures.

Brake fluid

21 The brake master cylinder is located on the left side of the engine compartment firewall and has two caps which must be removed to check the fluid level.

22 Before removing the caps, use a rag to clean all dirt, grease, etc. from the top of the reservoir. If any foreign matter enters the master cylinder with the caps removed, blockage in the brake system lines can occur. Make sure all painted surfaces around the master cylinder

4.24 The brake fluid level in the master cylinder reservoirs must be maintained at the split rings

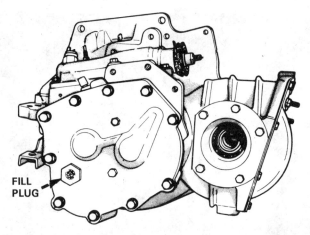

4.30 The manual transaxle fluid level is checked by removing the fill plug

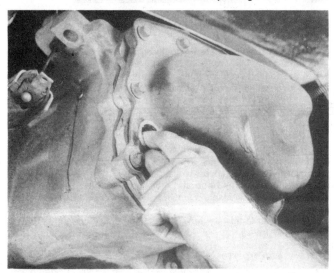

4.31 Use your finger as a dipstick to check that the manual transaxle fluid is at the upper edge of the hole

4.32 A funnel with a flexible spout or a syringe is necessary to add fluid to the manual transaxle

are covered, as brake fluid will ruin paint.
23 Unscrew the reservoir caps.
24 Observe the fluid level (see illustration). It should be at the bottom of the split rings.
25 If additional fluid is necessary to bring the level up to the proper height, carefully pour the specified brake fluid into the master cylinder. Be careful not to spill the fluid on painted surfaces. Be sure the specified fluid is used, as mixing different types of brake fluid can cause damage to the system. See *Recommended lubricants and fluids* or your owner's manual.
26 The fluid and the master cylinder should be inspected for contamination. Normally the braking system will not need periodic draining and refilling, but if rust deposits, dirt particles or water droplets are seen in the fluid, the system should be dismantled, drained and refilled with fresh fluid.
27 Reinstall the master cylinder caps.
28 The brake fluid in the master cylinder will drop slightly as the brake shoes or pads at each wheel wear during normal operation. If the master cylinder requires repeated replenishing to keep it at the proper level, this is an indication of leakage in the brake system which should be corrected immediately. Check all brake lines and connections, along with the wheel cylinders and booster (see Chapter 9 for more information).
29 If, upon checking the fluid level, you discover one or both reservoirs empty or nearly empty, the brake system should be bled (Chapter 9).

Manual transaxle lubricant

30 Manual shift transaxles do not have a dipstick. The fluid level is checked by removing a plug in the side of the transaxle case (see illustration). Locate the plug and use a rag to clean the plug and the area around it. It may be necessary to remove the left inner fender well cover for access.
31 With the engine cold, remove the plug. If fluid immediately starts leaking out, thread the plug back into the transaxle because the fluid level is correct. If there is no fluid leakage, completely remove the plug and place your finger inside the hole (see illustration). The fluid level should be just at the bottom of the plug hole.
32 If the transaxle needs more fluid, use a funnel with rubber filler tube or a syringe to pour or squeeze the appropriate lubricant into the plug hole to bring the fluid up to the proper level (see illustration). **Caution:** *Use only the specified transmission fluid; do not use gear oil.*
33 Thread the plug back into the transaxle and tighten it securely. Drive the vehicle and check for leaks around the plug.

Automatic transaxle fluid

34 The fluid inside the transaxle should be at normal operating temperature to get an accurate reading on the dipstick. This is done by driving the vehicle for several miles, making frequent starts and stops to allow the transaxle to shift through all gears.
35 Park the vehicle on a level surface, place the selector lever in Park and leave the engine running at an idle.
36 Remove the transaxle dipstick (see illustration) and wipe all the fluid from the end of the dipstick with a clean rag.
37 Push the dipstick back into the transaxle until the cap seats firmly on the dipstick tube. Remove the dipstick and note the fluid on the end. The fluid level should be in the crosshatched area marked *Hot* (between the two upper holes in the dipstick) (see illustration). If the

Chapter 1 Tune-up and routine maintenance

4.36 The automatic transaxle fluid level is checked by removing the dipstick — don't confuse it with the engine oil dipstick!

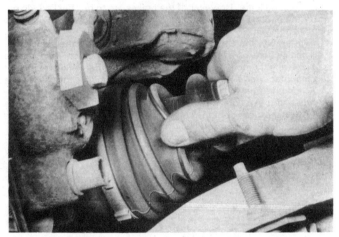

4.42 Carefully check the CV joint boots for cracks and other damage as well as leaks

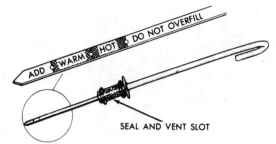

4.37 The automatic transaxle fluid level should be in the range indicated by the fluid temperature

4.49a The power steering fluid level should be checked with the engine off

4.49b The power steering dipstick has markings on opposite sides so the level can be checked with the fluid hot . . .

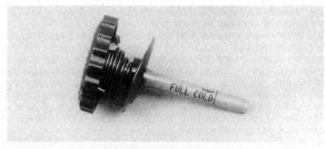

4.49c . . . or cold

fluid is not hot (temperature about 100°F), the level should be in the area marked *Warm* between the two lower holes.
38 If the fluid level is at or below the *Add* mark on the dipstick, add enough fluid to raise the level to within the marks indicated for the appropriate temperature. Fluid should be added directly into the dipstick hole, using a funnel to prevent spills.
39 It is important that the transaxle not be overfilled. Under no circumstances should the fluid level be above the upper hole on the dipstick, as this could cause internal damage to the transaxle. The best way to prevent overfilling is to add fluid a little at a time, driving the vehicle and checking the level between additions.
40 Use only transaxle fluid specified by the manufacturer. This information can be found in *Recommended lubricants and fluids* in the Specifications.
41 The condition of the fluid should be checked along with the level. If the fluid at the end of the dipstick is a dark reddish-brown color, or if it has a burned smell, the transaxle fluid should be changed. If you are in doubt about the condition of the fluid, purchase some new fluid and compare the two for color and smell.

Driveaxle lubricant

42 The driveaxle constant velocity joints are lubricated for life at the time of production. It is important to check the rubber boots for cracks, tears and holes (see illustration).
43 With the vehicle raised and supported securely, inspect the area around the boots for signs of grease splattering, indicating damage to the boots or retaining clamps.
44 Inspect around the inboard joint for signs of fluid leakage from the transaxle differential seal.

Power steering fluid

45 Unlike manual steering, the power steering system relies on fluid which may, over a period of time, require replenishing.
46 The reservoir for the power steering pump is located on the rear side of the engine.
47 The power steering fluid level can be checked with the engine cold.
48 With the engine shut off, use a rag to clean the reservoir cap and the area around the cap. This will help to prevent foreign material from falling into the reservoir when the cap is removed.
49 Twist off the reservoir cap (see illustration), which has a built-in dipstick attached to it. Pull off the cap and remove the fluid at the bot-

Chapter 1 Tune-up and routine maintenance

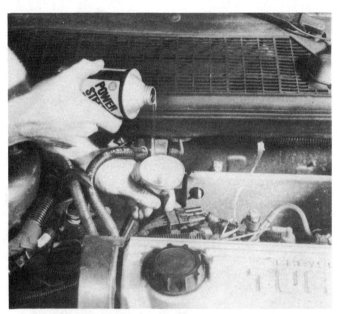

4.50 A funnel with a flexible spout or tube can be used to add fluid to the power steering reservoir

tom of the dipstick with a clean rag. Reinstall the cap/dipstick assembly to get a fluid level reading. Remove the dipstick and note the fluid level. It should be at the Full mark on the dipstick (see illustrations).

50 If additional fluid is required, pour the specified type directly into the reservoir using a funnel to prevent spills (see illustration).

51 If the reservoir requires frequent fluid additions, all power steering hoses, hose connections, the power steering pump and the steering box should be carefully checked for leaks.

5 Tire and tire pressure checks

1 Periodically inspecting the tires can prevent you from being stranded with a flat tire and can also give you clues as to possible problems with the steering and suspension systems before major damage occurs.

2 Proper tire inflation adds miles to the lifespan of the tires, allows the vehicle to achieve maximum miles per gallon figures, and contributes to overall ride quality.

3 When inspecting a tire, first check for wear at the tread. Irregularities in the tread pattern (cupping, flat spots, more wear on one side than the other) are indications of front end alignment and/or balance problems. If any of these conditions are found, take the vehicle to a repair shop which can correct the problem.

4 Check the tread area for cuts or punctures. Many times a nail or tack will embed itself into the tire tread and yet the tire will hold air pressure for a short time. In most cases, a repair shop or gas station can repair the punctured tire.

5 It is also important to check the sidewalls of the tire, both inside and outside. Check for deteriorated rubber, cuts and punctures. Also inspect the inboard side of the tire for signs of brake fluid, indicating a thorough brake inspection is needed immediately.

6 Incorrect tire pressure cannot be determined merely by looking at the tire. This is especially true for radial tires. A tire pressure gauge must be used. If you do not already have a reliable gauge, it is a good idea to purchase one and keep it in the glove compartment. Built-in pressure gauges at gas stations are often inaccurate. If you are in doubt as to the accuracy of your gauge, many repair shops have ''master'' pressure gauges which you can use for comparison purposes.

7 Always check tire inflation when the tires are cold. Cold, in this case, means the vehicle has not been driven more than one mile after sitting for three hours or more. It is normal for the pressure to increase 4 to 8 pounds when the tires are hot.

8 Unscrew the valve cap protruding from the wheel or hubcap and firmly press the gauge onto the valve stem. Note the reading on the gauge and check it against the recommended tire pressure listed on the tire placard located on the jamb of the driver's door.

9 Check all tires and add air as necessary to bring all tires up to the recommended pressure levels. Do not forget the spare tire. Be sure to reinstall the valve caps, which keep dirt and moisture out of the valve stem mechanism.

6 Tire rotation

Refer to illustration 6.2

1 The tires should be rotated at the specified intervals and whenever uneven wear is noticed. Since the vehicle will be raised and the tires removed anyway, this is a good time to check the brakes (Section 27) and repack the wheel bearings. Read over those Sections if this is to be done at the same time.

2 The rotation pattern depends on whether or not the spare is included in the rotation. The accompanying illustration shows the rotation pattern to be followed for four and five tire rotation procedures.

3 See the information in *Jacking and towing* at the front of this manual for the proper procedures to follow when raising the vehicle and changing a tire. However, if the brakes are to be checked, do not apply the parking brake as stated. Make sure the tires are blocked to prevent the vehicle from rolling.

4 Preferably, the entire vehicle should be raised at the same time. This can be done on a hoist or by jacking up each corner of the vehicle and lowering it onto jackstands placed under the frame rails. Note that the manufacturer *does not* indicate a lifting point for the rear of the vehicle with a floor jack in the *Jacking* diagram. Always use four jackstands and make sure the vehicle is securely supported.

5 After the tire rotation, check and adjust the tire pressures as necessary and be sure to check wheel lug nut tightness.

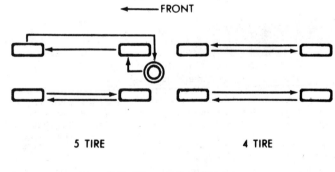

6.2 Tire rotation diagram

7 Wiper blade element removal and installation

Refer to illustrations 7.4 and 7.5a, 7.5b and 7.5c

1 The windshield and rear window wiper blade elements should be checked periodically for cracks and deterioration.

2 To gain access to the wiper blades, turn on the ignition switch and cycle the wipers to a position on the windshield or rear window where the work can be performed. Turn off the ignition to park the blades in the working position.

3 Lift the wiper blade assembly away from the glass.

4 Use a screwdriver to lift the release tab and remove the wiper blade assembly from the arm (see illustration).

5 Lift the lock tab on one of the end links and then squeeze the link to remove it from the center bridge. Slide the end link off the blade element and then slide the element from the claws of the other link (see illustrations).

6 Installation is the reverse of removal.

Chapter 1 Tune-up and routine maintenance

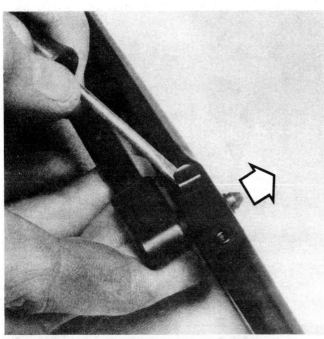

7.4 Lift the tab with a small screwdriver and slide the wiper blade off the pin in the direction shown

7.5a To remove the wiper element, lift the lock tabs at both ends ...

7.5b ... and squeeze the end links together, ...

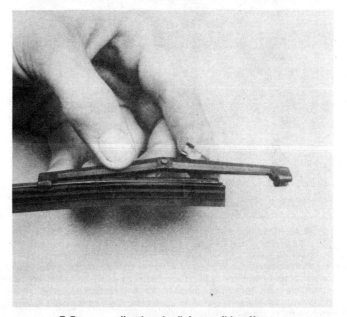

7.5c ... allowing the links to slide off

8 Battery check and maintenance

Refer to illustrations 8.4a, 8.4b, 8.4c and 8.4d

Warning: *Certain precautions must be followed when checking or servicing the battery. Hydrogen gas, which is highly flammable, is produced in the cells, so keep lighted tobacco, open flames, bare light bulbs and sparks away from the battery. The electrolyte inside the battery is dilute sulfuric acid, which can burn your skin and cause serious injury if splashed in the eyes (wear safety glasses). It will also ruin clothes and painted surfaces. Remove all metal jewelry which could contact the positive terminal and a ground, causing a short circuit.*

1 These models are equipped with a maintenance-free battery, which doesn't require the addition of water. These batteries have built-in test indicators which display different colors depending on battery condition. If the indicator shows *green*, the battery is properly charged. If it is *red* or *black*, charging is required. A light *yellow* indicator means the battery must be replaced with a new one. **Warning:** *Do not charge, test or jump start a battery with a yellow indicator visible.* If any doubt exists as to the battery state-of-charge, it should be tested by a dealer service department or service station.

2 The top of the battery should be kept clean and free from dirt and moisture so that the battery does not become partially discharged. Clean the top and sides of the battery with a baking soda and water solution, but make sure that it does not enter the battery. After it is clean, check the case for cracks and other damage.

3 Make sure the cable clamps are tight to ensure good electrical connections and check the cables for cracked insulation, frayed wires and corrosion.

Chapter 1 Tune-up and routine maintenance

8.4a Battery terminal corrosion usually appears as white, fluffy powder

8.4b Removing the cable from the battery terminal post — always remove the ground cable first and hook it up last!

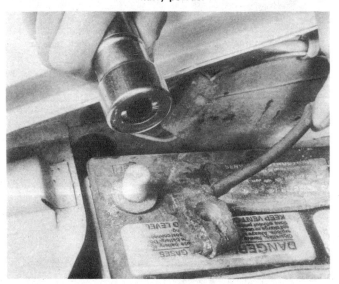

8.4c Cleaning the battery terminal post with a special tool

8.4d Cleaning the battery cable clamp

4 If the posts are corroded, remove the cables (negative first, then positive) and clean the clamps and battery posts with a battery terminal cleaning tool, then reinstall the cables (positive first, then negative) (see illustrations). Apply petroleum jelly to the cable clamps and posts to keep corrosion to a minimum.
5 Make sure that the battery carrier is in good condition and that the hold-down clamp bolts are tight. If the battery is removed, make sure that no parts remain in the bottom of the carrier when it is reinstalled. When reinstalling the clamp bolts do not overtighten them.
6 Corrosion on the carrier and hold-down components can be removed with a solution of baking soda and water. Rinse any treated areas with clean water, dry them thoroughly and apply zinc-based primer and paint.

9 Oil and oil filter change

Refer to illustrations 9.9, 9.14 and 9.19

1 Frequent oil changes may be the best form of preventive maintenance available to the home mechanic. When engine oil ages it gets diluted and contaminated, which leads to premature engine wear.
2 Although some sources recommend oil filter changes every other oil change, we feel that the minimal cost of an oil filter and the relative ease with which it is installed dictate that a new filter be used whenever the oil is changed.
3 The tools necessary for a routine oil and filter change include a wrench to fit the drain plug at the bottom of the oil pan, an oil filter wrench to remove the old filter, a container with at least a 5 quart capacity to drain the old oil into and a funnel or oil can spout to help pour fresh oil into the engine.
4 In addition, you should have plenty of clean rags and newspapers handy to mop up any spills. Access to the underside of the vehicle is greatly improved if it can be lifted on a hoist, driven onto ramps or supported by jackstands. **Warning:** *Do not work under a vehicle which is supported only by a bumper, hydraulic or scissors-type jack.*
5 If this is your first oil change on the vehicle, crawl underneath and familiarize yourself with the locations of the oil drain plug and the oil filter. Since the engine and exhaust components will be hot during the actual work, it is a good idea to figure out any potential problems in advance.
6 Allow the engine to warm to normal operating temperature. If the new oil or any tools are needed, use the warm-up time to gather everything necessary for the job.
7 With the engine oil warm (warm engine oil will drain better and more built-up sludge will be removed with the oil), raise the vehicle and support it securely on jackstands.
8 Move all necessary tools, rags and newspapers under the vehicle.

Chapter 1 Tune-up and routine maintenance

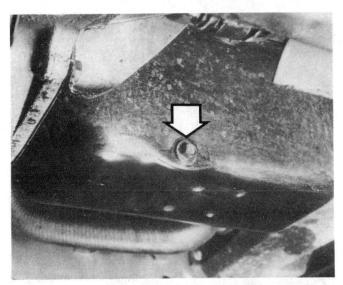

9.9 Use a six-point socket to remove the engine oil drain plug (it is usually very tight)

9.14 A strap-type wrench should be used to loosen the oil filter

9.19 Use your finger to apply a light coat of oil around the full circumference of the oil filter gasket

Position the drain pan under the drain plug. Keep in mind that the oil will initially flow from the engine with some force, so locate the pan accordingly.

9 Being careful not to touch any of the hot exhaust components, use the wrench to remove the drain plug at the bottom of the oil pan (see illustration). Depending on how hot the oil has become, you may want to wear gloves while unscrewing the plug the final few turns.

10 Allow the oil to drain into the pan. It may be necessary to move the pan further under the engine as the oil flow reduces to a trickle.

11 After all the oil has drained, clean the drain plug thoroughly with a rag. Small metal particles may cling to the plug and would immediately contaminate the new oil.

12 Clean the area around the drain plug opening and reinstall the plug.

13 Move the drain pan into position under the oil filter.

14 Use the filter wrench to loosen the oil filter (see illustration).

15 Sometimes the oil filter is on so tight it cannot be loosened, or it is positioned in an area which is inaccessible with a filter wrench. As a last resort, you can punch a metal bar or long screwdriver directly through the side of the canister and use it as a T-bar to turn the filter. If this must be done, be prepared for oil to spurt out of the canister as it is punctured.

16 Completely unscrew the old filter. Be careful, as it is full of oil. Empty the old oil inside the filter into the drain pan.

17 Compare the old filter with the new one to make sure they are identical.

18 Use a clean rag to remove all oil, dirt and sludge from the area where the oil filter mounts on the engine. Check the old filter to make sure the rubber gasket is not stuck to the engine mounting surface. If the gasket is stuck to the engine (use a flashlight if necessary to check), remove it.

19 Open one of the cans of new oil and apply a light coat of oil to the rubber gasket on the new oil filter (see illustration).

20 Attach the new filter to the engine following the tightening directions printed on the filter canister or packing box. Most filter manufacturers recommend against using a filter wrench due to possible over-tightening and damage to the sealing ring.

21 Remove all tools, rags, etc., from under the vehicle, being careful not to spill the oil in the drain pan. Lower the vehicle.

22 Move to the engine compartment and locate the oil filler cap on the engine.

23 If an oil spout if used, push the spout into the top of the oil can and pour the fresh oil through the filler opening. A funnel placed in the opening may also be used.

24 Pour the specified amount of fresh oil into the engine. Wait a few minutes to allow the oil to drain to the pan, then check the level on the oil dipstick (see Section 4 if necessary). If the oil level is at or above the lower Add mark, start the engine and allow the new oil to circulate.

25 Run the engine for only about a minute, then shut it off. Immediately look under the vehicle and check for leaks at the oil pan drain plug and around the oil filter. If either one is leaking, tighten with a bit more force.

26 With the new oil circulated and the filter now completely full, recheck the level on the dipstick and, if necessary, add enough oil to bring the level to the Full mark on the dipstick.

27 During the first few trips after an oil change, make it a point to check for leaks and keep a close watch on the oil level.

28 The old oil drained from the engine cannot be reused in its present state and should be disposed of. Oil reclamation centers, auto repair shops and gas stations will normally accept the oil, which can be refined and used again. After the oil has cooled, it can be drained into a suitable container (capped plastic jugs, topped bottles, milk cartons, etc.) for transport to one of these disposal sites.

10 Fuel filter replacement

Refer to illustrations 10.4 and 10.5

Warning: *Gasoline is extremely flammable, so extra precautions must be taken when working on any part of the fuel system. Do not smoke*

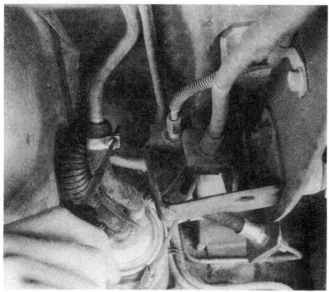

10.4 The new filter comes with the hose already mounted, so disconnect the old one at the fuel line after depressurizing the system

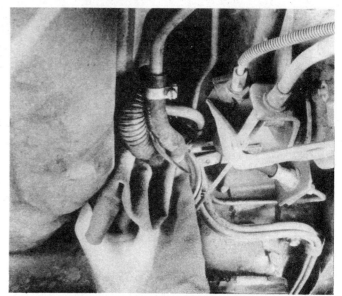

10.5 A cloth wrapped around the filter will keep the fuel from spurting out as the hose is detached

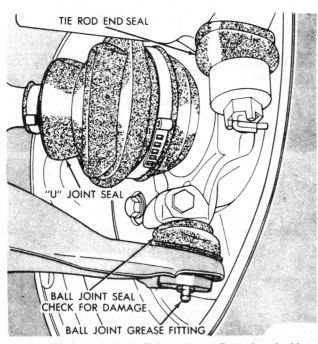

11.6a The suspension balljoints are usually equipped with grease fittings to ensure proper lubrication

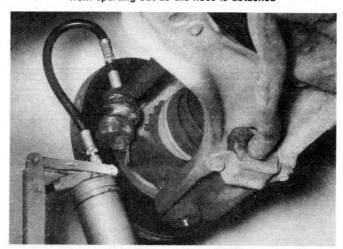

11.6b The steering system tie-rod end grease fittings are on the upper side and may be difficult to reach unless the grease gun is equipped with a flexible nozzle

or allow open flames or bare light bulbs near the vehicle. Also, do not work in a garage if a natural gas-type appliance with a pilot light is present.

1　The fuel filter is a disposable canister type and is located in the fuel line under the rear of the vehicle adjacent to the fuel tank.
2　**Warning:** *Because these models are equipped with Electronic Fuel Injection (EFI) which maintains pressure in the fuel system at all times, the system must be depressurized whenever it is worked on to avoid the possibility of spraying fuel. Refer to Chapter 4 for the depressurization procedure.*
3　Raise the rear of the vehicle and support it securely on jackstands.
4　Loosen the filter hose clamps (see illustration).
5　Wrap a cloth around the fuel filter to catch the residual fuel and disconnect the hoses (see illustration). It is a good idea to tie rags around your wrists to keep fuel from running down your arms.
6　Remove the bracket mounting bolt and lower the filter from the vehicle, holding your finger over the outlet to keep the residual fuel from running out.
7　Place the new filter in position and install the mounting bolt. Tighten the bolt securely.
8　Connect the filter hoses, using new clamps. Tighten the clamps securely.
9　Start the engine and check carefully for leaks at the hose connections.

11　Chassis lubrication

Refer to illustrations 11.6a and 11.6b

1　A grease gun and a cartridge filled with the proper grease (see *Recommended lubricants and fluids*) are usually the only items necessary to lubricate the chassis components. Occasionally, on later model vehicles, plugs will be installed rather than grease fittings. If so, grease fittings will have to be purchased and installed.
2　Look under the vehicle and see if grease fittings or plugs are installed. If there are plugs, remove them with a wrench and buy grease fittings which will thread into the component. A Chrysler dealer or auto parts store will be able to supply the correct fittings. Straight, as well as angled, fittings are available.
3　For easier access under the vehicle, raise it with a jack and place

Chapter 1 Tune-up and routine maintenance

12.5a Spark plug manufacturers recommend using a wire-type gauge when checking the gap — if the wire does not slide between the electrodes with a slight drag, adjustment is required

12.5b To change the gap, bend the *side* electrode only, as indicated by the arrows, and be very careful not to crack or chip the porcelain insulator surrounding the center electrode

jackstands under the frame. Make sure it is securely supported by the stands.
4 Before beginning, force a little grease out of the nozzle to remove any dirt from the end of the gun. Wipe the nozzle clean with a rag.
5 With the grease gun and plenty of clean rags, crawl under the vehicle and begin lubricating the components.
6 Wipe the balljoint grease fitting nipple clean and push the nozzle firmly over it. Squeeze the trigger on the grease gun to force grease into the component. The balljoints should be lubricated until the rubber seal is firm to the touch. Do not pump too much grease into the fittings as it could rupture the seal. For all other suspension and steering components, continue pumping grease into the fitting until it oozes out of the joint between the two components. If the escapes around the grease gun nozzle, the nipple is clogged or the nozzle is not completely seated on the fitting. Resecure the gun nozzle to the fitting and try again. If necessary, replace the fitting with a new one (see illustrations).
7 Wipe the excess grease from the components and the grease fitting. Repeat the procedure for the remaining fittings.
8 Lubricate the sliding contact and pivot points of the manual transaxle shift linkage with the specified grease. While you are under the vehicle, clean and lubricate the parking brake cable along with the cable guides and levers. This can be done by smearing some of the chassis grease onto the cable and its related parts with your fingers. Lubricate the clutch adjuster and cable, as well as the cable positioner, with a thin film of multi-purpose grease.
9 Lower the vehicle to the ground.
10 Open the hood and smear a little chassis grease on the hood latch mechanism. Have an assistant pull the inside hood release lever from inside the vehicle as you lubricate the cable at the latch.
11 Lubricate all the hinges (door, hood, etc.) with the recommended lubricant to keep them in proper working order.
12 The key lock cylinders can be lubricated with spray-on graphite, which is available at auto parts stores.
13 Lubricate the door weatherstripping with silicone spray. This will reduce chafing and retard wear.

12 Spark plug replacement

Refer to illustrations 12.5a and 12.5b

1 The spark plugs are located on the front side of the engine, facing the radiator grille. **Warning:** *Before beginning work, disconnect the negative battery cable to prevent the electric fan from coming on when working around the spark plugs.*
2 In most cases the tools necessary for spark plug replacement include a plug wrench or spark plug socket which fits onto a ratchet (this special socket will be insulated inside to protect the porcelain insulator) and a feeler gauge to check and adjust the spark plug gap. A special plug wire removal tool is available for separating the wire boot from the spark plug, but it is not absolutely necessary.
3 The best approach when replacing the spark plugs is to purchase the new spark plugs beforehand, adjust them to the proper gap and then replace each plug one at a time. When buying the new spark plugs it is important to obtain the correct plug for your specific engine. This information can be found on the *Emission Control Information label* located under the hood or in the owner's manual. If differences exist between the sources, purchase the spark plug type specified on the underhood label, as it was printed for your specific engine.
4 Allow the engine to cool completely before attempting to remove any of the plugs. During this cooling off time each of the new spark plugs can be inspected for defects and the gaps can be checked.
5 The gap is checked by inserting the proper thickness gauge between the electrodes at the tip of the plug (see illustration). The gap between the electrodes should be the same as that given in the Specifications or on the emissions label. The wire should touch each of the electrodes. If the gap is incorrect, use the notched adjuster on the thickness gauge body to bend the curved side electrode slightly until the proper gap is attained (see illustration). Check for cracks in the spark plug body. If any are found, the plug should not be used. If the side electrode is not exactly over the center one, use the notched adjuster to align the two.
6 Cover the front of the vehicle to prevent damage to the paint.
7 With the engine cool, remove the spark plug wire from one spark plug. Do this by pulling on the boot at the end of the wire, not the wire itself. Sometimes it is necessary to use a twisting motion while the boot and plug wire are pulled free.
8 If compressed air is available, use it to blow any dirt or foreign material away from the spark plug area. A common bicycle pump will also work. The idea is to eliminate the possibility of material falling into the cylinder after the spark plug is removed.
9 Place the spark plug wrench or socket over the plug and remove it from the engine by turning it in a counterclockwise direction.
10 Compare the spark plug to those shown in the accompanying photos to get an indication of the overall running condition of the engine.
11 Insert one of the new plugs into the hole, tightening it as much as possible by hand. The spark plug should thread easily into place. If it doesn't, change the angle of the spark plug slightly to match up the threads. **Caution:** *Be extremely careful, as these engines have an aluminum cylinder head, which means that the spark plug hole threads can be easily damaged.*
12 Attach the plug wire to the new spark plug, again using a twisting motion on the boot until it is firmly seated on the spark plug.
13 Follow the above procedures for the remaining spark plugs, replacing them one at a time to prevent mixing up the spark plug wires.

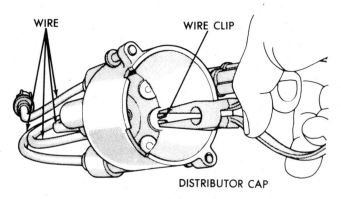

13.7 When replacing the spark plug wires, pliers must be used to compress the clips inside the distributor cap

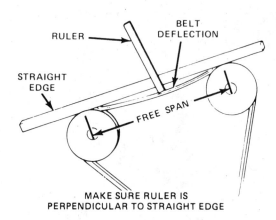

14.4 Checking drivebelt deflection with a straightedge and ruler

13 Spark plug wire, distributor cap and rotor check and replacement

Refer to illustration 13.7

1 The spark plug wires should be checked at the recommended intervals or whenever new spark plugs are installed.
2 The wires should be inspected one at a time to prevent mixing up the order, which is essential for proper engine operation.
3 Disconnect the plug wire from the spark plug. A removal tool can be used for this, or you can grab the rubber boot, twist slightly and then pull the wire free. *Do not pull on the wire itself, only on the rubber boot.*
4 Look inside the boot for corrosion, which will look like a white, crusty powder. Don't mistake the white dielectric grease used on some plug wire boots for corrosion.
5 Push the wire and boot back onto the end of the spark plug. It should be a tight fit on the plug end. If not, remove the wire and use pliers to carefully crimp the metal connector inside the wire boot until the fit is snug.
6 Clean each wire along its entire length. Remove all built-up dirt and grease. As this is done, inspect for burned areas, cracks and any other form of damage. Bend the wires in several places to ensure the conductive material inside has not hardened. Repeat the procedure for the remaining wires (don't forget the distributor cap-to-coil wire).
7 Remove the distributor cap splash shield and check the wires at the cap, making sure they are not loose and that the wires and boots are not cracked or damaged. **Caution:** *Do not attempt to pull the wires from the cap, as they are retained on the inside by wire clips. The manufacturer does not recommend removing the wires from the cap for inspection because this could damage the integrity of the boot seal. If the wires appear to be damaged, replace them with new ones. Remove the distributor cap (Chapter 5), release the wire clips with pliers and remove the wires (see illustration). Insert the new wires into the cap while squeezing the boots to release any trapped air as you push them into place. Continue pushing until you feel the wire electrodes snap into position.*
8 A visual check of the operation of the spark plug wires can also be made. In a darkened garage (make sure there is ventilation), start the engine and observe each plug wire. Be careful not to come into contact with any moving engine parts. If there is a break or fault in the wire you will be able to see arcing or a small spark at the damaged area.
9 Remove the distributor cap (see Chapter 5) with the wires attached and check the cap for cracks, carbon tracks and other damage. Examine the terminals inside the cap for corrosion. Slight corrosion can be removed with a pocket knife.
10 Check the rotor for cracks and a secure fit on the shaft. Make sure the terminals are not burned, corroded or pitted excessively. A small file can be used to restore the rotor terminals.
11 If new spark plug wires are needed, purchase a complete pre-cut set for your particular engine. The terminals and rubber boots should already be installed on the wires. Replace the wires one at a time to avoid mixing up the firing order and make sure the terminals are securely seated in the distributor cap and on the spark plugs.

14 Drivebelt check, adjustment and replacement

Refer to illustrations 14.4, 14.5a and 14.5b,

Warning: *The electric cooling fan on some models can activate at any time, even when the ignition switch is in the Off position. Disconnect the fan motor or negative battery cable when working in the vicinity of the fan.*

1 The drivebelts, or V-belts as they are sometimes called, at the front of the engine, play an important role in the overall operation of the vehicle and its components. Due to their function and material makeup, the belts are prone to failure after a period of time and should be inspected and adjusted periodically to prevent major damage.
2 The number of belts used on a particular engine depends on the accessories installed. Drivebelts are used to turn the alternator, air pump, power steering pump, water pump and air conditioning compressor. Depending on the pulley arrangement, a single belt may be used for more than one of these components.
3 With the engine off, open the hood and locate the various belts at the front of the engine. Using your fingers (and a flashlight if necessary), examine the belts, checking for cracks and separation of the plies. Also check for fraying and glazing, which gives the belt a shiny appearance. Both sides of each belt should be inspected, which means you will have to twist it to check the underside.
4 The tightness of each belt is checked by pushing on it at a distance halfway between the pulleys (see illustration). Apply about 10 pounds of force with your thumb and see how much the belt moves downward (deflects). Refer to the Specifications for the amount of deflection allowed in each belt.
5 If it is necessary to adjust the belt tension, to make it either tighter or looser, it is done by moving the belt driven accessory on the bracket (see illustrations).
6 For each component, there will be a locking bolt and a pivot bolt or nut. Both must be loosened slightly to enable you to move the component.
7 After the bolts have been loosened, move the component away from the engine to tighten the belt or toward the engine to loosen the belt. Many accessories are equipped with a square hole designed to accept a 3/8-inch or 1/2-inch drive breaker bar. The bar can be used to lever the component and tension the drivebelt. Hold the accessory in position and check the belt tension. If it is correct, tighten the two bolts until snug, then recheck the tension. If it is still correct, tighten the two bolts completely.
8 To adjust the alternator drivebelt, loosen the pivot nut and the locking screw or T-bolt locknut, then turn the adjusting bolt to tension the belt.
9 It will often be necessary to use some sort of pry bar to move the accessory while the belt is adjusted. If this must be done to gain the proper leverage, be very careful not to damage the component being moved, or the part being pried against.
10 Run the engine for about 15 minutes, then recheck the belt tension.

Chapter 1 Tune-up and routine maintenance

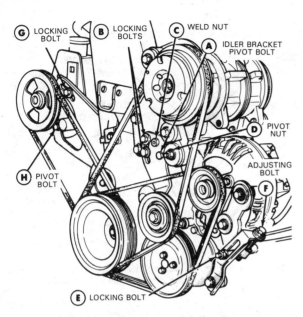

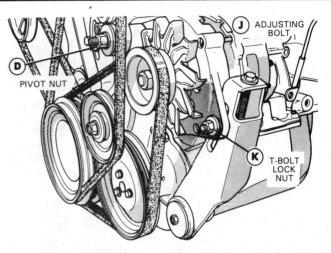

14.5a Typical engine drivebelt adjustment details, showing the 2.2/2.5L engine with the *Chrysler* alternator adjusting bracket

14.5b 2.2/2.5L engines equipped with the *Chrylser/Bosch* alternator have a slightly different adjusting bracket, although the adjusting procedure is essentially the same

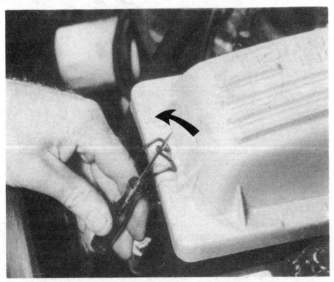

15.3a Use a small screwdriver to release the air cleaner housing cover clips

15.3b On turbocharged models, the filter element can be lifted out of the housing after completely removing the cover

15 Air filter element and PCV valve maintenance

Refer to illustrations 15.3a, 15.3b, 15.3c, 15.4, 15.11a and 15.11b

1 At the specified intervals, the air filter element and PCV valve should be replaced with new ones. In addition, the air filter and PCV valve should be inspected periodically.

2 The air filter element is located inside the housing, adjacent to the engine.

3 Loosen the clamps on the hose between the air cleaner housing and the throttle body. Release the air cleaner clips (see illustration), pull the hose off the throttle body and lift the cover and hose off the housing. On non-turbocharged models the cover can be rotated up without disconnecting the hose. Lift the air filter element out of the housing (see illustrations).

15.3c On non-turbocharged models, the cover can be rotated away from the filter after loosening the hose clamp (arrow)

15.4 The PCV filter must be replaced if it is dirty or clogged

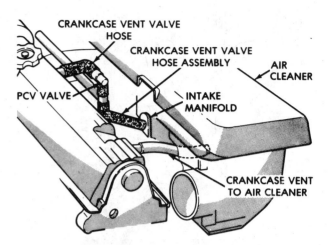

15.11a On non-turbocharged engines, the PCV system components are arranged as shown here

4   To check the filter, hold it up to sunlight or place a flashlight behind the element. If you can see light coming through the paper element, the filter is usable. Replace the PCV filter element if it is dirty (see illustration).
5   Clean the inside of the air cleaner housing with a rag.
6   Place the old filter (if in good condition) or the new filter (if the specified interval has elapsed) into the air cleaner housing with the screen facing up. Make sure it seats properly in the bottom of the housing.
7   If the hose was removed from the throttle body, loosely install the clamp and then push the hose onto the throttle body.
8   Align the upper and lower halves of the air cleaner housing and connect the clips.
9   Tighten the clamp around the hose at the throttle body securely.
10 The positive crankcase ventilation (PCV) valve should be replaced with a new one when the valve accumulates deposits which could cause it to stick. Refer to Chapter 6 for PCV valve checking procedures.
11 To replace the PCV valve, pull it from the hose and install a new one (see illustrations).
12 Inspect the PCV hose prior to installation to ensure that it isn't plugged or damaged. Compare the new PCV valve with the old one to make sure they are the same.
13 When replacing the PCV valve make sure the PCV filter is clean.

16 Underhood hose check and replacement

Warning: *Replacement of air conditioner hoses should be left to a dealer or air conditioning specialist who can relieve the pressure in the system and perform the work safely.*

1   The high temperatures present under the hood can cause deterioration of the numerous rubber and plastic hoses.
2   Periodic inspection should be made for cracks, loose clamps and leaks, since some of the hoses are part of the emissions system and can affect engine performance.
3   Remove the air cleaner if necessary and trace the entire length of each hose. Squeeze each hose to check for cracks and look for swelling, discoloration and leaks.
4   If the vehicle has a lot of miles or one or more of the hoses is deteriorated, it is a good idea to replace all of the hoses at one time.
5   Measure the length and inside diameter of each hose and obtain and cut the replacement to size. Original equipment hose clamps are often good for only one or two uses, so it is a good idea to replace them with screw-type clamps. The manufacturer recommends that any clamps involved with the fuel injection system be replaced with new ones designed for this specific purpose whenever they have been loosened or removed.

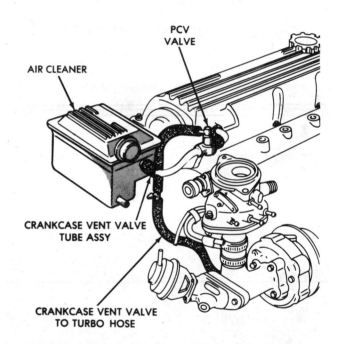

15.11b On turbocharged engines, the PCV system components are arranged like this

6   Replace each hose one at a time to eliminate the possibility of confusion. Hoses attached to the heater and choke contain coolant, so newspapers or rags should be kept handy to catch the spills when they are disconnected.
7   After installation, run the engine until it is up to operating temperature, shut it off and check for leaks. After the engine has cooled, retighten all of the screw-type clamps.

17 Throttle body mounting nut torque check

1   The Electronic Fuel Injection (EFI) throttle body is attached to the top of the intake manifold by four nuts. These fasteners can sometimes work loose during normal engine operation and cause a vacuum leak.
2   To properly tighten the throttle body mounting nuts, a torque wrench is necessary. If you do not own one, they can usually be rented on a daily basis.
3   Remove the air cleaner assembly-to-throttle body hose.

Chapter 1 Tune-up and routine maintenance

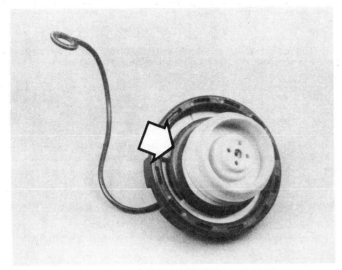

20.4 Check the fuel cap gasket (arrow) to make sure there is an even sealing imprint all the way around

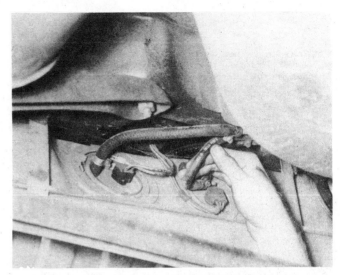

20.6 Move the fuel hoses back and forth to check for cracking

4   Locate the mounting nuts at the base of the throttle body. Decide what special tools or adaptors will be be necessary, if any, to tighten the nuts with a socket and the torque wrench.
5   Tighten the nuts to the specified torque. Do not overtighten the nuts, as the threads may strip.
6   If you suspect a vacuum leak exists at the bottom of the throttle body, obtain a short length of rubber hose. Start the engine and place one end of the hose next to your ear as you probe around the base of the throttle body with the other end. You will be able to hear a hissing sound if a leak exists.
7   If, after the nuts are properly tightened, a vacuum leak still exists, the throttle body must be removed and a new gasket installed. See Chapter 4 for more information.
8   After tightening the nuts, reinstall the air cleaner hose.

18 Combustion chamber conditioner application (Canadian models)

1   At the specified intervals combustion chamber conditioner (see *Recommended lubricants and fluids*) must be sprayed into the throttle body to help prevent the buildup of deposits in the combustion chamber and on the valves.
2   Remove the throttle body air inlet hose.
3   With the engine idling at normal operating temperature, the transaxle in Park (automatic) or Neutral (manual) with the parking brake applied, spray a can of the specified conditioner into the throttle body opening.
4   Attach the hose to the throttle body air inlet.

19 Heated inlet air system general check

1   Non-turbocharged models are equipped with a heated inlet air cleaner which draws air to the throttle body from different locations, depending upon engine temperature.
2   Remove the outside air duct and locate the vacuum flapper door in the air cleaner assembly. It will be located inside the "snorkel".
3   Check that the flexible heat duct is securely attached and not damaged.
4   The check should be done when the engine and outside air are cold. Start the engine and look through the snorkel at the flapper door, which should move to the Up or Heat On position. With the door up, air cannot enter through the end of the snorkel, but rather enters the air cleaner through the heat duct attached to the exhaust manifold.
5   As the engine warms to operating temperature, the door should move to the Down or Heat Off position to allow air through the snorkel end. Depending on ambient temperature, this may take 10 to 15 minutes. To speed up this check you can reconnect the outside air duct, drive the vehicle and then check that the door has moved down.
6   If the air cleaner is not operating properly, see Chapter 6 for more information.

20 Fuel system check

Refer to illustrations 20.4 and 20.6

Warning: *There are certain precautions to take when inspecting or servicing the fuel system components. Work in a well ventilated area and do not allow open flames (cigarettes, appliance pilot lights, etc.) near the vehicle. Mop up spills immediately and do not store fuel soaked rags where they could ignite.*

1   The fuel system on these models, which are equipped with Electronic Fuel Injection (EFI), is under pressure even when the engine is off. Consequently, the EFI system must be depressurized (Chapter 4) whenever the fuel system is worked on. Even after depressurization, if any fuel lines are disconnected for servicing, be prepared to catch the fuel as it spurts out. Plug all disconnected fuel lines immediately to prevent the tank from emptying itself.
2   The fuel system is most easily checked with the vehicle raised on a hoist where the components on the underside are readily visible and accessible.
3   If the smell of gasoline is noticed while driving, or after the vehicle has sat in the sun, the fuel system should be thoroughly inspected immediately.
4   Remove the gas tank cap and check for damage, corrosion and a proper sealing imprint on the gasket (see illustration). Replace the cap with a new one if necessary.
5   Inspect the gas tank and filler neck for punctures, cracks and other damage. The connection between the filler neck and the tank is especially critical. Sometimes a rubber filler neck will leak due to loose clamps or deteriorated rubber — problems a home mechanic can usually rectify. **Warning:** *Do not, under any circumstances, try to repair a fuel tank yourself except to replace rubber components. A welding torch or any open flame can easily cause the fuel vapors to explode if the proper precautions are not taken.*
6   Carefully check all rubber hoses and metal lines leading away from the fuel tank (see illustration). Check for loose connections, deteriorated hoses, crimped lines and damage of any kind. Follow the lines up to the front of the vehicle, carefully inspecting them all the way. Repair or replace damaged sections as necessary.

21 Cooling system check

Refer to illustration 21.4

Warning: *The electric cooling fan on some models can activate at any*

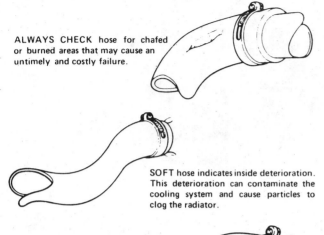

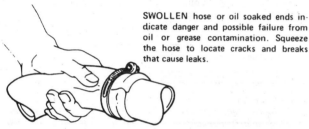

21.4 Radiator hose inspection details

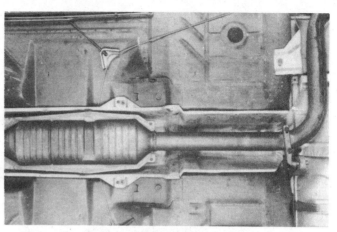

22.2a Check the entire length of the exhaust system for damaged pipes and . . .

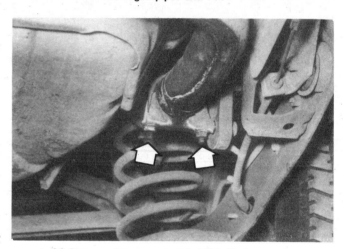

22.2b . . . loose or corroded U-bolt clamp nuts

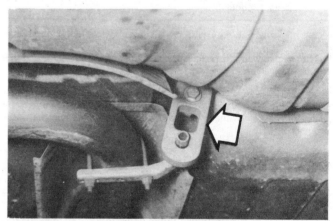

22.2c Check the rubber hangers to make sure they are secure and undamaged

time, even when the ignition switch is in the Off position. Disconnect the fan motor or the negative battery cable when working in the vicinity of the fan.

1   Many major engine failures can be attributed to a faulty cooling system. If the vehicle is equipped with an automatic transaxle, the cooling system is also used to cool the transaxle fluid.
2   The cooling system should be checked with the engine cold. Do this before the vehicle is driven for the day or after it has been shut off for two or three hours.
3   Remove the radiator cap and thoroughly clean the cap (inside and out) with clean water. Also clean the filler neck on the radiator. All traces of corrosion should be removed.
4   Carefully check the upper and lower radiator hoses along with the smaller diameter heater hoses. Inspect the entire length of each hose, replacing any that are cracked, swollen or show signs of deterioration. Cracks may become more apparent if the hose is squeezed (see illustration).
5   Also check that all hose connections are tight. A leak in the cooling system will usually show up as white or rust colored deposits on the areas adjoining the leak.
6   Use compressed air or a soft brush to remove bugs, leaves, etc. from the front of the radiator or air conditioning condenser. Be careful not to damage the delicate cooling fins, or cut yourself on them.
7   Finally, have the cap and system pressure tested. If you do not have a pressure tester, most gas stations and repair shops will do this for a minimal charge.

22 Exhaust system check

Refer to illustrations 22.2a, 22.2b and 22.2c

1   With the exhaust system cold (at least three hours after the vehicle has been driven), check the complete exhaust system from its starting point at the engine to the end of the tailpipe. This is best done on a hoist where full access is available.
2   Check the pipes and their connections for signs of leakage and/or corrosion indicating a potential failure. Make sure that all brackets and hangers are in good condition and tight (see illustrations).
3   Inspect the underside of the body for holes, corrosion, open seams, etc. which may allow exhaust gases to enter the passenger compartment. Seal all body openings with silicone or body putty.
4   Rattles and other noises can often be traced to the exhaust system, especially the mounts and hangers. Try to move the pipes, muffler and catalytic converter. If the components can come into contact with the body, secure the exhaust system with new mounts.

Common spark plug conditions

NORMAL
Symptoms: Brown to grayish-tan color and slight electrode wear. Correct heat range for engine and operating conditions.
Recommendation: When new spark plugs are installed, replace with plugs of the same heat range.

WORN
Symptoms: Rounded electrodes with a small amount of deposits on the firing end. Normal color. Causes hard starting in damp or cold weather and poor fuel economy.
Recommendation: Plugs have been left in the engine too long. Replace with new plugs of the same heat range. Follow the recommended maintenance schedule.

CARBON DEPOSITS
Symptoms: Dry sooty deposits indicate a rich mixture or weak ignition. Causes misfiring, hard starting and hesitation.
Recommendation: Make sure the plug has the correct heat range. Check for a clogged air filter or problem in the fuel system or engine management system. Also check for ignition system problems.

ASH DEPOSITS
Symptoms: Light brown deposits encrusted on the side or center electrodes or both. Derived from oil and/or fuel additives. Excessive amounts may mask the spark, causing misfiring and hesitation during acceleration.
Recommendation: If excessive deposits accumulate over a short time or low mileage, install new valve guide seals to prevent seepage of oil into the combustion chambers. Also try changing gasoline brands.

OIL DEPOSITS
Symptoms: Oily coating caused by poor oil control. Oil is leaking past worn valve guides or piston rings into the combustion chamber. Causes hard starting, misfiring and hesitation.
Recommendation: Correct the mechanical condition with necessary repairs and install new plugs.

GAP BRIDGING
Symptoms: Combustion deposits lodge between the electrodes. Heavy deposits accumulate and bridge the electrode gap. The plug ceases to fire, resulting in a dead cylinder.
Recommendation: Locate the faulty plug and remove the deposits from between the electrodes.

TOO HOT
Symptoms: Blistered, white insulator, eroded electrode and absence of deposits. Results in shortened plug life.
Recommendation: Check for the correct plug heat range, over-advanced ignition timing, lean fuel mixture, intake manifold vacuum leaks, sticking valves and insufficient engine cooling.

PREIGNITION
Symptoms: Melted electrodes. Insulators are white, but may be dirty due to misfiring or flying debris in the combustion chamber. Can lead to engine damage.
Recommendation: Check for the correct plug heat range, over-advanced ignition timing, lean fuel mixture, insufficient engine cooling and lack of lubrication.

HIGH SPEED GLAZING
Symptoms: Insulator has yellowish, glazed appearance. Indicates that combustion chamber temperatures have risen suddenly during hard acceleration. Normal deposits melt to form a conductive coating. Causes misfiring at high speeds.
Recommendation: Install new plugs. Consider using a colder plug if driving habits warrant.

DETONATION
Symptoms: Insulators may be cracked or chipped. Improper gap setting techniques can also result in a fractured insulator tip. Can lead to piston damage.
Recommendation: Make sure the fuel anti-knock values meet engine requirements. Use care when setting the gaps on new plugs. Avoid lugging the engine.

MECHANICAL DAMAGE
Symptoms: May be caused by a foreign object in the combustion chamber or the piston striking an incorrect reach (too long) plug. Causes a dead cylinder and could result in piston damage.
Recommendation: Repair the mechanical damage. Remove the foreign object from the engine and/or install the correct reach plug.

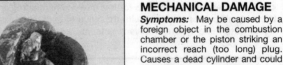

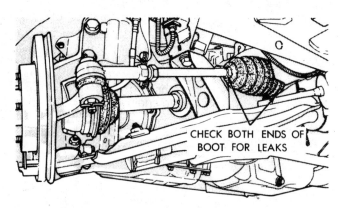

24.6 If the steering gear boots are leaking they must be replaced with new ones

5   Check the running condition of the engine by inspecting the end of the tailpipe. The exhaust deposits here are an indication of engine state-of-tune. If the pipe is black and sooty or coated with white deposits, the engine may be in need of a tune-up, including a thorough EFI system inspection and adjustment.

23 Clutch pedal free play check

There is no need for checking clutch pedal free play on these models because the clutch release system incorporates a self-adjuster. Excessive clutch pedal effort, failure of the clutch to disengage or noise from the adjuster indicates that a problem exists. Refer to Chapter 8 for further information on the clutch, adjuster and linkage.

24 Suspension and steering check

Refer to illustration 24.6

1   Whenever the front of the vehicle is raised for service visually check the suspension and steering components for wear.
2   Indications of a fault in these systems are excessive play in the steering wheel before the front wheels react, excessive sway around corners, body movement over rough roads or binding at some point as the steering wheel is turned.
3   Before the vehicle is raised for inspection, test the shock absorbers by pushing down to rock the vehicle at each corner. If it does not come back to a level position within one or two bounces, the shocks are worn and need to be replaced. As this is done, check for squeaks and strange noises from the suspension components. Information on shock absorber and suspension components can be found in Chapter 10.
4   Raise the front of the vehicle and support it securely with jackstands. Because of the work to be done, the vehicle must be stable.
5   Check the front wheel hub nut for looseness and make sure that it is properly locked in place.
6   Crawl under the vehicle and check for loose bolts, broken or disconnected parts and deteriorated rubber bushings on all suspension and steering components. Look for grease or fluid leaking from around the steering gear boots (see illustration). Check the power steering hoses and connections for leaks. Check the steering joints for wear.
7   Have an assistant turn the steering wheel from side-to-side and check the steering components for free movement, chafing and binding. If the steering does not react with the movement of the steering wheel, try to determine where the slack is located.

25 Steering shaft seal lubrication

1   The steering shaft seal protects the steering shaft at the point where it passes through the firewall. Lubricate the inner circumference of the seal with the specified lubricant if the shaft makes noise or sticks to the seal when it is turned.

27.5 The brake pads can be checked after removing the wheels by looking through the inspection hole at the edges of the pads (arrows)

2   Raise the vehicle and support it securely.
3   Peel back the upper edge of the seal and apply a light coat of grease all the way around the inner circumference where it contacts the steering shaft.
4   Lower the vehicle.

26 Wheel bearing check, adjustment and repacking

1   The front wheel bearings are adjusted and lubricated at the factory and normally only need to be checked for looseness, indicating bearing wear or an improperly tightened hub nut. Refer to Chapter 10 for checking and maintenance procedures for the front wheel bearings.
2   The rear wheel bearings should be checked at the specified intervals, or whenever the rear brake shoes are inspected. Adjustment, removal and installation and repacking procedures for the rear wheel bearings are also described in Chapter 10.

27 Brake checks

Refer to illustrations 27.5, 27.7 and 27.16

1   The brakes should be inspected every time the wheels are removed or whenever a defect is suspected. Indications of a potential brake system problem include the vehicle pulling to one side when the brake is applied, noises coming from the brakes, excessive brake pedal travel, pulsating pedal and leakage of fluid, usually seen on the inside of the tires or wheels.

Disc brakes

2   Disc brakes can be checked without removing any parts except the wheels.
3   Raise the vehicle and place it securely on jackstands. Remove the front wheels (see *Jacking and towing* at the front of this manual if necessary).
4   Now visible is the disc brake caliper which contains the pads. There is an outer brake pad and an inner pad. Both should be checked for wear.
5   Note the pad thickness by looking through the inspection hole in the caliper body (see illustration). If the combined thickness of the pad lining and metal shoe is 5/16-inch or less, the pads should be replaced.
6   Since it will be difficult, if not impossible, to measure the exact thickness of the pad, if you are in doubt as to the pad quality, remove them for further inspection or replacement. See Chapter 9 for disc brake pad replacement.
7   Before installing the wheels, check for leakage around the brake hose connections leading to the caliper and for damage (cracking, split-

Chapter 1 Tune-up and routine maintenance

27.7 Check the front brake hoses and connections for damage and leaks

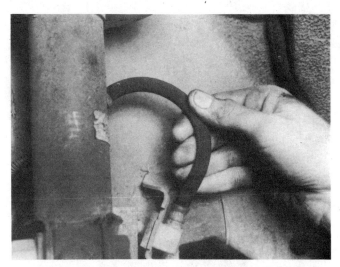

27.16 Check the rear brake hoses and connections for leaks and damage

ting etc.) to the brake hose (see illustration). Replace the hose or fittings as necessary, referring to Chapter 9.
8   Check the disc for scoring, wear and burned spots. If these conditions exist, the hub/rotor assembly should be removed for servicing (Chapter 9).

Drum brakes (rear)

9   Raise the vehicle and support it securely on jackstands. Note that the manufacturer *does not* indicate a lifting point for the rear of the vehicle with a floor jack in the *Jacking* diagram. Block the front tires to prevent the vehicle from rolling. Do not apply the parking brake as this will lock the drums in place.
10 Remove the wheels, referring to *Jacking and towing* at the front of this manual if necessary.
11 Mark the hub so it can be reinstalled in the same position. Use a scribe, chalk, etc. on the drum and hub.
12 Remove the brake drum as described in Chapter 10.
13 With the drum removed, carefully brush away any accumulations of dirt and dust. **Warning:** *Do not blow the dust out with compressed air. Make an effort not to inhale the dust as it contains asbestos and is harmful to your health.*
14 Note the thickness of the lining material on both front and rear brake shoes. If the material has worn to within 1/8-inch of the recessed rivets or metal backing, the shoes should be replaced. If the linings look worn, but you are unable to determine their exact thickness, compare them with a new set at the auto parts store. The shoes should also be replaced if they are cracked, glazed (shiny surface), or wet with brake fluid.
15 Check that all the brake assembly springs are connected and in good condition.
16 Check the brake components for any signs of fluid leakage. With your finger, carefully pry back the rubber cups on the wheel cylinder located at the top of the brake shoes. Any leakage here is an indication that the wheel cylinders should be overhauled immediately (Chapter 9). Check all hoses and connections for signs of leakage (see illustration).
17 Wipe the inside of the drum with a clean rag and denatured alcohol. Again, be careful not to breathe the dangerous asbestos dust.
18 Check the inside of the drum for cracks, scoring, deep scratches and hard spots, which will appear as small discolored areas. If these imperfections cannot be removed with fine emery cloth, the drum must be taken to a machine shop for resurfacing.
19 If, after the inspection process, all parts are in good working condition, reinstall the brake drum. Install the wheel and lower the vehicle to the ground.

Parking brake

20 The easiest way to check the operation of the parking brake is to park the vehicle on a steep hill with the parking brake set and the transmission in Neutral. If the parking brake cannot prevent the vehicle from rolling, it is in need of adjustment (see Chapter 9).

28 Cooling system servicing (draining, flushing and refilling)

Refer to illustration 28.6
Warning: *Because antifreeze is highly toxic, the radiator should always be drained into a container. Never allow the coolant to run onto the ground or driveway where a pet could drink it and be poisoned. The container should be capped and stored until it can be properly disposed of.*

1   The cooling system should be periodically drained, flushed and refilled to replenish the antifreeze mixture and prevent rust and corrosion, which can impair the performance of the cooling system and ultimately cause engine damage.
2   At the same time the cooling system is serviced, all hoses and the radiator cap should be inspected and replaced if faulty (see Section 21).
3   Antifreeze is a poisonous solution, so be careful not to spill any of it on your skin. If this happens, rinse immediately with plenty of clean water. Also, it is advisable to consult your local authorities about the dumping of antifreeze before draining the cooling system. In many areas reclamation centers have been set up to collect automobile oil and coolant mixtures rather than allowing these liquids to be added to the sewage system.
4   With the engine cold, remove the radiator cap and set the heater control to Heat (Max).
5   Move a large container under the radiator to catch the coolant mixture as it is drained.
6   Drain the radiator. Most models are equipped with a drain fitting (see illustration) at the bottom of the radiator. If the fitting has excessive

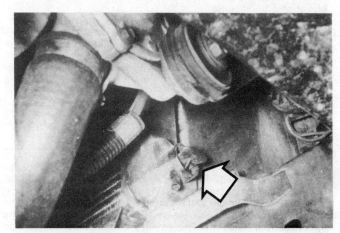

28.6 The radiator can be drained by opening the fitting (arrow) at the bottom

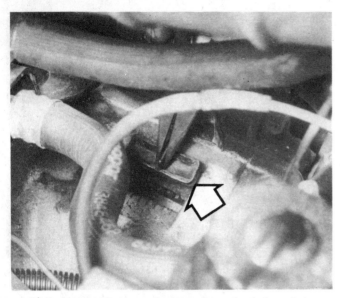

29.3 Use needle nose pliers to pull out the timing window cover (arrow)

29.5 Marking the timing notch (arrow) makes it easier to see during the timing adjustment procedure

corrosion and cannot be turned easily, or the radiator is not equipped with one, detach the lower radiator hose to allow the coolant to drain. Be careful that none of the solution is splashed on your skin or in your eyes. Remove the vacuum switch or plug from the top of the thermostat housing on the engine.

7 Disconnect the coolant reservoir hose, remove the reservoir and flush it with clean water.
8 Place a hose (a common garden hose is fine) in the radiator filler neck at the top of the radiator and flush the system until the water runs clear at all drain points.
9 In severe cases of contamination or clogging of the radiator, remove it (see Chapter 3) and reverse flush it. This involves inserting the hose in the bottom radiator outlet to allow the clean water to run against the normal flow, draining through the top. A radiator repair shop should be consulted if further cleaning or repair is necessary.
10 Where the coolant is regularly drained and the system refilled with the correct antifreeze mixture there should be no need to employ chemical cleaners or descalers.
11 Install the coolant reservoir, reconnect the hoses and close the drain fitting.
12 Add coolant to the radiator until it reaches the bottom of the threaded hole in the thermostat housing. Reinstall the vacuum switch or plug in the hole and tighten it to 15 ft-lbs. Continue adding coolant to the radiator until it reaches the radiator cap seat.
13 Add coolant to the reservoir until the level is between the *Min* and *Max* marks.
14 Run the engine until normal operating temperature is reached and, with the engine idling, add coolant up to the correct level.
15 Always refill the system with a mixture of antifreeze and water in the proportion called for on the antifreeze container or in your owner's manual. Chapter 3 also contains information on antifreeze mixtures.
16 Keep a close watch on the coolant level and the various cooling system hoses during the first few miles of driving. Tighten the hose clamps and add more coolant mixture as necessary.

29 Ignition timing check and adjustment

Refer to illustrations 29.3, 29.5, 29.8 and 29.10

1 All vehicles are equipped with an *Emissions Control Information* label inside the engine compartment. The label contains important ignition timing specifications and the proper procedures for your specific vehicle. If any information on the emissions label is different from the information provided in this Section, follow the procedure(s) given on the label.
2 Connect a timing light in accordance with the manufacturer's instructions. Usually, the light must be connected to the battery and the number 1 spark plug wire. The number 1 spark plug wire or terminal should be marked at the distributor. Trace it back to the spark plug and attach the timing light lead near the plug.
3 Remove the cover and locate the numbered timing tag in the timing window in the transaxle bellhousing (see illustration).
4 Locate the notched groove across the flywheel. It may be necessary to have an assistant temporarily turn the ignition on and off in short bursts without starting the engine in order to bring the groove into a position where it can easily be cleaned and marked. **Warning:** *Stay clear of all moving engine components when the engine is turned over in this manner.*
5 Use white soapstone, chalk or paint to mark the groove in the flywheel (see illustration). Also put a mark on the timing mark corresponding to the number of degrees specified on the *Emission Control Information label* in the engine compartment.
6 Connect a tachometer to the engine, setting the selector to the correct cylinder position.
7 Make sure that the wiring for the timing light is clear of all moving engine components, then start the engine and run it until normal operating temperature is reached.
8 Disconnect and then reconnect the water temperature sensor (located on the thermostat housing) connector by peeling back the boot with one hand while pulling the connector back with the other (see illustration). The Power Loss lamp (Chapter 5) should light up and stay on and the engine speed should now be as specified on the *Emissions Control Information label*.
9 Aim the timing light at the marks in the hole in the bellhousing, again being careful not to come into contact with moving parts. The marks you made should appear stationary. If the marks are in alignment, the timing is correct. If the marks are not aligned, turn off the engine.
10 Loosen the hold-down bolt or nut at the base of the distributor (see illustration). Loosen the bolt/nut only slightly, just enough to turn the distributor (see Chapter 5).
11 Restart the engine and turn the distributor until the timing marks are aligned.
12 Shut off the engine and tighten the distributor bolt/nut, being careful not to move the distributor.
13 Start the engine and recheck the timing to make sure the marks are still in alignment.
14 Turn the engine off, then disconnect and reconnect the positive battery cable quick disconnect. Start the vehicle and verify that the Power Loss lamp is off.
15 Shut the engine off and then turn the ignition On, Off, On, Off, On. The fault codes should clear with 88-51-55 (1984 and 1985 models) or 88-12-55 (1986 models) displayed.

Chapter 1 Tune-up and routine maintenance

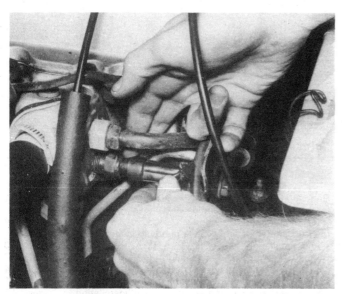

29.8 Peel back the water temperature sensor boot while pulling back on the connector to unplug it

29.10 Loosen the bolt and turn the distributor if the timing is not as specified

16 To keep "pinging" at a minimum, yet still allow you to operate the vehicle at the specified timing setting, it is advisable to use gasoline of the same octane at all times. Switching fuel brands and octane levels can decrease performance and economy, and may possibly damage the engine.

30 Compression check

1 A compression check will tell you a lot about the mechanical condition of your engine. For instance, it can tell you if compression is low because of leakage caused by worn piston rings, defective valves and seats or a blown head gasket.
2 Warm the engine to normal operating temperature, shut it off and allow it to sit for ten minutes to allow the catalytic converter temperature to drop.
3 Begin by cleaning the area around the spark plugs before you remove them. This will prevent dirt from falling into the cylinders while you are checking compression.
4 Remove the coil high tension lead from the distributor and ground it on the engine block. Depress the accelerator pedal all the way to the floor.
5 With the compression gauge in the number one cylinder spark plug hole, crank the engine over at least four compression strokes and observe the gauge. Compression should build up quickly in a healthy engine. Low compression on the first stroke, followed by gradually increasing pressure on successive strokes, indicates worn piston rings. A low compression reading on the first stroke, which does not build up during successive strokes, indicates leaking valves or a defective head gasket. Record the highest gauge reading that you obtained.
6 Repeat this procedure for the remaining cylinders and compare the results to the Specifications. Compression readings 10% above or below the specified amount can be considered normal.
7 Pour a couple of teaspoons of engine oil (a squirt can works well) into each cylinder through the spark plug hole and repeat the test.
8 If the compression increases after oil is added, the piston rings are worn. If the compression does not increase significantly, the leakage is occurring at the valves or head gasket.
9 If two adjacent cylinders have equally low compression, there is a strong possibility that the head gasket between them is blown. The appearance of coolant in the combustion chamber or the crankcase will verify this condition.
10 If the compression is higher than normal, the combustion chambers are probably coated with carbon deposits. If that is the case, the cylinder head should be removed and the carbon deposits removed.
11 If compression is very low, or varies greatly between cylinders, it is a good idea to have a leak-down test performed by an automotive repair shop. This test will pinpoint exactly where the leakage is occurring and how severe it is.

31 Manual transaxle fluid change

1 Under normal operating conditions the lubricant installed at the factory should last the life of the vehicle. If the vehicle is operated under severe conditions, or should some other cause require changing the lubricant, the front of the vehicle should be raised and supported securely on jackstands.
2 Remove the pan cover on the side of the differential and allow the fluid to drain into a container.
3 Attached to the inside of the pan cover is a small magnet, installed to trap metal particles before they can damage the transaxle bearings. Clean the magnet and the inside surface of the pan cover thoroughly.
4 Use RTV sealant to form a new gasket for the pan cover and install the cover on the differential. Tighten the bolts evenly and securely in a criss-cross pattern.
5 Fill the transaxle with Dexron II until the level is at the bottom edge of the filler plug. Drive the vehicle and check the pan cover seal for leakage.

Chapter 2 Part A 2.2L and 2.5L engines

Refer to Chapter 13 for Specifications on 1987 and later models

Contents

Automatic transaxle driveplate — removal and installation	3
Balance shaft assembly — removal and installation	9
Clutch and flywheel — removal and installation	4
Cylinder head and camshaft — installation	15
Cylinder head and camshaft — removal	8
Engine — installation	17
Engine — removal	2
External engine components — installation	16
External engine components — removal	5
Front oil seal and housing — removal and installation	12
General information	1
Intake and exhaust manifolds — removal, inspection and installation	See Chapter 4
Intermediate shaft and seal — removal, inspection and installation	10
Oil pan — removal and installation	7
Oil pump — reassembly and installation	14
Oil pump — removal, disassembly and inspection	11
Rear oil seal and housing — removal and installation	13
Timing belt and sprockets — removal, inspection and installation	6

Specifications

Camshaft
End play
  standard .... 0.005 to 0.013 in (0.13 to 0.33 mm)
  service limit .... 0.020 in (0.50 mm)

Oil Pump
Outer rotor-to-housing bore clearance
  standard .... 0.010 in (0.25 mm)
  service limit .... 0.014 in (0.35 mm)
Outer rotor thickness
  1984 and 1985
    standard .... 0.826 to 0.827 in (20.98 to 21.00 mm)
    service limit .... 0.825 in (20.96 mm)
  1986
    standard .... 0.944 to 0.946 in (23.97 to 24.00 mm)
    service limit .... 0.944 in (23.97 mm)
Inner rotor-to-outer rotor tip clearance
  1984 and 1985 .... 0.010 in (0.25 mm)
  1986
    standard .... 0.008 in (0.20 mm)
    service limit .... 0.010 in (0.25 mm)
Rotor-to-housing clearance
  standard .... 0.001 to 0.003 in (0.03 to 0.08 mm)
  service limit .... 0.004 in (0.10 mm)
Pump cover warpage
  standard .... 0.010 in (0.25 mm)
  service limit .... 0.015 in (0.38 mm)
Relief spring free length .... 1.95 in (49.5 mm)
Relief spring load .... 15 to 25 lbs @ 1.34 in (67 to 89 N @ 34 mm)

Firing order 1-3-4-2

Cylinder location and distributor rotation

Torque specifications

	Ft-lbs	Nm
Balance shaft carrier		
front chain cover bolt	9	12
chain tensioner adjustment bolt	9	12
chain tensioner pivot bolt	9	12
chain snubber stud	9	12
chain snubber nut	9	12
gear cover bolt	9	12
gear cover double-ended stud	9	12
gear and sprocket-to-balance shaft bolt	21	28
sprocket-to-crankshaft Torx drive bolt	11.1	15
rear cover bolt	9	12
carrier-to-cylinder block bolt	40	54

Torque specifications (continued)

	Ft-lbs	Nm
Cylinder head bolts (1984 and 1985)		
step 1	30	41
step 2	45	61
step 3	45	61
step 4	1/4-additional turn after reaching torque specified in step 3	
Cylinder head bolts (1986)		
step 1	45	61
step 2	65	88
step 3	65	88
step 4	1/4-additional turn after reaching torque specified in step 3	
Camshaft cover bolts	9	12
Camshaft sprocket bolt	65	88
Camshaft bearing cap nut	14	19
Crankshaft sprocket bolt	50	68
Main bearing cap bolt	30*	41*
Connecting rod bearing cap nut	40*	54*
Front crankshaft oil seal housing bolt	9	12
Rear crankshaft oil seal housing bolt	9	12
Intermediate shaft oil seal retainer bolt	9	12
Intermediate shaft sprocket bolt	65	88
Upper timimg belt cover bolt	3.5	4
Lower timing belt cover bolt	3.5	4
Exhaust manifold nut	17	23
Intake manifold bolt	17	23
Turbocharger-to-exhaust manifold nut	40	54
Turbocharger heat shield screw	9	12
Turbocharger fuel injection fuel rail screw	21	28
Turbocharger coolant tube nut	30	41
Turbocharger oil line tube nut	11	15
Thermostat housing bolt	21	28
Water pump housing bolt		
upper	21	28
lower	40	54
Oil pan bolt		
6 mm	9	12
8 mm	17	23
Oil pump mounting bolt	17	23
Oil pump cover bolt	9	12
Oil pump brace mounting bolt	9	12
Right engine bracket through-bolt	75	102
Right engine bracket insulator nut	75	102
Right engine bracket bolt	21	28
Transaxle-to-engine bolts	70	95
Turbocharger damper nut	16	22
Turbocharger damper mount bolt and nut	21	28
Left engine mount insulator bolt	40	54
Left engine mount through bolt	50	68
Anti-roll strut through-bolt	40	54
Anti-roll strut mount nut	21	28
Clutch pressure plate-to-flywheel bolts	21	28
Flywheel or driveplate-to-crankshaft bolts		
1984 and 1985	65	88
1986	70	95
Torque converter-to-driveplate bolts		
1984 and 1985	40	54
1986	55	74

Plus an additional 1/4-turn

1 General information

The 2.2 and 2.5 liter engines used in these models are virtually identical in design except that the 2.5 liter engine features a longer stroke (for increased displacement) and a chain driven, counter rotating balance shaft device mounted below the crankshaft (to reduce vibration). The majority of the 2.5 liter engine operations are the same as for the 2.2 liter engine except those affected by the balance shaft assembly.

The 2.2L/2.5L engine is an inline four with a belt-driven overhead camshaft. The belt also turns an intermediate shaft, mounted low in the block, which drives the oil pump and distributor.

The aluminum cylinder head has removable valve guides and valve seat inserts. The intake and exhaust ports are located on the rear (firewall) side with the spark plugs on the front (radiator) side of the cylinder head.

The crankshaft rides in five replaceable insert-type bearings. No vibration damper is used and a sintered iron timing belt sprocket is mounted on the front of the crankshaft.

The balance shaft and carrier assembly is bolted to the crankcase below the crankshaft and contains two counter-rotating shafts. The shafts are geared together and are turned at twice crankshaft speed by a crankshaft-driven roller chain.

The pistons have two compression rings and one oil control ring. The piston pins are semi-floating and press fit into the small end of the connecting rod. The big ends of the connecting rods are equipped with insert-type bearings.

The engine is liquid cooled. Coolant is circulated around the cylinders and combustion chambers and through the intake manifold by a centrifugal impeller-type pump which is driven by a belt from the crankshaft.

Lubrication is handled by an eccentric rotor-type oil pump mounted in the oil pan and driven by the intermediate shaft.

2 Engine — removal

Refer to illustrations 2.3, 2.6, 2.8, 2.12 and 2.17
Warning: *Before beginning any service procedure under the hood, always disconnect the negative cable at the battery. Place the cable out of the way so it cannot accidentally come in contact with the negative terminal of the battery, as this would once again allow power into the electrical system of the vehicle.*

1 Remove the hood as described in Chapter 11.
2 Drain the cooling system (Chapter 1).
3 Remove the radiator hoses, followed by the radiator and fan assembly (see illustration).
4 Remove the air cleaner assembly and hoses.
5 Detach the throttle cable from the linkage and bracket.
6 **Warning:** *Be sure to relieve the fuel system pressure as described in Chapter 4 before disconnecting any fuel lines or hoses.* Disconnect and plug the fuel and heater hoses at the engine (see illustration). Remove the oil filter.
7 Unbolt the air conditioning compressor and set it aside, but do not disconnect the hoses. **Caution:** *On vehicles equipped with an air conditioner, the system must be evacuated by a dealer service department or air conditioning repair shop before any refrigerant lines are disconnected. Do not attempt to do this yourself as injury could result.*
8 Disconnect all electrical wires from the engine, fuel injection and alternator, then remove the alternator. Be sure to tag all wires to ensure correct reinstallation (see illustration).
9 Remove the power steering pump with the hoses attached and set it aside.
10 Disconnect the heater hoses at the firewall fittings.
11 Disconnect the exhaust pipe at the manifold and remove the lower transaxle case cover.
12 On vehicles with a manual transaxle, disconnect the clutch cable (see illustration).
13 On vehicles with an automatic transaxle, mark the driveplate and torque converter so they can be mated correctly during installation, then remove the bolts. Attach a large C-clamp to the front edge of the torque converter housing to prevent the torque converter from falling out.
14 Support the transaxle with a jack or blocks, then remove the starter

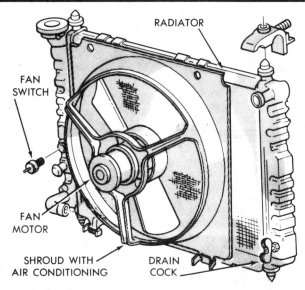

2.3 After detaching all cooling hoses, remove the radiator and fan assembly

2.6 Be sure to plug the fuel and vapor lines (arrows) as you remove them

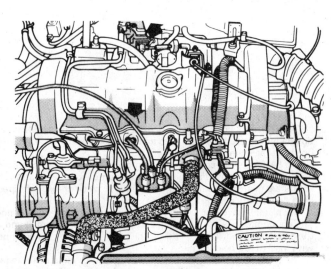

2.8 Alternator, carburetor and engine electrical connections (arrows) must be tagged for correct reassembly

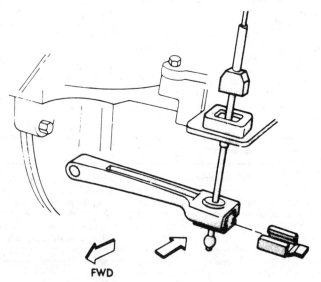

2.12 Remove the retainer to detach the clutch cable from the release arm

Chapter 2 Part A 2.2L and 2.5L engines

motor (Chapter 5) and the right inner splash shield.
15 Attach a chain to the engine lifting hooks and support the engine with a hoist.
16 Remove the engine ground strap between the cylinder head and the firewall.
17 Remove the through-bolt and nut and the insulator nut from the engine mount yoke at the timing belt end of the engine (see illustration). **Caution:** *Do not remove the insulator bolts from the body unless the insulator is marked to ensure installation in the exact same position.*
18 Remove the transaxle-to-engine bolts and the front engine mount through-bolt and nut.
19 If you have a manual transaxle model, remove the anti-roll strut or damper (see illustration 2.17).
20 Unbolt and remove the left side engine and transaxle mount (see illustration 2.17).
21 Lift the engine slowly and carefully up and out of the engine compartment.

3 Automatic transaxle driveplate — removal and installation

1 Remove the bolts and separate the driveplate from the crankshaft.
2 To install, hold the driveplate in position and thread the mounting bolts into place.
3 While locking the crankshaft so it won't turn, tighten the bolts (following a criss-cross pattern) to the specified torque.

4 Clutch and flywheel — removal and installation

1 Remove the clutch cover and clutch disc assembly (Chapter 8).
2 Remove the bolts and separate the flywheel from the crankshaft.
3 Hold the flywheel in position and install the bolts in the end of the crankshaft.
4 While holding the flywheel so that it doesn't turn, tighten the bolts (using a criss-cross pattern) to the specified torque.
5 Install the clutch disc and clutch cover assembly (Chapter 8).

5 External engine components — removal

Refer to illustrations 5.15 and 5.16

Note: *When removing the external components from the engine, pay close attention to details that may be helpful or important during installation. Note the locations of gaskets, seals, spacers, pins, washers, bolts and other small parts.*

1 It is much easier to dismantle and repair the engine if it is mounted on an engine stand. These stands can often be rented, for a reasonable fee, from an equipment rental yard.
2 If a stand isn't available you can dismantle the engine while it is blocked up on a sturdy workbench or on the floor. Be extra careful not to tip or drop the engine when working without a stand.
3 Remove the oxygen sensor from the exhaust manifold (Chapter 6).

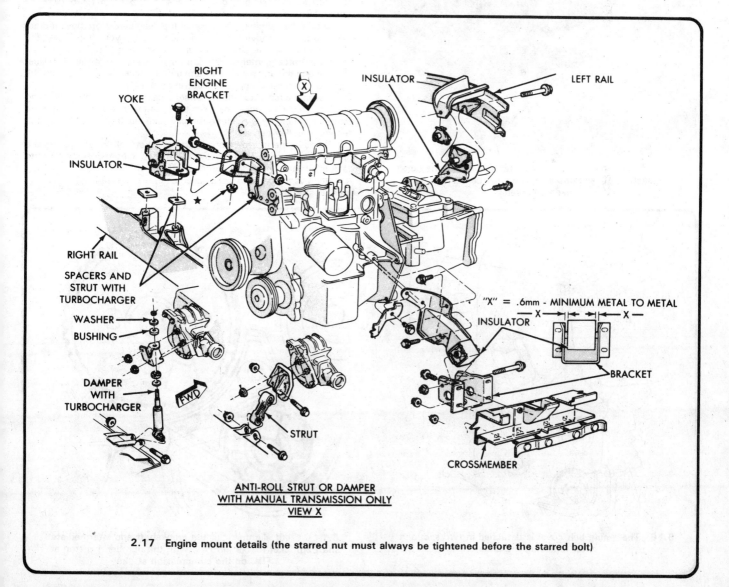

2.17 Engine mount details (the starred nut must always be tightened before the starred bolt)

4 Unscrew the oil pressure sending unit. Remove the dipstick, grasp the tube and pull it out of the engine block.
5 Disconnect the water hose from the water box and unscrew the adapter and coolant sensor.
6 Remove the thermostat and cover.
7 Remove the temperature vacuum switches (TVS) (if equipped) from the thermostat housing. Mark them for installation in the same locations.
8 Remove the two screws at the base and lift off the distributor cap shield.
9 Disconnect the spark plug wires and coil wire from the retainer on the valve cover.
10 Remove the distributor cap and distributor (Chapter 5).

11 Unscrew the coolant temperature sending unit from the cylinder head.
12 Disconnect the fuel lines from the fuel injection assembly.
13 Remove the spark plugs.
14 Remove the water pump (Chapter 3).
15 Remove the crankshaft pulley (see illustration).
16 Remove the upper and lower timing belt covers (see illustration).
17 Remove the right engine bracket.
18 Remove the ground strap from the intake manifold.
19 Remove the PCV valve and hose.
20 Remove the fuel injection throttle body (Chapter 4).
21 Remove the air cleaner heat tube from the exhaust manifold (if equipped).
22 Disconnect the EGR tube from the intake manifold (if equipped).
23 Unbolt and remove the EGR valve assembly (Chapter 6).
24 Remove the turbocharger assembly (Chapter 4).
25 Loosen the exhaust manifold bolts (work from the middle out and loosen them a little at a time).
26 Remove the exhaust manifold.
27 Remove the intake manifold, loosening the bolts in the same manner as for the exhaust manifold.
28 Remove the two engine mount brackets (if equipped) at the bottom edge of the block.

6 Timing belt and sprockets — removal, inspection and installation

Refer to illustrations 6.1, 6.2, 6.3, 6.4, 6.5, 6.11, 6.12, 6.14, 6.17 and 6.19

1 Locate the number one piston at top dead center on the compression stroke by removing the spark plug, placing your finger over the hole and turning the crankshaft until pressure is felt. The marks on the crankshaft and intermediate shaft sprocket will be aligned (see illustration) and the arrows on the camshaft sprocket will line up with the bearing cap parting line (see illustration 6.14).
2 Use one wrench to hold the offset tensioner pulley bolt while using another wrench or socket to loosen the center bolt, releasing the tension from the timing belt. Remove the belt (see illustration). Remove the tensioner pulley assembly.
3 Remove the intermediate shaft sprocket bolt while holding the sprocket with the special tool (no. C-4687) or a homemade substitute (see illustration). Pull the sprocket off the shaft.
4 Remove the bolt and use a puller to remove the crankshaft sprocket (see illustration).

5.15 A Torx head socket is required to remove the small crankshaft pulley bolts after the large center bolt has been removed

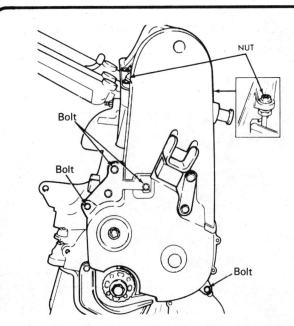

5.16 The timing belt cover is detached in two sections after removing the bolts

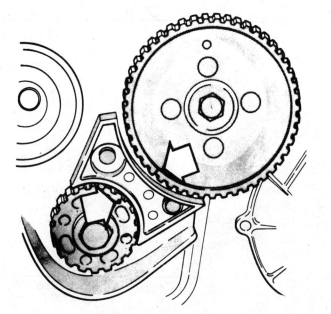

6.1 Correct alignment of the crankshaft and intermediate shaft sprocket marks (arrows) with the number 1 piston at TDC on the compression stroke

Chapter 2 Part A 2.2L and 2.5L engines

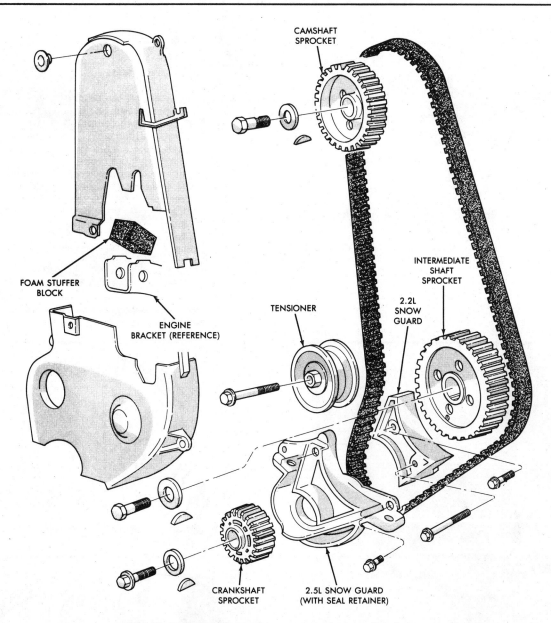

6.2 Timing belt and sprocket components — exploded view

6.3 Hold the intermediate shaft sprocket with a tool to keep it from turning as the bolt is loosened

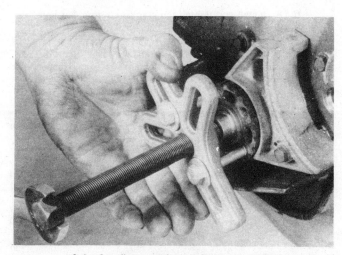

6.4 A puller must be used to remove the crankshaft sprocket

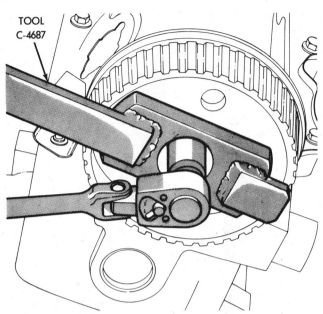

6.5 Removing the camshaft sprocket bolt using a special tool to keep the sprocket from turning

6.11 Use a straightedge to make sure the marks line up with the centers of the sprocket bolt holes

6.12 Install and tighten the crankshaft sprocket bolt

6.14 The small hole must be at the top and the arrows on the camshaft sprocket hub must be aligned with the bearing cap parting line when installing the timing belt

5 Hold the camshaft sprocket as described in Step 3 and remove the bolt (see illustration), then detach the sprocket from the cam.
6 Inspect the timing belt for wear, signs of stretching and damaged teeth. Check for signs of contamination by oil, gasoline, coolant and other liquids, which could cause the belt to break down and stretch. **Note:** *Unless the vehicle has very low mileage, it is a good idea to replace the timing belt with a new one any time it is removed.*
7 Inspect the tensioner pulley for damage, distortion and nicked or bent flanges. Replace the tensioner with a new one if necessary.
8 Inspect the camshaft, crankshaft and intermediate shaft sprockets for wear, damage, cracks, corrosion and rounding of the teeth. Replace them with new ones if necessary as damaged or worn sprockets could cause the belt to slip and alter camshaft timing (which could cause internal engine damage).
9 Inspect the crankshaft and intermediate shaft seals for signs of oil leakage and replace them with new ones if necessary (Sections 10 and 12).
10 When installing the timing belt, the keys on all shafts must be at the 12 o'clock position (facing up).
11 Install the crankshaft and intermediate shaft sprockets, then turn the shafts until the marks are aligned (see illustration).
12 Install the crankshaft sprocket bolt, lock the crankshaft to keep it from rotating and tighten the bolt to the specified torque (see illustration).
13 Install the intermediate shaft sprocket bolt and tighten it to the specified torque. Double-check to make sure the marks are aligned as shown in illustration 6.11.
14 Install the camshaft sprocket and bolt. Tighten the bolt to the specified torque. The arrows on the sprocket hub must align with the camshaft bearing cap parting line (see illustration).
15 Install the timing belt without turning any of the sprockets.
16 Install the tensioner pulley with the bolt finger tight.
17 With the help of an assistant, apply tension to the timing belt and temporarily tighten the tensioner bolt. Measure the deflection of the belt between the camshaft and tensioner pulley. Adjust the tensioner until belt deflection is approximately 5/16-inch (see illustration).
18 Turn the crankshaft two complete revolutions. This will align the belt on the pulleys. Recheck the belt deflection and tighten the tensioner pulley.

Chapter 2 Part A 2.2L and 2.5L engines

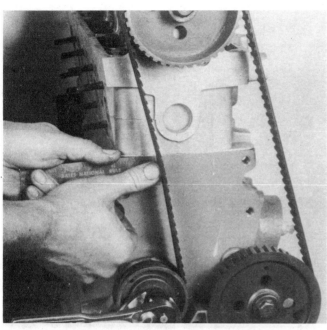

6.17 Use a ruler to measure timing belt deflection

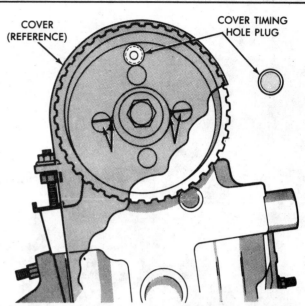

6.19 To check the cam timing with the cover in place, see if the small hole in the camshaft sprocket is aligned with the hole in the cover — this is done with the number 1 piston at TDC on the compression stroke

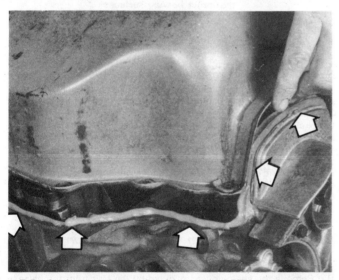

7.5 Apply a continuous bead (arrows) of RTV-type sealant to the block and end seals before installing the oil pan

8.3 Carefully slip a putty knife under the valve cover to break the seal (don't nick or otherwise damage the head)

19 Recheck the camshaft timing with the timing belt cover installed and the number one piston at TDC on the compression stroke. The small hole in the camshaft sprocket must be centered in the timing belt cover hole (see illustration).

7 Oil pan — removal and installation

Refer to illustration 7.5

1 Remove the bolts securing the oil pan to the engine block.
2 Tap on the pan with a soft-face hammer to break the gasket seal and lift the oil pan off the engine.
3 Using a gasket scraper, scrape off all traces of the old gasket from the engine block and oil pan. Remove the end seals from the oil seal retainers.
4 Clean the oil pan with solvent and dry it thoroughly. Check the gasket sealing surfaces for distortion.
5 Before installing the oil pan, install new end seals in the oil seal retainers. Apply a 3/16-inch bead of RTV sealant completely around the oil pan gasket surface of the engine block, including the end seals (see illustration).
6 Gently lay the oil pan in place.
7 Install the bolts and tighten them to the specified torque, starting with the bolts closest to the center of the pan and working out in a criss-cross pattern. *Do not overtighten them or leakage may occur.*

8 Cylinder head and camshaft — removal

Refer to illustrations 8.3, 8.5, 8.6 and 8.9

1 Remove the timing belt (Section 6).
2 Remove the camshaft sprocket (Section 6).
3 Remove the bolts and work carefully around the cover with a scraper or putty knife (see illustration) to release it from the sealant. Lift the cover off.
4 **Note:** *Removal of the cylinder head does not require removal of the camshaft, so you may skip the following steps pertaining to camshaft removal if you're just removing the head.* Loosen the camshaft bearing cap nuts, working from the ends toward the center, in 1/4-turn increments, until they can be removed by hand.

Chapter 2 Part A 2.2L and 2.5L engines

8.5 Tap the rear of the camshaft *lightly* with a soft-face hammer to dislodge the bearing caps

8.6 Lift straight up on the camshaft to avoid damage to the thrust bearing surfaces

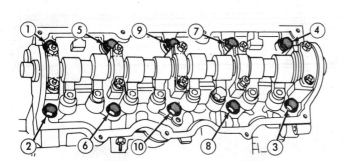

8.9 To avoid warping the head, *loosen* the head bolts a little at a time in the sequence shown until they can be removed

5 Lift off the cam bearing caps. It may be necessary to tap the camshaft lightly with a soft-faced hammer to loosen them (see illustration).
6 Remove the camshaft (see illustration). **Caution:** *If the camshaft is cocked as it is removed the thrust bearings could be damaged.*
7 Reinstall the caps temporarily to protect the studs and bearing surfaces.
8 Lift off the rocker arms and either mark them or place them in a marked container so they will be reinstalled in their original locations. Remove the lash adjusters.
9 Starting from the outside, loosen the head bolts, 1/8-turn at a time each in the sequence shown (see illustration). Remove the bolts and washers.
10 Use a soft-face hammer, if necessary, to tap the cylinder head and break the gasket seal. Do not pry between the cylinder head and the engine block. Remove the cylinder head.
11 Remove the gasket.

9 Balance shaft assembly — removal and installation

Refer to illustrations 9.1, 9.2, 9.3, 9.4, 9.5, 9.14, 9.17 and 9.23

1 Remove the oil pan, oil pickup, timing belt cover, timing belt, crankshaft sprocket and front oil seal retainer (see illustration).

Balance shafts — removal and installation

2 Remove the balance shaft assembly chain cover, guide and tensioner (see illustration). Remove the bolt from the end of the left shaft.
3 Remove the bolts retaining the balance shaft sprockets and remove the chain and sprockets as an assembly (see illustration). The bolts in the crankshaft sprocket may require a special tool for removal.
4 Remove the balance shaft drive gear cover double ended retaining stud and remove the cover and gears (see illustration).
5 Unbolt and remove the rear cover and slide the balance shafts out of the carrier (see illustration).
6 Remove the six bolts and separate the carrier from the crankcase.
7 Installation is the reverse of removal.

Carrier assembly — removal and installation

8 The carrier can be removed as an assembly with the gear cover, gears, balance shafts and rear cover intact.
9 Remove the chain cover, followed by the bolt from the balance shaft driven gear.
10 Loosen the tensioner pivot and adjusting bolts and push the driven balance shaft in and through the driven chain sprocket so the sprocket is hanging in the lower chain loop.
11 Remove the bolts and detach the carrier assembly from the crankcase.
12 Installation is the reverse of removal. Note that the crankshaft-to-balancer shaft timing must be established as described below.

Balance shaft timing

13 With the balance shafts installed in the carrier, place the carrier in position on the crankcase and install the bolts. Tighten the bolts to the specified torque.
14 Rotate the balance shafts until both keyways face up (in line with the centerline of the engine block) and install the gears. The short hub drive gear goes on the sprocket driven shaft and the long hub gear goes on the gear driven shaft. After installation the alignment dots must be directly opposite each other (see illustration).
15 Install the gear cover and tighten the double ended stud to the specified torque.
16 Install the crankshaft sprocket and tighten the bolts to the specified torque.

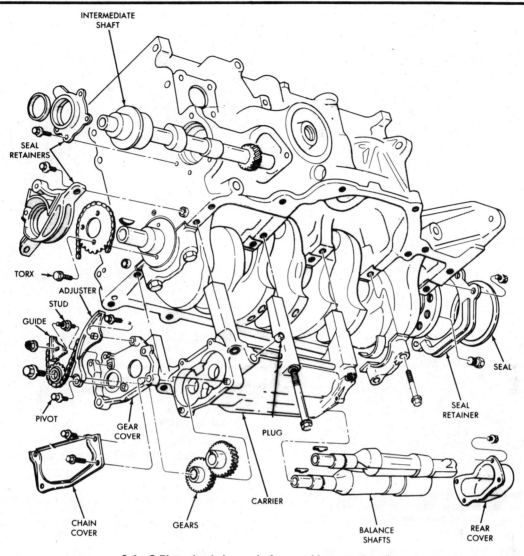

9.1 2.5L engine balance shaft assembly component layout

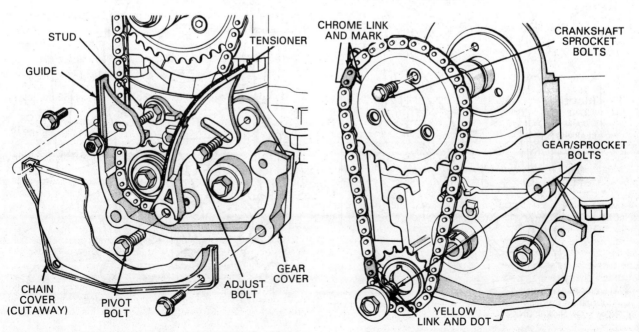

9.2 Balance shaft chain cover, guide and tensioner details

9.3 Balance shaft assembly drive chain and sprocket layout

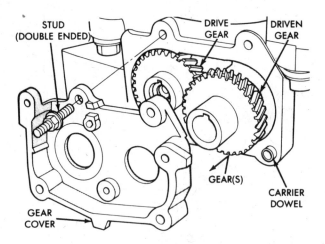

9.4 Balance shaft assembly drive gear and cover removal or installation details

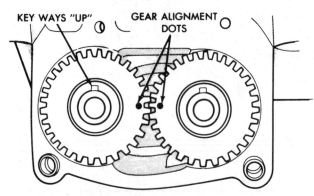

9.14 The balance shaft gear keyways must point directly up and the gear dots must be aligned for correct gear timing

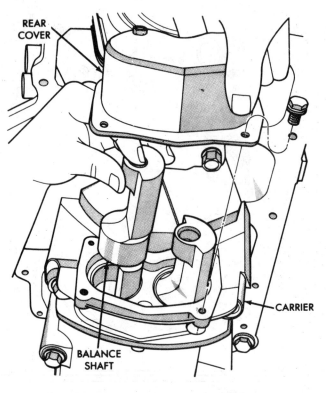

9.5 Removal/installation of the balance shafts

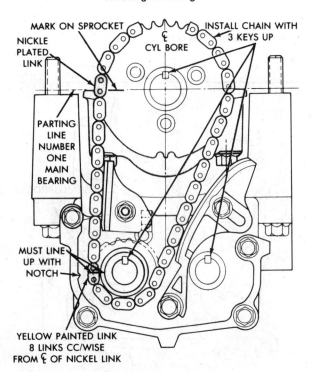

9.17 Balance shaft assembly shaft, gear and chain alignment details

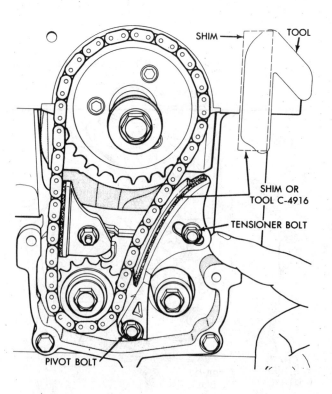

9.23 Balance shaft chain tension adjustment details

Chapter 2 Part A 2.2L and 2.5L engines

10.4 The retainer must be removed to withdraw the intermediate shaft

10.5 Be careful not to nick or gouge the bearings as the intermediate shaft is removed

10.7 With the retainer on a flat surface, carefully tap the new seal into place with a soft-face hammer

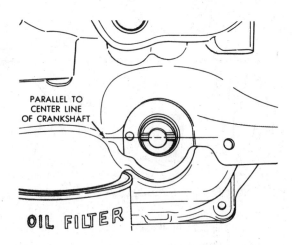

10.8 When viewed through the distributor hole, the oil pump shaft slot must be parallel to the crankshaft centerline

17 Rotate the crankshaft until the number one piston is at Top Dead Center (TDC). The timing marks on the chain sprocket should now be aligned with the parting line on the left side of the number one main bearing cap (see illustration). Place the timing chain over the crankshaft sprocket so that the nickel plated (bright) link is over the timing mark on the crankshaft sprocket as shown in the illustration.
18 Place the balance shaft sprocket in position in the timing chain, making sure the yellow timing mark on the sprocket mates with the yellow painted link on the chain. The mark on the sprocket should also line up with the notch in the cover (see illustration 9.17).
19 Make sure the balance shaft keyways are pointing straight up, then slide the sprocket onto the nose of the balance shaft. It may be necessary to push the balance shaft in slightly to obtain sufficient clearance. At this point the timing mark on the sprocket and the notch on the side of the gear cover must be aligned if the balance shafts are timed correctly.
20 Install the balance shaft bolts. Place a wood block between the crankshaft counterweight and the crankcase, to keep the crankshaft and gear from rotating, and tighten the bolts to the specified torque.

Balance shaft chain tensioning

21 Install the chain tensioner assembly with the bolts finger tight.
22 Place a piece of shim stock measuring 0.039-inch (1 mm) thick and 2.75-inch (70 mm) long or factory tool number C-4916 between the tensioner and the chain.
23 Push the tensioner and shim (or tool) up against the chain. Maintain pressure by pressing the tensioner directly behind the adjustment slot so that all slack is taken up as shown (see illustration).

24 With the tension maintained, tighten first the top and then the bottom tensioner bolts to the specified torque. Remove the shim.
25 Place the guide in position on the double-ended stud, making sure the tab on the guide fits into the slot on the gear cover. Install the nut and washer and tighten the nut to the specified torque.
26 Install the balance shaft assembly covers and tighten the bolts to the specified torque.

10 Intermediate shaft and seal — removal, inspection and installation

Refer to illustrations 10.4, 10.5, 10.7 and 10.8

1 Remove the timing belt (Section 6).
2 Remove the sprocket (Section 6).
3 Inspect the sprocket as described in Section 6.
4 Unbolt and remove the shaft retainer (see illustration).
5 Grasp the shaft and carefully withdraw it from the engine (see illustration).
6 Clean the shaft thoroughly with solvent and inspect the gear, bearing surfaces and lobes for wear and damage.
7 Use a punch to drive the old oil seal out of the retainer. Apply a thin coat of RTV sealant to the inner surface of the retainer and tap the new seal into place with a soft-face hammer (see illustration).
8 Lightly lubricate the gear, lobes and bearing surfaces with engine assembly lubricant and carefully insert the shaft into place. After installation, make sure the shaft is securely in place in the oil pump. The oil pump slot must be parallel to the crankshaft centerline and the inter-

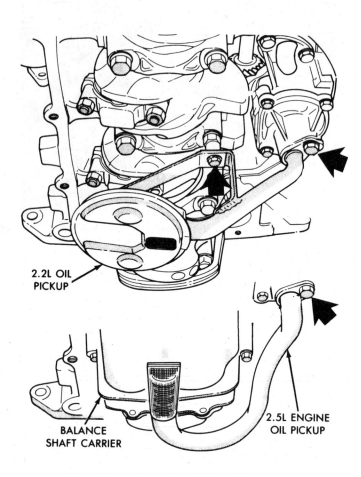

11.1 The 2.2L (top) and 2.5L oil pickup bolts (arrows)

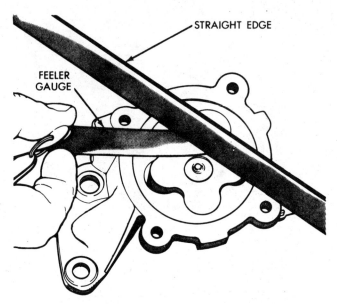

11.4 Checking oil pump rotor end play with a straightedge and feeler gauge

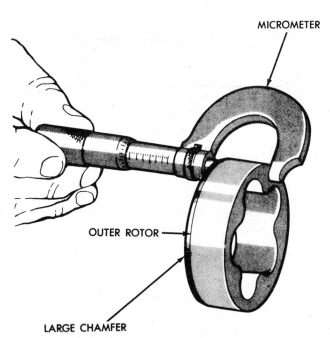

11.5 Measuring the oil pump rotor thickness with a micrometer

mediate shaft keyway must be in the 12 o'clock position (see illustration).
9 Apply a 1 mm wide bead of *anaerobic* sealant to the contact surface of the retainer and place it in position over the end of the shaft. Install the retaining bolts and tighten them to the specified torque.
10 Install the sprocket as described in Section 6.
11 Install the timing belt (Section 6).

11 Oil pump — removal, disassembly and inspection

Refer to illustrations 11.1, 11.4, 11.5, 11.6, 11.7, 11.8 and 11.9

1 Unbolt and remove the oil pickup (see illustration).
2 Unbolt and remove the oil pump.
3 Remove the retaining bolts and lift off the oil pump cover.
4 Check the end play of the rotors, using a feeler gauge and a straightedge (see illustration). Compare the measurement to the Specifications.
5 Remove the outer rotor and measure its thickness (see illustration). **Caution:** *Install the rotor with the large chamfered edge facing the pump body.*
6 Check the clearance between the rotors with a feeler gauge (see illustration) and compare the results to the Specifications.
7 Measure the outer rotor-to-body clearance (see illustration) and compare it to the Specifications.
8 Check the oil pump cover for warpage with a feeler gauge and a straightedge to make sure it is as specified (see illustration).
9 Measure the oil pressure relief valve spring to ensure that it is the specified length (see illustration).
10 If any components are worn beyond the specified limit, the oil pump will have to be replaced with a new one.

12 Front oil seal and housing — removal and installation

Refer to illustrations 12.1 and 12.3

1 With the timing belt and sprockets, the oil pan and the intermediate shaft sprocket removed for access, unbolt and remove the oil seal housing (see illustration).

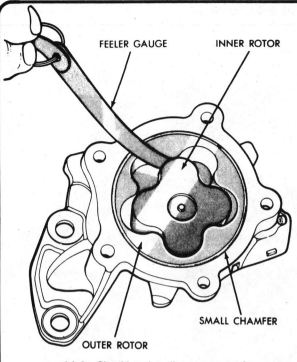

11.6 Checking the oil pump rotor clearance

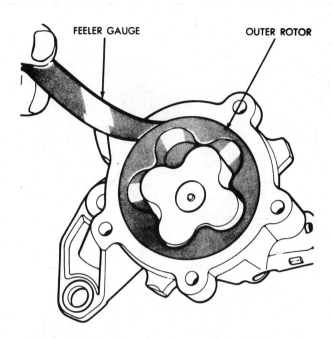

11.7 Checking the outer rotor-to-pump body clearance

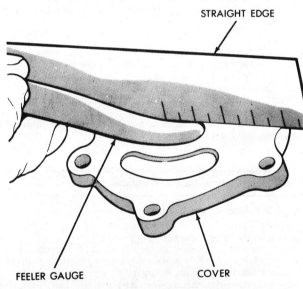

11.8 Checking the oil pump cover for warpage with a straightedge and feeler gauge

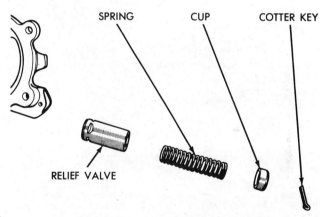

11.9 Oil pressure relief valve components — exploded view

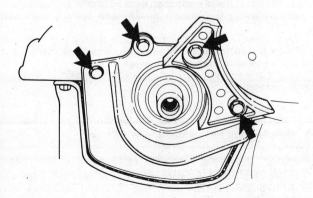
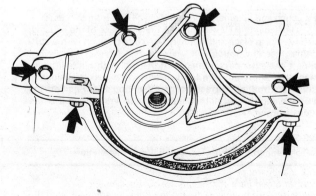

12.1 2.2L (top) and 2.5L front oil seal housing retaining bolts (arrows)

12.3 Tap the new front crankshaft oil seal into place with a soft-face hammer

13.1 Pull the rear crankshaft oil seal housing away from the crankshaft and detach it from the engine

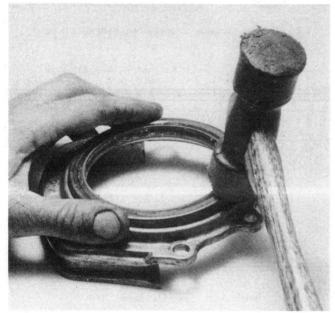

13.4 Tap the new rear oil seal squarely into place with a soft-face hammer

2 Use a punch and hammer to drive the old oil seal from the housing.
3 Apply a thin coat of RTV sealant to the inner surface of the housing, place the new seal in place and carefully tap it into position with a soft-face hammer (see illustration).
4 Lubricate the inner circumference of the seal with white lithium-base grease and apply a 1 mm bead of *anaerobic* gasket sealant to the engine block mating surfaces of the seal housing. Position the housing on the engine. Install the retaining bolts and tighten them to the specified torque.

13 Rear oil seal and housing — removal and installation

Refer to illustrations 13.1 and 13.4

1 Remove the four bolts and detach the housing and seal from the rear of the engine block (see illustration).
2 Drive the old seal from the housing.
3 Clean the seal surface thoroughly with solvent and inspect it for nicks and other damage.
4 Apply a thin coat of RTV sealant to the inner circumference of the housing, lay the new seal in place and tap it squarely into position with a soft-face hammer (see illustration).
5 Lubricate the seal inner surface with white lithium-base grease.
6 Apply a 1mm bead of anaerobic gasket sealant to the engine block mating surfaces of the seal housing.
7 Place the assembly in position, install the bolts and tighten them to the specified torque.

14 Oil pump — reassembly and installation

Refer to illustrations 14.4 and 14.6

1 Install the rotor (large chamfered edge toward the pump body) and oil pressure relief valve and spring assembly.
2 Install the pump cover and tighten the bolts to the specified torque.
3 Apply a thin coat of RTV sealant to the contact surface of the pump and lower it fully into position. Rotate it back and forth slightly to make sure there is positive contact between the pump and the cylinder block. While holding the pump securely in place, coat the threads of the retaining bolts with sealant and install and tighten them to the specified torque. The slot in the oil pump shaft must be parallel to the crankshaft centerline when viewed through the distributor opening (see illustration 10.8).
4 Install a new O-ring in the oil pump pickup opening (see illustration).
5 Carefully work the pickup into the pump, install the retaining bolts and tighten them to the specified torque.
6 On 2.2 liter engines, install the brace bolt and tighten it securely (see illustration).

15 Cylinder head and camshaft — installation

Refer to illustrations 15.5, 15.7, 15.8, 15.12, 15.14, 15.15 and 15.18

1 Before installing the cylinder head, check to make sure the number one (front) piston is at top dead center, the timing belt sprockets are properly aligned (Section 6) and the oil pump shaft slot (viewed through the distributor installation hole) is parallel to the crankshaft centerline.
2 Place the head gasket in position on the engine block and press it into place over the alignment dowels.

Chapter 2 Part A 2.2L and 2.5L engines

14.4 Lubricate the new O-ring and install it in the oil pump before attaching the pickup assembly

14.6 Don't forget to tighten the pickup brace bolt on 2.2L engines

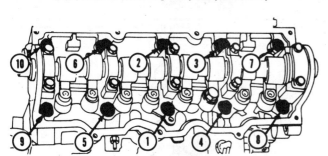

15.5 Cylinder head bolt *tightening* sequence

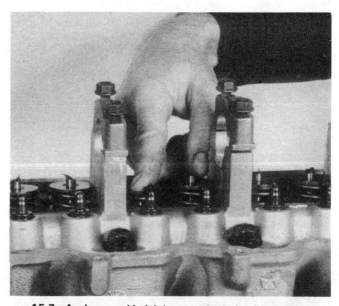

15.7 Apply assembly lubricant to the lash adjusters and the ends of the valve stems

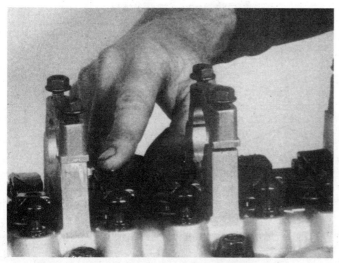

15.8 Seat the rocker arms on the valves and adjusters, then apply assembly lubricant to the cam lobe contact surfaces

3 Place the cylinder head in position.
4 Apply a thin coat of sealant to the threads of the head bolts and install the bolts finger tight. **Caution:** *Head bolts used in 1986 models are 11 mm in diameter and have an 11 on the bolt head. The 10 mm bolts used in previous years will thread into the 11 mm holes, but will strip the threads when they are tightened.*
5 Tighten the head bolts to the specified torque, following the sequence shown (see illustration). **Caution:** *Bolt torque after the final 1/4-turn (Step 4) must be over 90 ft-lbs on 1986 models. If it isn't, replace the bolt(s).*
6 Lubricate the valve lash adjusters with engine oil and insert them into the bores.
7 Lightly lubricate the contact points of the valve stems and lash adjusters with assembly lube (see illustration).
8 Install the rocker arms (see illustration).
9 Lubricate the contact surfaces on the top side of the rocker arms with assembly lube.
10 Lubricate the camshaft bearing surfaces in the head with assembly lube. Wipe the camshaft carefully with a clean, dry, lint-free cloth.
11 Lubricate the camshaft lobes with assembly lubricant and lower it into position with the sprocket keyway facing up.
12 Apply a thin coat of assembly lubricant to the camshaft bearing caps and install them on their respective pedestals (the arrows must

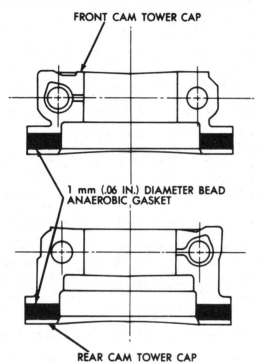

15.12 Apply anaerobic-type gasket sealant to the dark areas of the front and rear camshaft bearing caps (do not get it in the oil passages)

15.14 Use a small screwdriver to remove excess sealant from the bearing cap seal surfaces

15.15 Tap the new camshaft seal into place with the broad end of a punch and a hammer

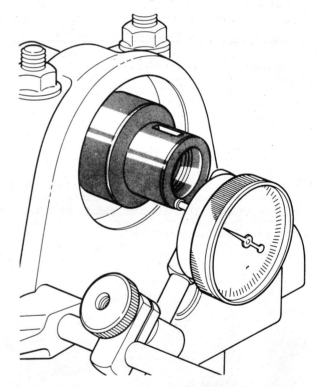

15.18 Measuring camshaft end play with a dial indicator

point toward the timing belt). Apply *anaerobic-type* sealant to the contact surfaces of the two end bearing caps (see illustration).
13 Install the bearing cap nuts and tighten them with your fingers until they are snug. Tighten the nuts evenly, in 1/4-turn increments, to the specified torque.
14 Remove any excess sealant from the two end bearing caps (see illustration).
15 Press the camshaft seals into the end bearing caps and seat them in place with a hammer and the large end of a drift punch (see illustration).
16 The camshaft end play can be checked with a dial indicator set or a feeler gauge.

17 If a feeler gauge is used, gently pry the camshaft all the way toward the front of the engine. Slip a feeler gauge between the flange at the front of the camshaft and the front bearing cap. Compare the measured end play to the Specifications.
18 If a dial indicator is used, mount it at the front of the engine with the indicator stem touching the end of the camshaft. Carefully pry the camshaft all the way toward the front of the engine, then zero the indicator. Gently pry the camshaft as far as possible in the opposite direction and observe the needle movement on the dial indicator, which will indicate the amount of end play. Compare the reading to the Specifications (see illustration).

Chapter 2 Part A 2.2L and 2.5L engines

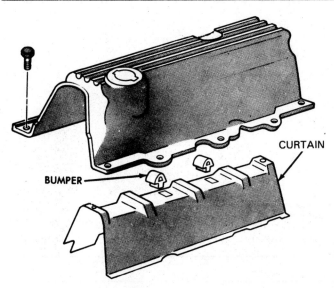

16.4 Camshaft cover installation details

16 External engine components — installation

Refer to illustration 16.4

1 Install the camshaft drivebelt sprocket, making sure it is properly aligned (Section 6). Install the retaining bolt finger tight.
2 Lock the camshaft timing belt sprocket with a suitable tool such as Chrysler tool C-4687 or equivalent to keep it from turning, install the bolt and tighten it to the specified torque.
3 Install and adjust the timing belt (Section 6).
4 Attach the rubber bumpers in the camshaft cover curtain and place the curtain in position on the cylinder head with the clearance cutouts facing the manifold (see illustration).
5 Apply a 1/8-inch bead of RTV sealant around the sealing surface of the camshaft cover.
6 Place the cover in position and press down to seat it. Install the retaining bolts and tighten them to the specified torque.
7 Place the exhaust manifold and gasket in position, lubricate the mounting stud threads with white lithium-base grease and install the nuts. Tighten them to the specified torque in a criss-cross pattern.
8 Install the intake manifold and bolts. Tighten the bolts to the specified torque in increments, working from the center to the ends.
9 Install the turbocharger assembly (Chapter 4).
10 Install the EGR valve (Chapter 6).
11 Apply anti-seize compound to the threads of the oxygen sensor and install it in the exhaust manifold.
12 Install the fuel injection throttle body (Chapter 4)
13 Install the PCV valve and hose assembly.
14 Attach the engine mount to the front of the engine.
15 Install the lower timing belt cover, followed by the upper cover.
16 Install the water pump and gasket (Chapter 3).
17 Lubricate the O-ring on the distributor shaft with white lithium-base grease and install the distributor (Chapter 5).
18 Install the thermostat, gasket and cover (Chapter 3).
19 Install the heater water outlet.
20 Install the coolant temperature switch(es) in the thermostat housing.
21 Install the oil pressure switch.
22 Install the water temperature sensor in the cylinder head.

17 Engine — installation

1 Attach the hoist to the engine, carefully lower the engine into the engine compartment and line up the transaxle and engine bolt holes.
2 Align the engine mounts and install all bolts/nuts finger tight. *Do not tighten any of the bolts/nuts until all of them have been installed.*
3 Install the transaxle-to-engine bolts and tighten them securely. Remove the hoist and the transaxle supports.
4 Attach the engine ground strap and install the right engine splash shield.
5 Install the starter and connect the exhaust pipe to the manifold.
6 On vehicles with a manual transaxle, install the lower transmission case cover and hook up the clutch cable.
7 On vehicles with an automatic transaxle, remove the C-clamp from the housing and align the driveplate and torque converter. Install the bolts and tighten them to the specified torque, then install the lower transmission case cover.
8 Install the power steering pump and drivebelt.
9 Install the alternator and drivebelt.
10 Hook up the fuel and vapor lines, the throttle cable and the heater hoses.
11 Reattach the wires to the engine, fuel injection and alternator.
12 Install a new oil filter and fill the crankcase with the specified oil (Chapter 1).
13 Install the air conditioning compressor and drivebelt.
14 Install the radiator and fan assembly and hook up the radiator hoses, then refill the cooling system (Chapter 1).
15 Install the air cleaner assembly and the hoses.
16 Install the battery and connect the cables (positive first, then negative).
17 Install the hood (Chapter 11).

Chapter 2 Part B
General engine overhaul procedures

Refer to Chapter 13 for Specifications on 1987 and later models

Contents

Crankshaft — inspection	16
Crankshaft — installation and main bearing oil clearance check	19
Crankshaft — removal	12
Cylinder head — cleaning and inspection	8
Cylinder head — disassembly	7
Cylinder head — reassembly	10
Engine block — cleaning	13
Engine block — inspection	14
Engine overhaul — disassembly and reassembly sequence	6
Engine overhaul — general information	3
Engine rebuilding alternatives	4
Engine removal — methods and precautions	5
General information	1
Initial start-up and break-in after overhaul	21
Main and connecting rod bearings — inspection	17
Piston/connecting rod assembly — inspection	15
Piston/connecting rod assembly — installation and bearing oil clearance check	20
Piston/connecting rod assembly — removal	11
Piston rings — installation	18
Repair operations possible with the engine in the vehicle	2
Valves — servicing	9

Specifications

General
Displacement
  2.2L engine .... 135 cu in (2.2 liters)
  2.5L engine .... 153 cu in (2.5 liters)
Bore and stroke
  2.2L engine .... 3.44 x 3.62 in (87.5 x 92.0 mm)
  2.5L engine .... 3.44 x 4.09 in (87.5 x 104 mm)
Compression pressure .... See Chapter 1

Valve timing
2.2L non-turbocharged engine
  intake valve
    opens .... 16° BTDC
    closes .... 48° ABDC
  exhaust valve
    opens .... 52° BBDC
    closes .... 12° ATDC
2.2L turbocharged engine
  intake valve
    opens .... 10° BTDC
    closes .... 50° ATDC
  exhaust valve
    opens .... 50° BBDC
    closes .... 10° ATDC
2.5L engine
  intake valve
    opens .... 12° BTDC
    closes .... 52° ATDC
  exhaust valve
    opens .... 48° BBDC
    closes .... 16° ATDC
Oil pressure
  1984 (at 2000 rpm) .... 40 psi (276 kPa)
  1985 (at 3000 rpm) .... 25 to 90 psi (172 to 620 kPa)
  1986 (at 3000 rpm) .... 25 to 80 psi (172 to 552 kPa)

Chapter 2 Part B General engine overhaul procedures

Engine block
Cylinder bore diameter .... 3.44 in (87.4 mm)
Taper limit
   1984 .... 0.010 in (0.25 mm)
   1985 and 1986 .... 0.005 in (0.125 mm)
Out-of-round limit
   1984 .... 0.005 in (0.125 mm)
   1985 and 1986 .... 0.002 in (0.050 mm)

Pistons and rings
Piston diameter
   non-turbocharged engine .... 3.443 to 3.445 in (87.442 to 87.507 mm)
   turbocharged engine .... 3.4416 to 3.4441 in (87.416 mto 87.481 mm)
Piston ring-to-groove clearance
   top ring
      standard .... 0.0015 to 0.0031 in (0.038 to 0.078 mm)
      service limit .... 0.004 in (0.10 mm)
   2nd ring
      standard .... 0.0015 to 0.0037 in (0.038 to 0.093 mm)
      service limit .... 0.004 in (0.10 mm)
   oil ring .... 0.008 in (0.20 mm)
Piston ring end gap (non-turbocharged engine)
   top ring
      standard .... 0.011 to 0.021 in (0.28 to 0.53 mm)
      service limit .... 0.039 in (1.0 mm)
   2nd ring
      standard .... 0.011 to 0.021 in (0.28 to 0.53 mm)
      service limit .... 0.039 in (1.0 mm)
   Oil ring
      standard .... 0.015 to 0.055 in (0.38 to 1.40 mm)
      service limit .... 0.074 in (1.88 mm)
Piston ring end gap (turbocharged engine)
   top ring
      standard .... 0.010 to 0.020 in (0.25 to 0.51 mm)
      service limit .... 0.038 in (0.98 mm)
   2nd ring
      standard .... 0.009 to 0.019 in (0.23 to 0.48 mm)
      service limit .... 0.037 in (0.95 mm)
   oil ring
      standard .... 0.015 to 0.55 in (0.38 to 1.40 mm)
      service limit .... 0.074 in (1.88 mm)

Crankshaft and flywheel
Main journal diameter
   non-turbocharged engine .... 2.362 to 2.363 in (59.987 to 60.013 mm)
   turbocharged engine .... 2.3622 to 2.3627 in (60.000 to 60.013 mm)
   taper limit .... 0.0004 in (0.010 mm)
   out-of-round limit .... 0.012 in (0.03 mm)
Main bearing oil clearance (non-turbocharged engine)
   standard .... 0.0003 to 0.0031 in (0.007 to 0.080 mm)
   service limit .... 0.004 in (0.10 mm)
Main bearing oil clearance (turbocharged engine) .... 0.0004 to 0.0023 in (0.011 to 0.54 mm)
Connecting rod journal diameter .... 1.968 to 1.969 in (49.979 to 50.005 mm)
Connecting rod bearing oil clearance (non-turbocharged engine)
   standard .... 0.0008 to 0.0034 in (0.019 to 0.087 mm)
   service limit .... 0.004 in (0.10 mm)
Connecting rod bearing oil clearance
   (turbocharged engine) .... 0.0.008 to 0.0031 in (0.019 to 0.079 mm)
Connecting rod side clearance .... 0.005 to 0.013 in (0.13 to 0.32 mm)
Crankshaft end play
   standard .... 0.002 to 0.007 in (0.05 to 0.18 mm)
   service limit .... 0.014 in (0.35 mm)

Camshaft
End play
   standard .... 0.005 to 0.013 in (0.13 to 0.33 mm)
   service limit .... 0.020 in (0.50 mm)
Bearing journal diameter .... 1.375 to 1.376 in (34.939 to 34.960 mm)
Camshaft lobe wear limit
   1984 .... 0.005 in (0.13 mm)
   1985 and 1986 .... 0.010 in (0.25 mm)

Cylinder head and valve train
Head warpage limit .... 0.004 in (0.1 mm)
Valve seat angle .... 45°

Cylinder head and valve train (continued)

Valve seat width
- intake .... 0.079 in (2.00 mm)
- exhaust .... 0.068 in (1.75 mm)

Valve face angle.. 45°

Valve margin width
- intake
  - standard .... 0.06 in (1.5 mm)
  - service limit .... 0.03 in (0.793 mm)
- exhaust
  - standard .... 0.06 in (1.5 mm)
  - service limit .... 0.05 in (1.19 mm)

Valve stem diameter
- intake .... 0.3124 in (7.935 mm)
- exhaust .... 0.3103 in (7.881 mm)

Valve head diameter
- intake .... 1.60 in (40.6 mm)
- exhaust .... 1.39 in (35.4 mm)

Valve stem-to-guide clearance
- intake .... 0.0009 to 0.0026 in (0.022 to 0.065 mm)
- exhaust .... 0.0030 to 0.0047 in (0.076 to 0.119 mm)

Valve spring free length
- 1984 .... 2.28 in (57.9 mm)
- 1985 and 1986
  - non-turbocharged engine .... 2.39 in (60.8 mm)
  - turbocharged engine........................... 2.28 in (57.9 mm)

Valve spring installed height
- intake .... 1.62 to 1.68 in (41.2 to 42.7 mm)
- exhaust .... 1.62 to 1.68 in (41.2 to 42.7 mm)

Collapsed tappet gap .... 0.024 to 0.060 in (0.62 to 1.52 mm)
Spring seat-to-valve tip dimension*............ 1.960 to 2.009 in (49.76 to 51.04 mm)
Rocker arm-to-spring retainer .... 0.050 in (1.25 mm) minimum

Must be checked if valve faces or seats are reground

Intermediate shaft

Journal diameter
- large .... 1.6812 to 1.6822 in (42.67 to 42.695 mm)
- small .... 0.775 to 0.776 in (19.67 to 19.695 mm)

Bearing inside diameter
- large .... 1.6836 to 1.6844 in (42.73 to 42.75 mm)
- small .... 0.777 to 0.778 in (19.72 to 19.75 mm)

Oil pump

Outer rotor-to-housing bore clearance
- standard .... 0.010 in (0.25 mm)
- service limit .... 0.014 in (0.035 mm)

Outer rotor thickness
- 1984 and 1985
  - standard .... 0.826 to 0.827 in (20.98 to 21.00 mm)
  - service limit .... 0.825 in (20.96 mm)
- 1986
  - standard .... 0.944 to 0.946 in (23.97 to 24.00 mm)
  - service limit .... 0.825 in (20.96 mm)

Inner rotor-to-outer rotor tip clearance
- 1984 and 1985 .... 0.010 in (0.25 mm)
- 1986
  - standard .... 0.008 in (0.20 mm)
  - service limit .... 0.010 in (0.25 mm)

Rotor-to-housing clearance
- standard .... 0.001 to 0.003 in (0.03 to 0.08 mm)
- service limit .... 0.004 in (0.10 mm)

Pump cover warpage
- standard .... 0.010 in (0.025 mm)
- service limit .... 0.015 in (0.038 mm)

Relief spring free length .... 1.95 in (49.5 mm)
Relief spring load .... 15 to 25 lbs @ 1.34 in (67 to 89 N @ 34 mm)

Torque specifications

	Ft-lbs	Nm
Main bearing cap bolt	30*	41*
Connecting rod bearing cap nut	40*	54*

Plus 1/4-turn

Chapter 2 Part B General engine overhaul procedures

1 General information

Included in this part of Chapter 2 are the general overhaul procedures for the cylinder head and internal engine components. This information ranges from advice about preparing for an overhaul and the purchase of replacement parts to detailed, step-by-step procedures covering removal and installation of internal engine components and the inspection of parts.

In the following Sections, it is assumed that the engine has been removed from the vehicle. For information concerning removal and installation of the engine and its external components, see Part A of this Chapter and Section 2 of this Part.

The specifications included here in Part B are only those necessary for the inspection and overhaul procedures which follow. Refer to Part A for additional specifications.

2 Repair operations possible with the engine in the vehicle

Many major repair operations can be accomplished without removing the engine from the vehicle.

It is a very good idea to clean the engine compartment and the exterior of the engine with some type of pressure washer before any work is begun. A clean engine will make the job easier and will prevent the possibility of getting dirt into internal areas of the engine.

Remove the hood (Chapter 11) and cover the fenders to provide as much working room as possible and to prevent damage to the painted surfaces.

If oil or coolant leaks develop, indicating a need for gasket or seal replacement, the repairs can generally be made with the engine in the vehicle. The oil pan gasket, the cylinder head gasket, intake and exhaust manifold gaskets (non-turbocharged engines), timing cover gaskets, and the front crankshaft oil seal are accessible with the engine in place.

Exterior engine components, such as the water pump, the starter motor, the alternator, the distributor, the fuel injection and turbocharger assemblies, as well as the intake and exhaust manifolds (non-turbocharged engines), are quite easily removed for repair with the engine in place.

Since the cylinder head can be removed without pulling the engine, valve component servicing can also be accomplished with the engine in the vehicle. On turbocharged engines, the intake and exhaust manifolds can be removed only after the cylinder head has been removed (Chapter 4).

Replacement of, repairs to or inspection of the timing sprockets and belt, the balance shaft and chain assembly, the oil pump and front cover seals are all possible with the engine in place.

In extreme cases caused by a lack of necessary equipment, repair or replacement of piston rings, pistons, connecting rods and rod bearings and reconditioning of the cylinder bores is possible with the engine in the vehicle. However, this practice is not recommended because of the cleaning and preparation work that must be done to the components involved.

Detailed removal, inspection, repair and installation procedures for the above mentioned components can be found in the appropriate Part of Chapter 2 or the other Chapters in this manual.

3 Engine overhaul — general information

It is not always easy to determine when, or if, an engine should be completely overhauled, as a number of factors must be considered.

High mileage is not necessarily an indication that an overhaul is needed, while low mileage does not preclude the need for an overhaul. Frequency of servicing is probably the most important consideration. An engine which has had regular and frequent oil and filter changes, as well as other required maintenance, will most likely give many thousands of miles of reliable service. Conversely, a neglected engine may require an overhaul very early in its life.

Excessive oil consumption is an indication that piston rings and/or valve guides are in need of attention. Make sure that oil leaks are not responsible before deciding that the rings and guides are bad. Have a cylinder compression or leakdown test performed by an experienced tune-up mechanic to determine the extent of the work required.

If the engine is making obvious knocking or rumbling noises, the connecting rod and/or main bearings are probably at fault. Check the oil pressure with a gauge installed in place of the oil pressure sending unit and compare it to the Specifications. If it is extremely low, the bearings and/or oil pump are probably worn out.

Loss of power, rough running, excessive valve train noise and high fuel consumption rates may also point to the need for an overhaul, especially if they are all present at the same time. If a complete tune-up does not remedy the situation, major mechanical work is the only solution.

An engine overhaul involves restoring the internal parts to the specifications of a new engine. During an overhaul, the piston rings are replaced and the cylinder walls are reconditioned (rebored and/or honed). If a rebore is done, new pistons are required. The main and connecting rod bearings are replaced with new ones and, if necessary, the crankshaft may be reground to restore the journals. Generally, the valves are serviced as well, since they are usually in less-than-perfect condition at this point. While the engine is being overhauled, other components, such as the fuel injection, distributor, starter and alternator, can be rebuilt as well. The end result should be a like new engine that will give many trouble free miles.

Before beginning the engine overhaul, read through the entire procedure to familiarize yourself with the scope and requirements of the job. Overhauling an engine is not difficult, but it is time consuming. Plan on the vehicle being tied up for a minimum of two weeks, especially if parts must be taken to an automotive machine shop for repair or reconditioning. Check on availability of parts and make sure that any necessary special tools and equipment are obtained in advance. Most work can be done with typical hand tools, although a number of precision measuring tools are required for inspecting parts to determine if they must be replaced. Often an automotive machine shop will handle the inspection of parts and offer advice concerning reconditioning and replacement. **Note:** *Always wait until the engine has been completely disassembled and all components, especially the engine block, have been inspected before deciding what service and repair operations must be performed by an automotive machine shop.* Since the condition of the block will be the major factor to consider when determining whether to overhaul the original engine or buy a rebuilt one, never purchase parts or have machine work done on other components until the block has been thoroughly inspected. As a general rule, time is the primary cost of an overhaul, so it does not pay to install worn or substandard parts.

As a final note, to ensure maximum life and minimum trouble from a rebuilt engine, everything must be assembled with care in a spotlessly clean environment.

4 Engine rebuilding alternatives

The do-it-yourselfer is faced with a number of options when performing an engine overhaul. The decision to replace the engine block, piston/connecting rod assemblies and crankshaft depends on a number of factors, especially the condition of the block. Other considerations are cost, access to machine shop facilities, parts availability, time required to complete the project and experience.

Some of the rebuilding alternatives include:

Individual parts — If the inspection procedures reveal that the engine block and most engine components are in reusable condition, purchasing individual parts may be the most economical alternative. The block, crankshaft and piston/connecting rod assemblies should all be inspected carefully. Even if the block shows little wear, the cylinder bores should receive a finish hone.

Crankshaft kit — This rebuild package consists of a reground crankshaft and a matched set of pistons and connecting rods. The pistons will already be installed on the connecting rods. Piston rings and the necessary bearings will be included in the kit. These kits are commonly available for standard cylinder bores, as well as for engine blocks which have been bored to a regular oversize.

Short block — A short block consists of an engine block with a crankshaft and piston/connecting rod assemblies already installed. All new bearings are incorporated and all clearances will be correct. The existing camshaft, valve train components, cylinder head and external parts can be bolted to the short block with little or no machine shop work necessary.

Long block — A long block consists of a short block plus an oil pump, oil pan, cylinder head, cylinder head cover, camshaft and valve train

Chapter 2 Part B General engine overhaul procedures

components, timing sprockets and belt/chain and timing cover. All components are installed with new bearings, seals and gaskets incorporated throughout. The installation of manifolds and external parts is all that is necessary.

Give careful thought to which alternative is best for you and discuss the situation with local automotive machine shops, auto parts dealers or parts store countermen before ordering or purchasing replacement parts.

5 Engine removal — methods and precautions

If it has been decided that an engine must be removed for overhaul or major repair work, certain preliminary steps should be taken.

Locating a suitable work area is extremely important. A shop is, of course, the most desirable place to work. Adequate work space, along with storage space for the vehicle, is very important. If a shop or garage is not available, use a flat, level, clean work surface made of concrete or asphalt.

Cleaning the engine compartment and engine prior to removal will help keep tools clean and organized.

An engine hoist or A-frame will also be necessary. Make sure that the equipment is rated in excess of the combined weight of the engine and its accessories. Safety is of primary importance, considering the potential hazards involved in lifting the engine out of the vehicle.

If the engine is being removed by a novice, a helper should be available. Advice and aid from someone more experienced would also be helpful. There are many instances in which one person cannot simultaneously perform all of the operations necessary to lift the engine out of the vehicle.

Plan the operation ahead of time. Obtain all of the tools and equipment you will need — or know where to get them — prior to beginning the job. Some of the equipment necessary for safe and easy engine removal and installation are an engine hoist, a heavy duty floor jack, complete sets of wrenches and sockets as described in the front of this manual, wooden blocks and plenty of rags and cleaning solvent for mopping up the inevitable spills. If you plan to rent a hoist, arrange for it in advance and perform all of the operations possible without it beforehand. This will save you money and time.

Plan for the vehicle to be out of use for a considerable period of time. A machine shop is a must for that work which you cannot accomplish without special skills and equipment. Machine shops often have busy schedules, so it's a good idea to coordinate your plans for overhaul work with your local shop. Careful planning will result in less down time for your vehicle.

Always use extreme caution when removing and installing the engine. Serious injury can result from careless actions. Plan ahead. If you take your time and think out each step before you proceed, even a job of this magnitude can be accomplished successfully.

6 Engine overhaul — disassembly and reassembly sequence

1   It is much easier to disassemble and reassemble on the engine if it is mounted on a portable engine stand, which can usually be rented for a reasonable fee from an equipment rental yard. Before the engine is mounted on a stand, the flywheel/driveplate should be removed from the engine (refer to Part A).
2   If you can't find a stand suitable for your engine, you can block it up on a sturdy workbench or on the floor. If you elect to work on the engine in this manner, be extra careful not to tip or drop it.
3   If you are performing a complete engine rebuild yourself, the following external components will have to come off. Even if you are going to obtain a rebuilt engine, the same components must still be removed from the old engine so they can be installed on the rebuilt. In general, this includes:

   Alternator and brackets
   Emissions control components
   Distributor, spark plug wires and spark plugs
   Thermostat and housing
   Water pump
   Fuel injection assembly
   Turbocharger assembly
   Intake/exhaust manifolds
   Engine mounts
   Flywheel/driveplate

Note: *When removing the external components from the engine, pay close attention to details that may be helpful or important during installation. Note the installed position of gaskets, seals, spacers, pins, washers, bolts and other small items.*

4   If you are obtaining a short block, which consists of the engine block, crankshaft, pistons and connecting rods all assembled, then the cylinder head, oil pan and oil pump will also have to be removed from the old engine. See Section 4 for additional information regarding the alternatives.
5   If you are planning a complete overhaul, the engine must be disassembled in the following order:

   *External engine components
   Oil pan
   Cylinder head cover
   Timing cover
   Balance shaft and carrier assembly (2.5L engine)
   Timing belt and sprockets
   Intermediate shaft and sprocket
   Cylinder head and camshaft
   Oil pump
   Piston/connecting rod assemblies
   Front oil seal housing
   Rear oil seal housing
   Crankshaft*

6   Before beginning the disassembly and overhaul procedures, make sure the following items are available:

   *Common hand tools
   Small cardboard boxes or plastic bags for storing parts
   Gasket scraper
   Ridge reamer
   Vibration damper puller
   Micrometers
   Telescoping gauges
   Dial indicator set
   Valve spring compressor
   Cylinder surfacing hone
   Piston ring groove cleaning tool
   Electric drill motor
   Tap and die set
   Wire brushes
   Cleaning solvent*

7   Before beginning engine reassembly, make sure you have all the necessary new parts, gaskets and seals as well as the following items on hand:

   *Common hand tools
   A 1/2-inch drive torque wrench
   Piston ring installation tool
   Piston ring compressor
   Short lengths of rubber or plastic hose to fit over connecting
      rod bolts
   Plastigage
   Feeler gauges
   A fine-tooth file
   New engine oil
   Engine assembly lube or moly-base grease
   RTV gasket sealant
   Anaerobic gasket sealant
   Thread locking compound*

8   In order to save time and avoid problems, engine reassembly should be done in the following order.

   *Crankshaft and main bearings
   Piston rings
   Piston/connecting rod assemblies
   Balance shaft assembly (2.5L engine)
   Oil pump and oil strainer
   Front oil seal housing
   Rear oil seal housing
   Oil pan
   Flywheel/driveplate
   Cylinder head
   Camshaft/rocker arm assembly
   Timing belt, sprockets and tensioner
   Timing cover
   Cylinder head cover
   External components*

Chapter 2 Part B General engine overhaul procedures

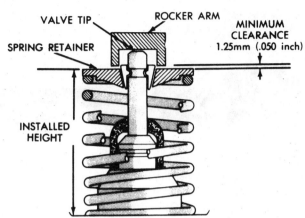

7.1 The valve spring installed height is measured from the upper edge of the retainer (note the minimum clearance required between the rocker arm and retainer — if the valves are serviced, the clearance may not be correct)

7 Cylinder head — disassembly

Refer to illustrations 7.1 and 7.2

Note: *Cylinder head service involves removal and disassembly of the intake and exhaust valves and their related components.*

Because some specialized tools are necessary for disassembly and inspection and because replacement parts are not always readily available, it may be more practical and economical for the home mechanic to purchase a replacement head rather than taking the time to disassemble, inspect and recondition the original head. New and rebuilt cylinder heads for most engines are usually available at dealerships and auto parts stores.

1 Before removing the valves, arrange to label and store them, along with their related components, so they can be stored separately and reinstalled in the same valve guides from which they are removed. Measure the valve spring installed height of each valve and compare it to the Specifications (see illustration). If it is greater than specified, the valve seats and faces need attention.

2 Compress the valve springs on the first valve with a spring compressor and remove the keepers (see illustration). Carefully release the valve spring compressor and remove the retainer, the springs, the valve stem seal, the spring seat and the valve from the head. If the valve binds in the guide and won't come out, push it back into the head and deburr the area around the keeper groove with a fine file or whetstone.

3 Repeat this procedure for each of the remaining valves. Remember to keep all the parts for each valve together so they can be reinstalled in the same locations.

4 Once the valves have been removed and safely stored, the head should be thoroughly cleaned and inspected. If a complete engine overhaul is being done, finish the engine disassembly procedures before beginning the cylinder head cleaning and inspection process.

8 Cylinder head — cleaning and inspection

Refer to illustrations 8.12, 8.14, 8.16, 8.18, 8.20a, 8.20b, 8.21a and 8.21b

1 Thorough cleaning of the cylinder head and related valve train components, followed by a detailed inspection, will enable you to decide how much valve service work must be done during the engine overhaul.

Cleaning

2 Scrape away all traces of old gasket material and sealing compound from the head gasket, intake manifold and exhaust manifold sealing surfaces. **Caution:** *Do not gouge the cylinder head.* Special gasket removal solvents which dissolve the gasket, making removal much easier, are available at auto parts stores.

3 Remove any built up scale around the coolant passages.

4 Run a stiff wire brush through the oil holes and EGR gas port passages to remove any deposits that may have formed in them.

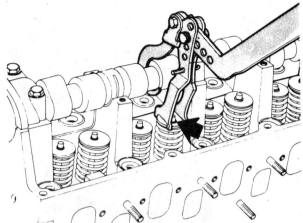

7.2 Use a valve spring compressor to compress the valve springs, then remove the keepers (arrow) from the valve stem (a magnet or needle-nose pliers may be needed)

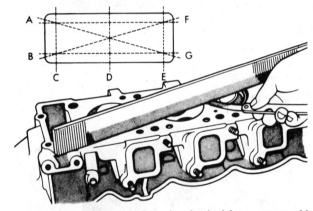

8.12 The cylinder head can be checked for warpage with a straightedge and feeler gauges

5 Run an appropriate size tap into each of the threaded holes to remove any corrosion and thread sealant that may be present. If compressed air is available, use it to clear the holes of debris produced by this operation. **Warning:** *Always wear safety goggles when using compressed air to blow away debris.*

6 Clean the exhaust and intake manifold stud threads with an appropriate size die. Clean the rocker arm pivot bolt or stud threads with a wire brush.

7 Clean the cylinder head with solvent and dry it thoroughly. Compressed air will speed the drying process and ensure that all holes and recessed areas are clean. **Note:** *Decarbonizing chemicals may prove helpful for cleaning cylinder heads and valve train components. They are very caustic and should be used with caution. Be sure to follow the instructions on the container.*

8 Clean the rocker arms and hydraulic lash adjusters with solvent and dry them thoroughly.

9 Clean all the valve springs, keepers and retainers with solvent and dry them thoroughly. Clean these assemblies one at a time to avoid mixing up the parts.

10 Scrape off any heavy deposits that may have formed on the valves, then use a motorized wire brush to remove the remaining deposits from the valve heads and stems. Again, do not mix up the valves.

Inspection

Cylinder head

11 Inspect the head very carefully for cracks, evidence of coolant leakage and other damage. If cracks are discovered, a new cylinder head must be obtained. Check the camshaft bearing surfaces in the head and the bearing caps. If there is evidence of excessive cam bearing galling or scoring, the head must be replaced. Failure to do so can lead to camshaft seizure.

12 Using a straightedge and feeler gauge, check the head gasket mating surface for warpage (see illustration). If the warpage exceeds

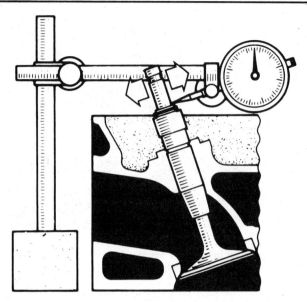

8.14 A dial indicator can be used to check for excessive valve stem-to-guide clearance

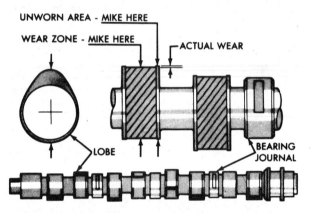

8.18 Cam lobe wear can be determined by measuring each lobe at the edge and in the center and subtracting the two measurements

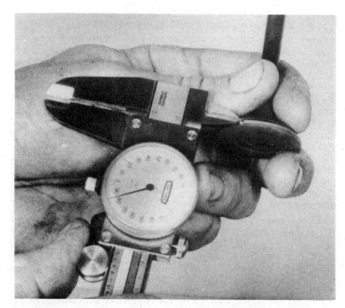

8.20a Measuring valve margin width with a dial caliper

8.16 If the camshaft bearing journals are worn, scored or pitted, a new camshaft is required

the specified amount, the head should be resurfaced at an automotive machine shop.

13 Examine the valve seats in each of the combustion chambers. If they are pitted, cracked or burned, take the head to an automotive machine shop for a valve job. This procedure is beyond the scope of the home mechanic.

14 Check the valve stem-to-valve guide clearance. Use a dial indicator to measure the lateral movement of each valve stem with the valve in the guide and raised off the seat slightly (see illustration). If there is still some doubt regarding the condition of the valve guides after this check, the exact clearance and condition of the guides can be checked by an automotive machine shop.

Rocker arms

15 The rocker arms ride below the camshaft and contact the valve stem at one end and the hydraulic lash adjuster at the other end. Check each rocker arm for wear, galling and pitting of the contact surfaces. Inspect the lash adjuster contact surfaces for pitting and wear as well.

Camshaft

16 Inspect the camshaft bearing journals for excessive wear and evidence of galling, scoring or seizure (see illustration). If the journals are damaged, the bearing surfaces in the head and bearing caps are probably damaged as well. Both the camshaft and cylinder head will have to be replaced.

17 Check the cam lobes for grooves, flaking, pitting and scoring. Measure the cam lobe height and compare it to the Specifications. If the lobe height is less than the minimum specified, and/or the lobes are damaged, get a new camshaft.

18 To determine the extent of cam lobe wear, measure the height of each lobe at the edge, in the unworn area, and in the center, where the rocker arm contacts the lobe (see illustration). Subtract the center measurement from the edge measurement to obtain the wear. Compare the results to the Specifications.

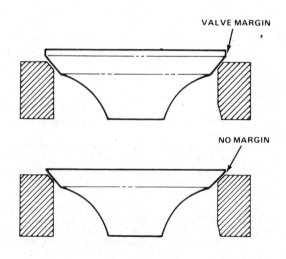

8.20b The margin width on each valve must be as specified (if no margin exists, the valve cannot be reused)

Chapter 2 Part B General engine overhaul procedures

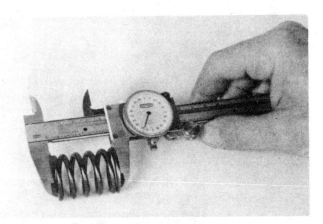

8.21a Measure the free length of each valve spring with a dial or vernier caliper

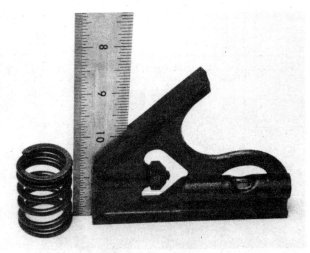

8.21b Check each valve spring for squareness

Valves

19 Carefully inspect each valve face for cracks, pits and burned spots. Check the valve stem and neck for cracks. Rotate the valve and check for any obvious indication that it is bent. Check the end of the stem for pits and excessive wear. The presence of any of these conditions indicates the need for valve service by an automotive machine shop.
20 Measure the width of the valve margin on each valve (see illustrations) and compare it to Specifications. Any valve with a margin narrower than specified will have to be replaced with a new one.

Valve components

21 Check each valve spring for wear and pitting. Measure the free length and compare it to the Specifications (see illustration). If a spring is shorter than specified, it has sagged and should not be reused. Stand the spring on a flat surface and check it for squareness (see illustration).
22 Check the spring retainers and keepers for obvious wear and cracks. Any questionable parts should be replaced with new ones. In the event that a retainer or keeper should fail during operation of the engine, extensive damage will occur.
23 If the inspection process indicates that the valve components are in generally poor condition and worn beyond the limits specified, which is often the case in an engine being overhauled, reassemble the valves in the cylinder head and refer to Section 9 for valve servicing recommendations.
24 If the inspection turns up no excessively worn parts, and if the valve faces and seats are in good condition, the valve train components can be reinstalled in the cylinder head without major servicing. Refer to the appropriate Section for cylinder head reassembly procedures.

9 Valves — servicing

1 Because of the complex nature of the job and the special tools and equipment needed, servicing of the valves, the valve seats and the valve guides, commonly known as a "valve job," is best left to a professional.
2 The home mechanic can remove and disassemble the head, do the initial cleaning and inspection, then reassemble and deliver the head to a dealer service department or an automotive machine shop for the actual valve servicing.
3 The dealer service department or automotive machine shop will remove the valves and springs, recondition or replace the valves and valve seats, recondition the valve guides, check and replace the valve springs, spring retainers and keepers (as necessary), replace the valve seals with new ones, reassemble the valve components and make sure the installed spring height is correct. The cylinder head gasket surface will also be resurfaced if it is warped.
4 After the valve job has been performed, the head will be in like new condition. When the head is returned, be sure to clean it again — with compressed air, if available — to remove any metal particles and abrasive grit that may still be present from the valve service or head resurfacing operations. If you have compressed air, use it blow out all the oil holes and passages.

10 Cylinder head — reassembly

1 Regardless of whether or not the head was sent to an automotive machine shop for valve servicing, make sure it is clean before beginning reassembly.
2 If the head was sent out for valve servicing, the valves and related components will already be in place.
3 Lay all of the spring seats in position, then install new seals on each of the valve guides. Push the seals down firmly over the valve guides.
4 Install the valves, taking care not to damage the new valve stem oil seals, the valve springs and the retainers. Coat the valve stems with engine assembly lube or moly-base grease before slipping them into the guides and install the springs with the painted side next to the retainer.
5 Compress the spring with a valve compressor tool and install the keepers. Do not allow the retainer to touch the seal. Release the compressor, making sure the keepers are seated properly in the valve stem groove(s). If necessary, grease can be used to hold the keepers in place until the compressor is released.
6 Double check the installed valve spring height. If it was correct before disassembly, it should still be within the specified limits. **Note:** *The spring height is measured from the bottom of the spring to the upper edge of the spring retainer.*

11 Piston/connecting rod assembly — removal

Refer to illustrations 11.1, 11.5 and 11.7

1 Using a ridge reamer, completely remove the ridge at the top of each cylinder (see illustration). Follow the manufacturer's instructions

11.1 A cutting tool is required to remove the ridge from the top of each cylinder

Chapter 2 Part B General engine overhaul procedures

11.5 Checking connecting rod end play with a feeler gauge

11.7 To prevent damage to the crankshaft journals and cylinder walls, slip sections of hose over the rod bolts before removing the pistons

12.1 Checking crankshaft end play with a dial indicator

12.3 Checking crankshaft end play with a feeler gauge

provided with the ridge reaming tool. **Caution:** *Failure to remove the ridge before attempting to remove the piston/connecting rod assemblies will result in piston breakage.*

2   With the engine in the upside-down position, remove the oil pickup tube and screen assembly from the bottom of the engine block. It is held in place with one bolt (plus the brace bolt on the 2.2L engine).
3   Before the connecting rod caps are removed, check the rod end play. Mount a dial indicator with its stem in line with the crankshaft and touching the side of the number one connecting rod cap.
4   Push the connecting rod backward, as far as possible, and zero the dial indicator. Next, push the connecting rod all the way to the front and check the reading on the dial indicator. The distance that it moves is the end play. If the end play exceeds the service limit, a new connecting rod will be required. Repeat the procedure for the remaining connecting rods.
5   An alternative method is to slip feeler gauges between the connecting rod and the crankshaft throw until the play is removed (see illustration). The end play is equal to the thickness of the feeler gauge(s).
6   Check the connecting rods and connecting rod caps for identification marks. If they are not plainly marked, use a small punch or scribe to label them correctly so that they will be reinstalled to the same cylinder from which they were removed.
7   Loosen each of the connecting rod cap nuts 1/2-turn. Remove the number one connecting rod cap and bearing insert. Do not drop the bearing insert out of the cap. Slip a short length of plastic or rubber hose over each connecting rod cap bolt to protect the crankshaft journal and cylinder wall when the piston is removed (see illustration) and push the connecting rod/piston assembly out through the top of the engine. Use a wooden tool to push on the upper bearing insert in the connecting rod. If resistance is felt, double-check to make sure that the ridge has been completely removed from the cylinder.
8   Repeat this procedure for each of the remaining cylinders. After removal, reattach the connecting rod caps and bearing inserts to their respective connecting rods and install the cap nuts finger-tight. Leaving the old bearing inserts in place until reassembly will help prevent the connecting rod bearing surfaces from being accidentally nicked or gouged.

12 Crankshaft — removal

Refer to illustrations 12.1, 12.3 and 12.4
Note: *The front crankshaft oil seal housing, the rear crankshaft oil seal housing and the oil pickup tube and screen assembly must be removed before the crankshaft is removed.*

1   Before the crankshaft is removed, check the end play. Mount a dial indicator with the stem in line with the crankshaft and just touching one of the crank throws (see illustration).

Chapter 2 Part B General engine overhaul procedures

12.4 Mark the main bearing caps with a center punch before removing them

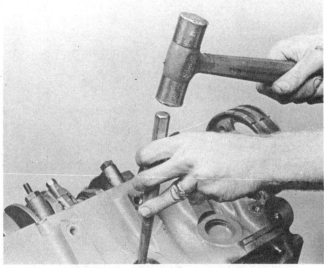

13.1a Drive each soft plug into the block with a large punch and hammer, . . .

5 Gently tap the caps with a soft-face hammer, then separate them from the engine block. If necessary, use the main bearing cap bolts as levers to remove the caps. Sometimes the bearing inserts come out with the caps. If they do, don't drop them.
6 Carefully lift the crankshaft out of the engine. It is a good idea to have an assistant available, since the crankshaft is quite heavy. With the bearing inserts in place in the engine block and in the main bearing caps, return the caps to their respective locations on the engine block and tighten the bolts finger tight.

13 Engine block — cleaning

Refer to illustrations 13.1a, 13.1b and 13.10

1 Remove the soft plugs from the engine block. To do this, knock the plugs into the block, using a hammer and punch, then grasp them with large pliers and pull them back through the holes (see illustrations).
2 Using a gasket scraper, remove all traces of gasket material from the engine block. Be very careful not to nick or gouge the gasket sealing surfaces.
3 Remove the main bearing caps and separate the bearing inserts from the caps and the engine block. Tag the bearings to indicate the cap or saddle from which they were removed and the cylinder to which they must be returned. Set them aside.
4 Using an allen wrench of the correct size, remove any threaded oil gallery plugs from the block.
5 If the engine is extremely dirty it should be taken to an automotive machine shop to be steam cleaned or hot tanked.
6 After the block is returned, clean all oil holes and oil galleries one more time. Brushes for cleaning oil holes and galleries are available at most auto parts stores. Flush the passages with warm water until the water runs clear, dry the block thoroughly and wipe all machined surfaces with a light, rust preventive oil. If you have access to compressed air, use it to speed the drying process and to blow out all the oil holes and galleries.
7 If the block is not extremely dirty or sludged up, you can do an adequate cleaning job with warm soapy water and a stiff brush. Take plenty of time and do a thorough job. Regardless of the method used, thoroughly clean all oil holes and galleries, dry the block completely and coat all machined surfaces with light oil.
8 The threaded holes in the block must be clean to ensure accurate torque readings during reassembly. Run the proper size tap into each of the holes to remove any rust, corrosion, thread sealant or sludge and to restore any damaged threads. If possible, use compressed air to clear the holes of debris produced by this operation. Thoroughly clean the threads on the head bolts and the main bearing cap bolts as well.

13.1b . . . then grip it with a pair of pliers and lever it out of the hole

2 Push the crankshaft all the way to the rear and zero the dial indicator. Next, pry the crankshaft to the front as far as possible and check the reading on the dial indicator. The distance that it moves is the end play. If it is greater than specified, check the crankshaft thrust surfaces for wear. If no wear is apparent, new main bearings should correct the end play.
3 If a dial indicator is not available, feeler gauges can be used. Gently pry or push the crankshaft all the way to the front of the engine. Slip feeler gauges between the crankshaft and the front face of the thrust main bearing to determine the clearance (see illustration).
4 Loosen each of the main bearing cap bolts 1/4-turn at a time, until they can be removed by hand. Check the main bearing caps to see if they are marked correctly with respect to their locations. They are usually numbered consecutively from the front of the engine to the rear and may have arrows which point to the front of the engine. If they are not marked, label them with number stamping dies or a center punch (see illustration).

Chapter 2 Part B General engine overhaul procedures

13.10 A large socket, mounted on an extension, can be used to drive the new soft plugs into the block

14.4a Use a telescoping gauge to determine the cylinder bore size, . . .

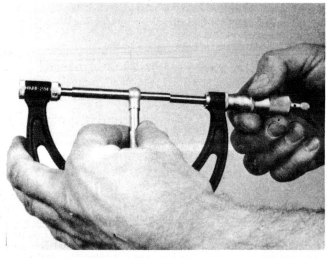

14.4b . . . then measure the gauge with a micrometer to obtain the diameter

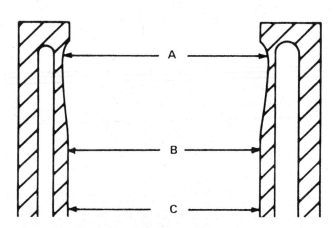

14.4c Measure the diameter of each cylinder just under the wear ridge (A), at the center (B) and at the bottom (C)

clean the threads on the head bolts and the main bearing cap bolts as well.
9   Reinstall the main bearing caps and tighten the bolts finger tight.
10 After coating the sealing surfaces of the new soft plugs with a good quality gasket sealer, install them in the engine block (see illustration). Make sure they are driven in straight and seated properly or leakage could result. Special tools are available for this job, but equally good results can be obtained with a hammer and large socket. The outside diameter of the socket should just slip into the soft plug.
11 If the engine is not going to be reassembled right away, cover it with a large plastic trash bag to keep it clean.

14 Engine block — inspection

Refer to illustrations 14.4a, 14.4b, 14.4c, 14.7a and 14.7b

1   Thoroughly clean the engine block as described in Section 13 and double-check to make sure that the ridge at the top of each cylinder has been completely removed.
2   Visually check the block for cracks, rust and corrosion. Look for stripped threads in the threaded holes. It is also a good idea to have the block checked for hidden cracks by an automotive machine shop that has the special equipment to do this type of work. If defects are found, have the block repaired. If this isn't possible, replace it.
3   Check the cylinder bores for scuffing and scoring.
4   Measure each cylinder's diameter at the top (just under the ridge), center and bottom of the cylinder bore, parallel to the crankshaft axis (see illustrations). Next, measure each cylinder's diameter at the same three locations across the crankshaft axis. Compare the results to the Specifications. If the cylinder walls are badly scuffed or scored, or if they are out of round or tapered beyond the specified limits, have the engine block rebored and honed at an automotive machine shop. If the block is rebored, you will have to obtain correctly oversized pistons and rings.
5   If the cylinders are in reasonably good condition and not worn to the outside of the limits, and if the piston-to-cylinder clearances can be maintained properly, then they do not have to be rebored. But they still must be honed.
6   Before honing the cylinders, install the main bearing caps and tighten the bolts to the specified torque.
7   To perform the honing operation you will need the proper size flexible hone and fine stones, plenty of light oil or honing oil, some rags and an electric drill motor. Mount the hone in the drill motor, compress the stones and slip the hone into the first cylinder (see illustration). Lubricate the cylinder thoroughly, turn on the drill and move the hone up and down in the cylinder at a pace which will produce a fine crosshatch pattern on the cylinder walls with the crosshatch lines intersecting at approximately a 60° angle (see illustration). Be sure to use

Chapter 2 Part B General engine overhaul procedures

14.7a A surfacing hone should be used to prepare the cylinders for the new rings

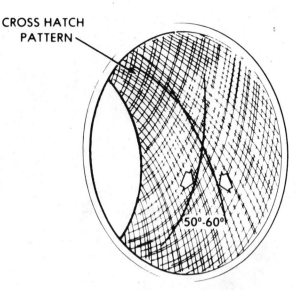

14.7b The cylinder hone should leave a cross hatch pattern with the lines intersecting at approximately a 60° angle

15.4a If available, a ring groove cleaning tool can be used to remove carbon from the ring grooves

15.4b If the tool is not available, use a piece of broken piston ring to carefully remove carbon deposits from the ring grooves (use caution not to remove any of the piston material)

plenty of lubricant. Do not withdraw the hone from the cylinder while it is running. Instead, shut off the drill and continue moving the hone up and down in the cylinder until it comes to a complete stop, then compress the stones and withdraw the hone. Wipe the oil out of the cylinder and repeat the procedure on the remaining cylinders. If you do not have the tools or do not desire to perform the honing operation, most automotive machine shops will do it for a reasonable fee.
8 After the honing job is complete, chamfer the top edges of the cylinder bores with a small file so the rings will not catch when the pistons are installed.
9 The entire engine block must be thoroughly washed again with warm, soapy water to remove all traces of the abrasive grit produced during the honing operation. Be sure to run a brush through all oil holes and galleries and flush them with running water. After rinsing, dry the block and apply a coat of light rust preventative oil to all machined surfaces. Wrap the block in a plastic trash bag to keep it clean and set it aside until reassembly.

15 Piston/connecting rod assembly — inspection

Refer to illustrations 15.4a, 15.4b, 15.10, 15.11a, 15.11b and 15.11c
1 Before the inspection process can be carried out, the piston/connecting rod assemblies must be cleaned and the original piston rings removed from the pistons. **Note:** *Always use new piston rings when the engine is reassembled.*
2 Using a piston ring installation tool, carefully remove the rings from the pistons. Do not nick or gouge the pistons in the process.
3 Scrape all traces of carbon from the top (or crown) of the piston. A hand-held wire brush or a piece of fine emery cloth can be used once the majority of the deposits have been scraped away. **Caution:** *Do not, under any circumstances, use a wire brush mounted in a drill motor to remove deposits from the pistons. The piston material is soft and will be eroded away by the wire brush.*
4 Use a piston ring groove cleaning tool to remove any carbon deposits from the ring grooves (see illustration). If a tool is not available, a broken piece of the old ring will do the job (see illustration). Be very careful to remove only the carbon deposits. Do not remove any metal and do not nick or scratch the sides of the ring grooves.
5 Once the deposits have been removed, clean the piston/rod

Chapter 2 Part B General engine overhaul procedures

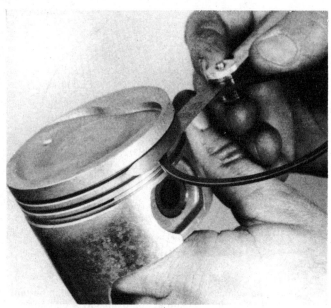

15.10 Checking the piston ring side clearance with a feeler gauge

15.11a Measure the piston diameter at the point indicated in the text

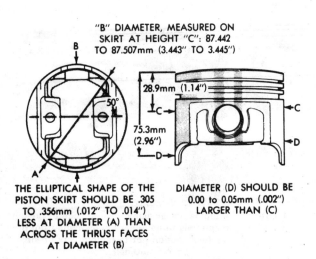

15.11b Non-turbocharged engine piston measurement details

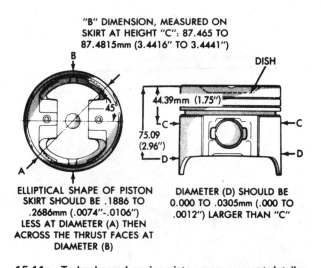

15.11c Turbocharged engine piston measurement details

assemblies with solvent and dry them thoroughly. Make sure that the oil return holes in the back sides of the lower ring grooves are clear.

6 Normal piston wear is indicated by even vertical lines on the piston thrust surfaces and slight looseness of the top ring in its groove. If the pistons are not damaged or worn excessively and if the engine block is not rebored, new pistons will not be necessary. However, new piston rings should always be used when an engine is rebuilt.

7 Carefully inspect each piston for cracks around the skirt, at the pin bosses and at the ring lands.

8 Look for scoring and scuffing on the thrust faces of the skirt, holes in the piston crown and burned areas at the edge of the crown. If the skirt is scored or scuffed, the engine may have been suffering from overheating and/or abnormal combustion which caused excessively high operating temperatures. The cooling and lubrication systems should be checked thoroughly. A hole in the piston crown is an indication that abnormal combustion (preignition) was occurring. Burned areas at the edge of the piston crown are usually evidence of spark knock (detonation). If any of the above problems exist, the causes must be corrected or the damage will occur again.

9 Corrosion of the piston, evidenced by pitting, indicates that coolant is leaking into the combustion chamber and/or the crankcase. Again, the cause must be corrected or the problem will persist in the rebuilt engine.

10 Measure the piston ring side clearance by laying a new piston ring in each ring groove and slipping a feeler gauge between the ring and the edge of the ring groove (see illustration). Check the clearance at three or four locations around each groove. Be sure to use the correct ring for each groove; they are different. If the side clearance is greater than specified, new pistons will have to be used.

11 Check the piston-to-bore clearance by measuring the bore (see Section 14) and the piston diameter (see illustrations). Make sure that the pistons and bores are correctly matched. Measure the piston across the skirt, on the thrust faces (at a 90° angle to the piston pin). Take the measurement 1.140-inches (28.9 mm) (non-turbocharged engine) or 1.750-inches (44.39 mm) (turbocharged engines) down from the edge of the crown. Subtract the piston diameter from the bore diameter to obtain the clearance. If it is greater than specified, the block will have to be rebored and new pistons and rings installed. Check the

Chapter 2 Part B General engine overhaul procedures

16.2 Measure the diameter of each crankshaft journal at several points to detect taper and out-of-round conditions

piston-to-rod clearance by twisting the piston and rod in opposite directions. Any noticeable play indicates that there is excessive wear, which must be corrected. The piston/connecting rod assemblies should be taken to an automotive machine shop to have the pistons and connecting rods rebored and new pins installed.

12 If the pistons must be removed from the connecting rods — for instance, when new pistons must be installed — or if the piston pins have too much play in them, they should be taken to an automotive machine shop. Have the rods checked for bend and twist at the same time. Automotive machine shops have special equipment for this purpose. Unless new pistons or connecting rods must be installed, do not disassemble the pistons from the connecting rods.

13 The connecting rod bolts on 2.2L/2.5L engines are designed to stretch a certain amount during tightening of the nuts. If the bolts are stretched ("necked down") beyond specifications they must be replaced with new ones. The connecting rod bolts can be checked to make sure they are not stretched by running a 3/8 in x 24 nut all the way down the threads. If the nut does not run down the threads smoothly, replace the bolts with new ones.

14 Check the connecting rods for cracks and other damage. Temporarily remove the rod caps, lift out the old bearing inserts, wipe the rod and cap bearing surfaces clean and inspect them for nicks, gouges and scratches. After checking the rods, replace the old bearings, slip the caps into place and tighten the nuts finger tight.

16 Crankshaft — inspection

Refer to illustration 16.2

1 Clean the crankshaft with solvent and dry it thoroughly. Be sure to clean the oil holes with a stiff brush and flush them with solvent. Check the main and connecting rod bearing journals for uneven wear, scoring, pitting and cracks. Check the remainder of the crankshaft for cracks and damage.

2 Measure the diameter of the main and connecting rod journals with a micrometer (see illustration) and compare the results to the Specifications. By measuring the diameter at a number of points around the journal's circumference, you will be able to determine whether or not the journal is out-of-round. Take the measurement at each end of the journal, near the crank counterweights, to determine whether the journal is tapered.

3 If the crankshaft journals are damaged, tapered, out-of-round or worn beyond the limits given in the Specifications, have the crankshaft reground by an automotive machine shop. Be sure to use the correct size bearing inserts if the crankshaft is reconditioned.

4 Refer to Section 17 and examine the main and rod bearing inserts.

17 Main and connecting rod bearings — inspection

1 The main and connecting rod bearings should always be replaced with new ones when the engine is overhauled. Don't discard the old bearings. They can reveal valuable information about the condition of the engine.

2 Bearing failure occurs because of lack of lubrication, the presence of dirt or other foreign particles, overloading the engine and corrosion. Regardless of the cause of bearing failure, it must be corrected before the engine is reassembled to prevent it from happening again.

3 When examining the bearings, remove them from the engine block, the main bearing caps, the connecting rods and the rod caps and lay them out on a clean surface in the same order and location which they occupied in the engine. This is the only way you can match a bearing problem to its corresponding crankshaft journal.

4 Dirt and other foreign particles get into the engine in a number of ways. Sometimes it isn't removed from the engine during assembly. Or it enters through filters or breathers. Either way, it gets into the oil and then into the bearings. Metal chips from machining operations and normal engine wear invade the oil as well. Abrasives are sometimes left in engine components after reconditioning, especially when parts are not thoroughly cleaned using the proper cleaning methods. No matter where it comes from, this stuff usually finds its way into bearing clearances, where it embeds itself into soft bearing material. This kind of problem is easy to identify. Larger particles, however, will not embed in the bearing. They will score or gouge the bearing and shaft. So even though the cause itself may not be visible, the effect will be just as easy to see as smaller embedded particles. The best, and really the only, prevention for either of these causes of bearing failure is to clean all parts thoroughly and keep everything spotlessly clean during engine assembly. Frequent and regular engine oil and filter changes are also recommended.

5 Lack of lubrication (or lubrication breakdown) has a number of interrelated causes. Excessive engine operation heat thins the oil. Overloading between journals and bearing surfaces sometimes squeezes the thin film of oil from the bearing face. Oil leakage or throw off from excessive bearing clearances, worn oil pump or high engine speeds all contribute to lubrication breakdown. Blocked oil passages, which usually are the result of misaligned oil holes in a bearing shell, will also oil starve a bearing and destroy it. When lack of lubrication is the cause of bearing failure, the bearing material is wiped or extruded from the steel backing of the bearing. Temperatures may increase to the point where the steel backing turns blue from overheating.

6 Driving habits can have a definite effect on bearing life. Full throttle low speed operation, lugging the engine, puts very high loads on bearings, which tends to squeeze out the oil film. These loads cause the bearings to flex, which produces fine cracks in the bearing face (fatigue failure). Eventually the bearing material will loosen in pieces and tear away from the steel backing. Short trip driving leads to corrosion of bearings because insufficient engine heat is produced to drive off the condensed water and corrosive gases. These products collect in the engine oil, forming acid and sludge. As the oil is carried to the engine bearings, the acid attacks and corrodes the bearing material.

7 Incorrect bearing installation during engine assembly will lead to bearing failure as well. Tight fitting bearings leave insufficient bearing oil clearance and will result in oil starvation. Dirt or foreign particles trapped behind a bearing insert result in high spots on the bearing which lead to failure.

18 Piston rings — installation

Refer to illustrations 18.3a, 18.3b, 18.9a, 18.9b, 18.10 and 18.12

1 Before installing the new piston rings, the ring end gaps must be checked. It is assumed that the piston ring side clearance has been measured and verified to be correct (Section 15).

2 Lay out the piston/connecting rod assemblies and the new ring sets so the ring sets will be matched with the same piston and cylinder during the end gap measurement and engine assembly.

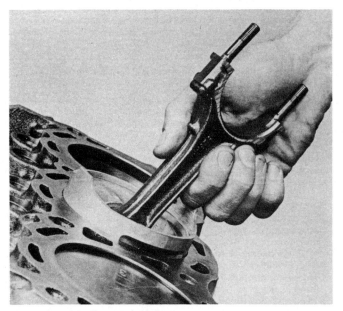

18.3a Use the piston to square up the ring in the cylinder prior to checking the ring end gap

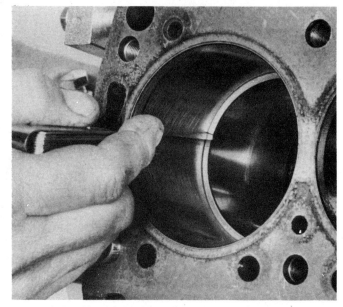

18.3b Measure the ring end gap with a feeler gauge

18.9a Installing the spacer/expander in the oil control ring groove

18.9b *Do not* use a piston ring tool when installing the oil ring side rails

3 Insert the top (number one) ring into the first cylinder and square it up with the cylinder walls by pushing it in with the top of the piston (see illustration). The ring should be near the bottom of the cylinder at the lower limit of ring travel. To measure the end gap, slip a feeler gauge between the ends of the ring (see illustration). Compare the measurement to the Specifications.

4 If the gap is larger or smaller than specified, double-check to make sure that you have the correct rings before proceeding.

5 If the gap is too small, it must be enlarged or the ring ends may come in contact with each other during engine operation, which can cause serious damage to the engine. The end gap can be increased by filing the ring ends very carefully with a fine file. Mount the file in a vise equipped with soft jaws, slip the ring over the file with the ends contacting the file face and slowly move the ring to remove material from the ends. When performing this operation, file only from the outside in.

6 Excess end gap is not critical unless it is greater than 0.040-inch (1 mm). Again, double-check to make sure you have the correct rings for your engine.

7 Repeat the procedure for each ring that will be installed in the first cylinder and for each ring in the remaining cylinders. Remember to keep rings, pistons and cylinders matched up.

8 Once the ring end gaps have been checked and corrected, the rings can be installed on the pistons.

9 The oil control ring (lowest one on the piston) is installed first. It is composed of three separate components. Slip the spacer/expander into the groove (see illustration), then install the lower side rail with the size mark and manufacturer's stamp facing up. Do not use a piston ring installation tool on the oil ring side rails, as they may be damaged. Instead, place one end of the side rail into the groove between the spacer/expander and the ring land, hold it firmly in place and slide a finger around the piston while pushing the rail into the groove (see illustration). Next, install the upper side rail in the same manner.

10 After the three oil ring components have been installed, check to make sure that both the upper and lower side rails can be turned smoothly in the ring groove. Position the end gaps correctly (see illustration).

11 The number two (middle) ring is installed next. It is stamped with a mark which should face toward the top of the piston. **Note:** *Always follow the instructions printed on the ring package or box. Different manufacturers may specify slight variations in method of installation. Do not mix up the top and middle rings — they have different cross*

Chapter 2 Part B General engine overhaul procedures

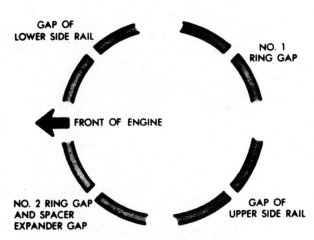

18.10 Position the ring end gaps as shown here before installing the pistons in the block

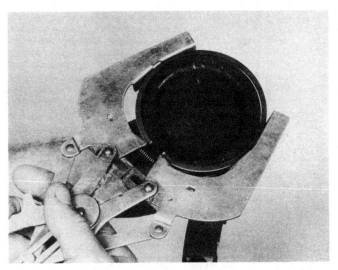

18.12 Install the compression rings with a ring expander

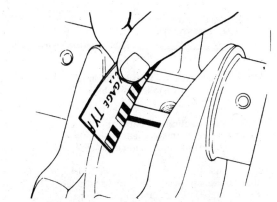

19.14 Compare the width of the crushed Plastigage (at its widest point) to the scale on the container to obtain the main bearing oil clearance

sections.
12 Use a piston ring installation tool and make sure that the identification mark is facing the top of the piston, then slip the ring into the middle groove on the piston (see illustration). Do not expand the ring any more than is necessary to slide it over the piston.
13 Install the number one (top) ring in the same manner. Make sure the identifying mark is facing up. Be careful not to confuse the number one and number two rings. Refer to the illustration for ring gap positioning.
14 Repeat this procedure for the remaining pistons and rings.

19 Crankshaft — installation and main bearing oil clearance check

Refer to illustration 19.14

1 Crankshaft installation is generally one of the first steps in engine reassembly. It is assumed at this point that the engine block and crankshaft have been cleaned, inspected and repaired or reconditioned.
2 Position the engine with the bottom facing up.
3 Remove the main bearing cap bolts and lift out the caps. Lay them out in the proper order to ensure that they are installed correctly.
4 If they are still in place, remove the old bearing inserts from the block and the main bearing caps. Wipe the main bearing surfaces of the block and caps with a clean, lint-free cloth. They must be kept spotlessly clean.
5 Clean the back sides of the new main bearing inserts and lay one bearing half in each main bearing saddle in the block. Lay the other bearing half from each bearing set in the corresponding main bearing cap. Make sure the tab on the bearing insert fits into the recess in the block or cap. Also, the oil holes in the block must line up with the oil holes in the bearing insert. Do not hammer the bearing into place and do not nick or gouge the bearing faces. No lubrication should be used at this time.
6 The flanged thrust bearing must be installed in the number three (center) cap and saddle.
7 Clean the faces of the bearings in the block and the crankshaft main bearing journals with a clean, lint-free cloth. Check or clean the oil holes in the crankshaft, as any dirt here can go only one way — straight through the new bearings.
8 Once you are certain that the crankshaft is clean, carefully lay it in position (an assistant would be very helpful here) in the main bearings.
9 Before the crankshaft can be permanently installed, the main bearing oil clearance must be checked.
10 Trim several pieces of the appropriate size of Plastigage, so they are slightly shorter than the width of the main bearings, and place one piece on each crankshaft main bearing journal, parallel with the journal axis.
11 Clean the faces of the bearings in the caps and install the caps in their respective positions (do not mix them up) with any arrows pointing toward the front of the engine. Do not disturb the Plastigage.
12 Starting with the center main and working out toward the ends, tighten the main bearing cap bolts, in three steps, to the specified torque. **Note:** *Do not rotate the crankshaft at any time during this operation.*
13 Remove the bolts and carefully lift off the main bearing caps. Keep them in order. Do not disturb the Plastigage or rotate the crankshaft. If any of the main bearing caps are difficult to remove, tap them gently from side-to-side with a soft-face hammer to loosen them.
14 Compare the width of the crushed Plastigage on each journal to the scale printed on the Plastigage wrapper to obtain the main bearing oil clearance (see illustration). Check the Specifications to make sure your measurement is correct.
15 If the clearance is not correct, double-check to make sure you have the right size bearing inserts. Also, make sure that no dirt or oil was between the bearing inserts and the main bearing caps or the block when the clearance was measured.
16 Carefully scrape all traces of the Plastigage material off the main bearing journals and/or the bearing faces. Do not nick or scratch the bearing faces.
17 Carefully lift the crankshaft out of the engine. Clean the bearing faces in the block, then apply a thin, uniform layer of clean, high quality moly-base grease or engine assembly lube to each of the bearing surfaces. Be sure to coat the thrust faces as well as the journal face of the thrust bearing.
18 Make sure the crankshaft journals are clean, then lay the crankshaft back in place in the block. Clean the faces of the bearings in the caps, then apply a thin layer of clean, moly-base grease or engine assembly lube to each of the bearing faces. Install the caps in their respective positions with the arrows pointing toward the front of the engine. Install

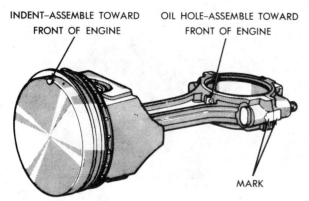

20.8a On 1984 and 1985 2.2L engines, the indent on the piston and the oil hole in the rod must face the *front* of the engine

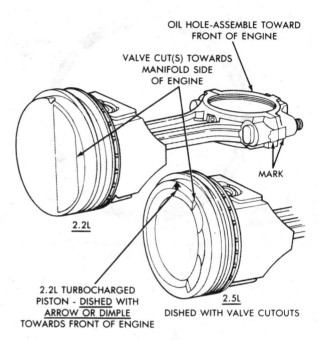

20.8b 1986 model 2.2L and 2.5L engine piston and rod installation details

20.8c Leave the piston (arrow) protruding about 1/4-inch out of the bottom of the ring compressor to align it in the bore

the bolts and tighten them to the specified torque, starting with the center main and working out toward the ends. Work up to the final torque in three steps.
19 On manual transaxle models, install a new pilot bearing in the end of the crankshaft. Lubricate the crankshaft cavity and the outer circumference of the bearing with clean engine oil and place the bearing in position. Tap it fully and evenly into the cavity using a section of pipe and a hammer. Lubricate the inside of the bearing with grease.
20 Rotate the crankshaft a number of times by hand to check for any obvious binding.
21 The final step is to check crankshaft end play with a feeler gauge or a dial indicator (Section 12).

20 Piston/connecting rod assembly – installation and bearing oil clearance check

Refer to illustrations 20.8a, 20.8b, 20.8c, 20.9, 20.11 and 20.13

1 Before installing the piston/connecting rod assemblies, the cylinder walls must be perfectly clean, the top edge of each cylinder must be chamfered, and the crankshaft must be in place.
2 Remove the connecting rod cap from the end of the number one connecting rod. Remove the old bearing inserts and wipe the bearing surfaces of the connecting rod and cap with a clean, lint free cloth. Everything must be spotlessly clean.
3 Clean the back side of the new upper bearing half, then lay it in place in the connecting rod. Make sure that the tab on the bearing fits into the recess in the rod. Do not hammer the bearing insert into place and be very careful not to nick or gouge the bearing face. Do not lubricate the bearing at this time.
4 Clean the back side of the other bearing insert and install it in the rod cap. Again, make sure the tab on the bearing fits into the recess in the cap, and do not apply any lubricant. It is critically important that the mating surfaces of the bearing and connecting rod are perfectly clean and oil free when they are assembled.
5 Position the piston ring gaps as shown (see illustration 18.10), then slip a section of plastic or rubber hose over the connecting rod cap bolts.
6 Lubricate the piston and rings with clean engine oil and attach a piston ring compressor to the piston. Leave the skirt protruding about 1/4-inch to guide the piston into the cylinder. The rings must be compressed as far as possible.
7 Rotate the crankshaft until the number one connecting rod journal is as far from the number one cylinder as possible (bottom dead center), and apply a coat of engine oil to the cylinder walls.
8 The indentation on the piston and oil hole in the connecting rod big end must face the front of the engine. On 1986 models, the valve relief on the piston crown must be on the manifold side of the engine (see illustrations). Gently insert the piston/connecting rod assembly into the number one cylinder bore (see illustration) and rest the bottom edge of the ring compressor on the engine block. Tap the top edge of the ring compressor to make sure it contacts the block around its entire circumference.
9 Carefully tap on the top of the piston with the end of a wooden hammer handle (see illustration) while guiding the end of the connecting rod into place on the crankshaft journal. The piston rings may try to pop out of the ring compressor just before entering the cylinder bore, so keep some downward pressure on the ring compressor. Work slowly, and if any resistance is felt as the piston enters the cylinder, stop immediately. Find out what is hanging up and fix it before proceeding. **Caution:** *Do not, for any reason, force the piston into the cylinder. If you force it, you will break a ring and/or the piston.*
10 Once the piston/connecting rod assembly is installed, the connecting rod bearing oil clearance must be checked before the rod cap is permanently bolted in place.

Chapter 2 Part B General engine overhaul procedures

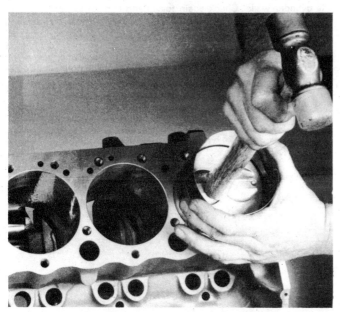

20.9 If resistance is encountered when tapping the piston into the block, *stop immediately* and make sure the rings are fully compressed

20.11 Carefully lay the Plastigage on the crankshaft rod journal

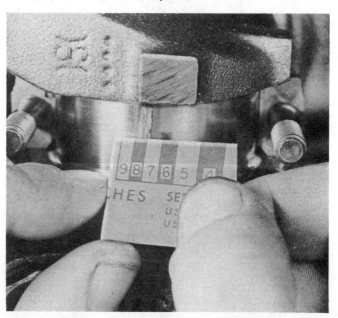

20.13 The crushed Plastigage is compared to the scale printed on the container to obtain the rod bearing oil clearance

11 Cut a piece of the appropriate size Plastigage slightly shorter than the width of the connecting rod bearing and lay it in place on the number one connecting rod journal, parallel with the journal axis (see illustration). It should not cross the oil hole in the journal.
12 Clean the connecting rod cap bearing face, remove the protective hoses from the connecting rod bolts and install the rod cap. Make sure the mating mark on the cap is on the same side as the mark on the connecting rod. Lubricate the rod bolt threads, install the nuts finger tight and then tighten them alternately to the specified torque. **Note:** *Do not rotate the crankshaft at any time during this operation.*
13 Remove the rod cap, being very careful not to disturb the Plastigage. Compare the width of the crushed Plastigage to the scale printed on the Plastigage container to obtain the oil clearance (see illustration). Compare it to the Specifications to make sure the clearance is correct. If the clearance is not correct, double-check to make sure that you have the correct size bearing inserts. Also, recheck the crankshaft connecting rod journal diameter and make sure that no dirt or oil was between the bearing inserts and the connecting rod or cap when the clearance was measured.
14 Carefully scrape all traces of the Plastigage material off the rod journal and bearing face. Be very careful not to scratch the bearing — use your fingernail or a piece of hardwood. Make sure the bearing faces are perfectly clean, then apply a uniform layer of clean, high quality moly-base grease or engine assembly lube to both of them. You will have to push the piston into the cylinder to expose the face of the bearing insert in the connecting rod. Be sure to slip the protective hoses over the rod bolts first.
15 Slide the connecting rod back into place on the journal, remove the protective hoses from the rod cap bolts, install the rod cap and tighten the nuts to the specified torque. Again, work up to the torque alternately.
16 Repeat the entire procedure for the remaining piston/connecting rod assemblies. Keep the back sides of the bearing inserts and the inside of the connecting rod and cap perfectly clean when assembling them. Make sure you have the piston matched to the correct cylinder. Use plenty of oil to lubricate the piston before installing the ring compressor. Also, when installing the rod caps for the final time, be sure to lubricate the bearing faces adequately.
17 After all the piston/connecting rod assemblies have been properly installed, rotate the crankshaft a number of times by hand to check for any obvious binding.
18 As a final step, the connecting rod end play must be checked (Section 11). Compare the measured end play to the Specifications to make sure it is correct.

21 Initial start-up and break-in after overhaul

1 Once the engine has been properly installed in the vehicle, double-check the engine oil and coolant levels.
2 With the spark plugs out of the engine and the coil high tension lead grounded to the engine block, crank the engine until oil pressure registers on the gauge (if so equipped) or until the oil light goes off.
3 Install the spark plugs, hook up the plug wires and the coil high tension lead.
4 It may take a few moments for the gasoline to reach the fuel injectors, but the engine should start without a great deal of effort.
5 As soon as the engine starts it should be allowed to warm up to normal operating temperature. While the engine is warming up, make

a thorough check for oil and coolant leaks.

6   After the engine reaches normal operating temperature, shut it off and recheck the engine oil and coolant levels. Restart the engine and check the ignition timing and the engine idle speed (refer to Chapter 1). Make any necessary adjustments.

7   Drive the vehicle to an area with minimum traffic, accelerate at full throttle from 30 to 50 mph, then allow the vehicle to slow to 30 mph with the throttle closed. Repeat the procedure 10 or 12 times. This will load the piston rings and cause them to seat properly against the cylinder walls. Check again for oil and coolant leaks.

8   Drive the vehicle gently for the first 500 miles (no sustained high speeds) and keep a constant check on the oil level. It is not unusual for an engine to use oil during the break-in period.

9   At approximately 500 to 600 miles, change the oil and filter again.

10 For the next few hundred miles, drive the vehicle normally. Do not pamper it or abuse it.

11 After 2000 miles, change the oil and filter again and consider the engine fully broken in.

Chapter 3
Cooling, heating and air conditioning systems

Contents

Air conditioning system — description and testing	15
Antifreeze — general information	2
Cooling system check	See Chapter 1
Cooling system servicing (draining, flushing and refilling)	See Chapter 1
Drivebelts — inspection, replacement and adjustment	7
Fan motor and shroud assembly — disassembly, inspection and reassembly	9
Fan motor and shroud assembly — removal and installation	8
General information	1
Heater core, evaporator core and blower motor — removal and installation	13
Heater and air conditioner control — removal and installation	11
Heater and air conditioner evaporator assembly — removal and installation	12
Radiator — inspection	6
Radiator — removal and installation	5
Thermostat — removal and installation	3
Turbocharger coolant hose and tube assembly — checking and replacement	10
Underhood hose check and replacement	See Chapter 1
Water pump — removal and installation	4

Specifications

General
Radiator pressure cap rating .... 14 to 18 psi

Thermostat
Rating .... 195°F (91°C)
Initial opening temperature .... Same as thermostat rating
Fully open temperature .... 219°F (104°C)

Torque specifications	Ft-lbs	Nm
Fan shroud bolts	8.9	12.1
Radiator hose clamp	3	4
Turbocharger coolant tube connections	30	41
Turbocharger coolant hose clamp	1.5	2
Thermostat housing		
bolt		
1984 and 1985	14	19
1986	20	27
nut		
1984 and 1985	17	23
1986	20	27
Water pump mounting bolts		
upper three bolts	21	28
lower bolt		
1984 and 1985	50	68
1986	40	54
Water pump pulley nuts and bolts	10.4	14

1 General information

Warning: *When working in the vicinity of the fan, always make sure the ignition is turned off and the negative battery cable is disconnected.*

The cooling system on all models consists of a radiator, an electrically-driven fan mounted in the radiator shroud, a thermostat, a water pump and a coolant reserve tank.

Coolant is circulated through the radiator tubes and is cooled by air passing through the cooling fins. The coolant is circulated by a pump mounted on the engine and driven by a belt.

A thermostat allows the engine to warm up by remaining closed until the coolant in the engine is at operating temperature. The thermostat then opens, allowing full circulation of the coolant throughout the cooling system.

The electric fan is actuated by the on-board computer, when the air conditioning turns on, and on some models, by a thermal switch. This aids cooling by drawing air through the radiator.

The radiator cap contains a vent valve which allows coolant to escape through a tube to the reserve tank. When the engine cools, vacuum in the radiator draws the coolant back from the tank so the coolant level remains constant.

On turbocharged models, coolant is circulated from the cylinder head through the turbocharger housing (to cool the center bearing) and back to the cylinder head water box.

The heating system operates by directing air through the heater core mounted in the dash and then to the interior of the vehicle by a system of ducts. Temperature is controlled by mixing heated air with fresh air, using a system of flapper doors in the ducts and a heater motor.

Some models are equipped with an air conditioner/heater system consisting of an evaporator core and ducts in the dash and a compressor in the engine compartment.

Chapter 3 Cooling, heating and air conditioning systems

2 Antifreeze — general information

Warning: *Do not allow antifreeze to contact your skin or the painted surfaces of the vehicle. Flush contacted areas immediately with water. Antifreeze can be fatal to children and pets (they like its sweet taste). Wipe up garage floor and drip pan coolant spills immediately. Keep antifreeze containers covered and repair leaks in vehicle cooling systems as quickly as possible.*

The cooling system should be filled with a water/ethylene glycol-based antifreeze solution, which will give protection down to at least −20°F at all times. It also provides protection against corrosion and increases the coolant boiling point.

The cooling system should be drained, flushed and refilled at least every other year. The use of antifreeze solutions for periods longer than two years is likely to cause damage and encourage the formation of rust and scale in the system.

Before adding antifreeze to the system, check all hose connections (antifreeze tends to search out and leak through very minute openings).

The exact mixture of antifreeze-to-water which you should use depends upon the relative weather conditions. The mixture should contain at least 50 percent antifreeze, but should never contain more than 70 percent antifreeze.

3 Thermostat — removal and installation

Refer to illustrations 3.5a, 3.5b, 3.6a, 3.6b and 3.9

Warning: *The engine must be completely cool before beginning this procedure.*

1 A faulty thermostat is indicated by failure of the engine to reach operating temperature or requiring longer than normal time to do so, or by overheating.
2 Disconnect the negative cable at the battery. Place the cable out of the way so it cannot accidentally come in contact with the negative terminal of the battery, as this would once again allow power into the electrical system of the vehicle.
3 Drain the coolant (see Chapter 1).
4 Remove the upper radiator hose from the thermostat housing.
5 Remove the thermostat housing upper bolt and lower nut, pull the

3.5a After removing the nut from the housing stud, pull the dipstick bracket away to remove the stud

3.5b Grasp the thermostat housing and rock it back and forth to break the gasket seal

3.6a Dislodge the thermostat by inserting a screwdriver behind it

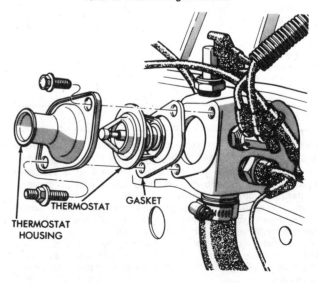

3.6b 2.2L thermostat — exploded view

Chapter 3 Cooling, heating and air conditioning systems

dipstick bracket away and remove the stud (see illustration). Grasp the thermostat housing securely, break it loose from the gasket by rocking it back and forth and remove it (see illustration).
6   Use a screwdriver to dislodge it and lift the thermostat out of the engine (see illustrations).

3.9 Make sure the spring on the thermostat (arrow) is installed facing *into* the engine

7   Clean all traces of gasket from the housing and engine mating surfaces with a scraper, taking care not to gouge or nick the metal.
8   Coat both sides of the new gasket with RTV sealant, then position the gasket on the engine.
9   Install the new thermostat in the engine (see illustration). **Note:** *Make sure the spring side faces into the engine and center the thermostat in the gasket.*
10 Install the housing and bolts. Tighten the bolts to the specified torque.
11 Install the hose, refill the radiator with the specified coolant and connect the negative battery cable.
12 Start the engine and check for coolant leaks around the thermostat housing.

4   Water pump — removal and installation

Refer to illustrations 4.7, 4.8, 4.9, 4.10, 4.12, 4.13a and 4.13b
Warning: *The engine must be completely cool before beginning this procedure.*

1   Disconnect the negative cable at the battery. Place the cable out of the way so it cannot accidentally come in contact with the negative terminal of the battery, as this would once again allow power into the electrical system of the vehicle.
2   Drain the coolant (see Chapter 1) and remove the upper radiator hose.
3   On air conditioner-equipped models, loosen the compressor idler pulley bolt and release the drivebelt tension. Unplug the air conditioner electrical connector, unbolt the compressor and secure it out of the way. **Warning:** *Do not disconnect or kink the hoses, as injury could result.*
4   Remove the wires from the alternator.
5   Loosen the alternator drivebelt adjuster and slip off the drivebelt.
6   Remove the alternator through-bolt and detach the alternator.
7   Unbolt and remove the air conditioner compressor bracket (if equipped) (see illustration).

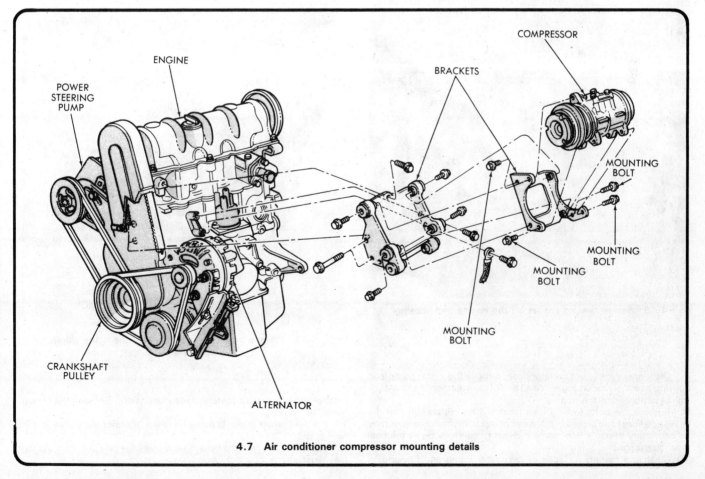

4.7 Air conditioner compressor mounting details

4.8 Water pump mounting bolt locations

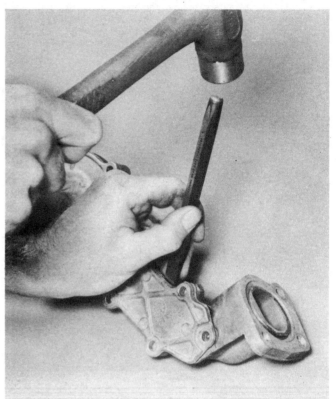

4.9 Separate the water pump and housing using a hammer and chisel

4.10 Scrape the old gasket off the engine and housing

4.12 Press a *new* O-ring evenly into the water pump housing groove

8 Disconnect the lower radiator hose, remove the water pump and housing assembly mounting nuts and bolts (see illustration) and detach the assembly from the engine.
9 Remove the bolts (noting their locations for reinstallation in the same positions) and separate the water pump from the housing. It may be necessary to use a hammer and chisel to carefully break the seal (see illustration).
10 Clean the mating surfaces of the water pump and housing to remove the old sealant material (see illustration). Remove the O-ring from the housing.
11 If a new water pump is being installed, transfer the pulley to the new pump.
12 Clean the groove in the housing and press the new O-ring into place (see illustration).

Chapter 3 Cooling, heating and air conditioning systems

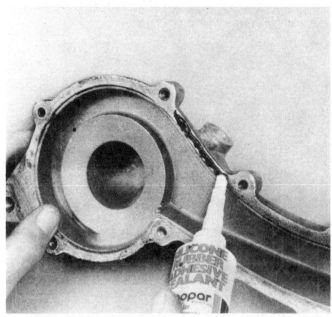

4.13a Carefully apply a bead of RTV-type sealant to the contact surface of the water pump

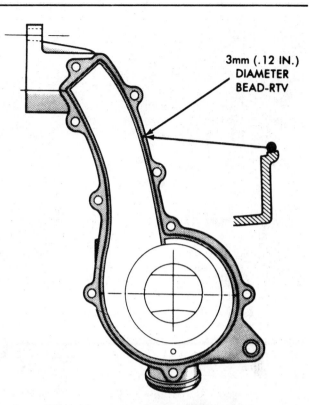

4.13b Water pump sealant application diagram

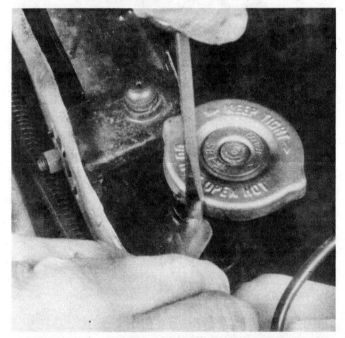

5.3a Use a screwdriver to pry the overflow hose off the radiator fitting . . .

5.3b . . . and pull the hose off

13 Apply a bead of RTV-type sealant to the housing mating surface and attach the pump (see illustrations). Install the bolts and tighten them to the specified torque.
14 Make sure the O-ring is in place, then attach the pump and housing assembly to the engine. Tighten the bolts to the specified torque. Install the radiator and heater hoses.
15 Install the air conditioner bracket and alternator and adjust the drivebelt.
16 Connect the wires to the alternator.
17 Install the air conditioner compressor (if removed) and adjust the drivebelt.
18 Refill the cooling system with the specified coolant (see Chapter 1).
19 Connect the negative battery cable.
20 Start the engine and check for coolant leaks at the water pump.

5 Radiator — removal and installation

Refer to illustrations 5.3a, 5.3b, 5.4a, 5.4b, 5.6a, 5.6b, and 5.10
Warning: *The engine must be completely cool before beginning this procedure.*

1 Disconnect the negative cable at the battery. Place the cable out of the way so it cannot accidentally come in contact with the negative terminal of the battery, as this would once again allow power into the electrical system of the vehicle.
2 Drain the cooling system (Chapter 1).
3 Remove the radiator hoses and the coolant reservoir hose (see illustrations).

Chapter 3 Cooling, heating and air conditioning systems

5.4a Pull back the radiator fan switch cover . . .

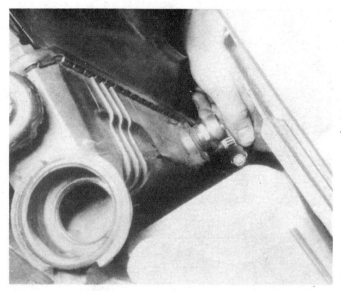

5.4b . . . and unplug the switch

5.6a Radiator mounting nuts (arrows)

5.6b Rotate the retaining brackets up and out of the mounts

5.10 Make sure the radiator tabs seat securely in the rubber mount (arrow)

4 Unplug the radiator fan switch connector (see illustrations).
5 Remove the fan motor and shroud assembly (Section 8).
6 Remove the two upper mounting nuts and the radiator retaining brackets (see illustrations).
7 Grasp the radiator securely and work it from side to side while lifting to disengage the mounting tabs from the rubber grommets at the bottom. Once it is disengaged, the radiator can be lifted straight up, out of the engine compartment.
8 Inspect the radiator for leaks, bent fins, damaged tubes, cracks around the tanks and signs of corrosion.
9 Lubricate all of the hose fittings lightly with white lithium grease to ease installation.
10 Lower the radiator into place and push down to seat the tabs in the rubber grommets (see illustration).
11 Install the bolts and tighten them securely.
12 Install the radiator, heater and reservoir hoses. Because the radiator tanks are plastic, be sure not to overtighten the hose clamps. Also, make sure that the radiator hoses are securely mounted in any mounting clips.
13 Install the fan motor and shroud assembly.
14 Refill the cooling system with the specified coolant.
15 Connect the negative battery cable.
16 Start the engine and check for coolant leaks at the hose fittings.

Chapter 3 Cooling, heating and air conditioning systems

8.2 Unplug the fan motor at the connector nearest the motor

8.3 Fan shroud retaining bolts (arrows)

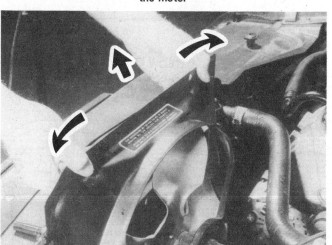

8.4 Rock the fan shroud from side to side while pulling up to disengage the clips

8.5 Lower the motor and shroud assembly into the clips until the bolt holes (arrows) line up

6 Radiator — inspection

1 The radiator should be kept free of obstructions such as leaves, paper, insects, mud and other debris which could affect the flow of air through it.
2 Periodically inspect the radiator for bent cooling fins or tubes, signs of coolant leakage and cracks around the upper and lower tanks.
3 Check the filler neck sealing surface for dents which could affect the radiator cap sealing effectiveness.

7 Drivebelts — inspection, replacement and adjustment

1 The drivebelts should be inspected periodically for wear, cuts and contamination by oil, gasoline or coolant as well as for signs of glazing, indicating improper adjustment.
2 To replace a drivebelt, loosen the bolts and push the pivoting component away from the belt until it can be removed. Do not pry on the pulley surface as this could cause nicks or gouges which will damage the new belt.
3 Install the new belt and adjust it as described in Chapter 1.

8 Fan motor and shroud assembly — removal and installation

Refer to illustrations 8.2, 8.3, 8.4 and 8.5

1 Disconnect the negative cable at the battery. Place the cable out of the way so it cannot accidentally come in contact with the negative terminal of the battery, as this would once again allow power into the electrical system of the vehicle.
2 Unplug the fan motor wiring connector (see illustration).
3 Remove the two fan shroud retaining bolts (see illustration).
4 Pull the assembly up, working it from side to side to disengage the shroud from the clips at the bottom of the radiator and lift it from the engine compartment (see illustration).
5 To install, lower the motor and shroud assembly into position, line up the bolt holes and push down to seat it in the clips (see illustration).
6 Install the bolts and tighten them securely.
7 Plug in the fan motor wiring connector and reattach the negative battery cable.

Chapter 3 Cooling, heating and air conditioning systems

9.2 The fan retaining clip can be removed with needle nose pliers

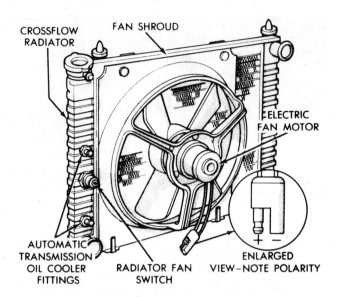

9.3 Cooling fan connector terminal polarity

9 Fan motor and shroud assembly — disassembly, inspection and reassembly

Refer to illustration 9.2 and 9.3

1 Remove the assembly from the vehicle (Section 8) and place it on a workbench. Be very careful not to bend the fan blades.
2 Remove the clip and slide the fan off the motor shaft (see illustration).
3 Inspect the motor for a bent shaft, damage and worn wiring insulation. Check the motor by inserting two 14 gauge wires into the connector and attaching them to the battery posts (see illustration). Replace the motor with a new one if it does not run and the wiring and connector are in good condition. Since the motor cannot be removed from the shroud, the entire motor/shroud assembly must be replaced as a unit. Inspect the fan for warping, cracks or damage. Replace it with a new one of the same design if necessary.
4 Slide the fan onto the shaft and retain it with the clip.
5 Install the assembly.

10 Turbocharger coolant hose and tube assembly — checking and replacement

Refer to illustration 10.2

1 Check the hoses, tubes and connections for cracks, damage and leaks. Replace any damaged components.
2 When installing new components, apply thread sealant the full length of all screw-in connections which thread into water passages and tighten to the specified torque (see illustration).

11 Heater and air conditioner control — removal and installation

Refer to illustrations 11.4a, 11.4b and 11.4c

1 Disconnect the negative cable at the battery. Place the cable out of the way so it cannot accidentally come in contact with the negative terminal of the battery, as this would once again allow power into the electrical system of the vehicle.
2 Remove the two screws at the base of the heater control bezel, grasp the base of the bezel and pull sharply outward to disengage the clips and remove the bezel.
3 Remove the screws that hold the control base to the instrument panel.

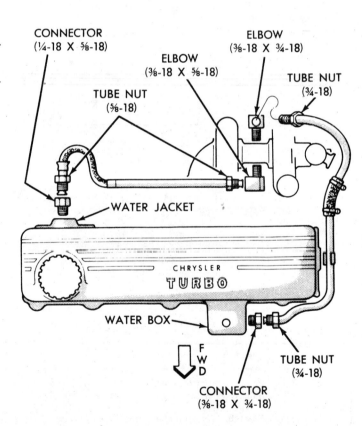

10.2 Turbocharger coolant hose and tube details

4 Withdraw the assembly from the dash and disengage the control cable (see illustrations). Unplug the electrical connectors. Some electrical connectors are retained by clips which can be released using a small screwdriver (see illustration).
5 Remove the control assembly from the vehicle.
6 To install, place the assembly in position and connect the vacuum and electrical connectors. Hook up the control cable.
7 Install the control base screws and bezel and connect the negative battery cable.

Chapter 3 Cooling, heating and air conditioning systems

11.4a Squeeze the control cable flag tab to release it from the receiver, lift it up . . .

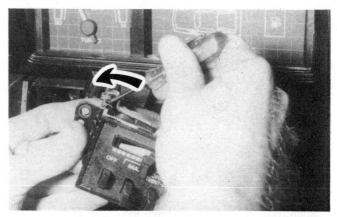

11.4b . . . and pry the clip off the end of the cable with a screwdriver

11.4c Use a small screwdriver to lift the electrical connector locking tab

12 Heater and air conditioner evaporator assembly — removal and installation

Refer to illustrations 12.3, 12.8, 12.9, 12.11 and 12.14

Warning: *On vehicles equipped with an air conditioner, the system must be evacuated by a dealer service department or air conditioning repair shop before any refrigerant lines are disconnected.*

1 Refer to Chapter 1 and drain the cooling system.
2 Disconnect the battery cables (negative first, then positive). Place the cables out of the way so they cannot accidentally come in contact with the terminals of the battery, as this would once again allow power into the electrical system of the vehicle.
3 Disconnect the heater hoses at the core fittings. Plug the core fittings to prevent coolant from spilling out when the assembly is removed (see illustration).

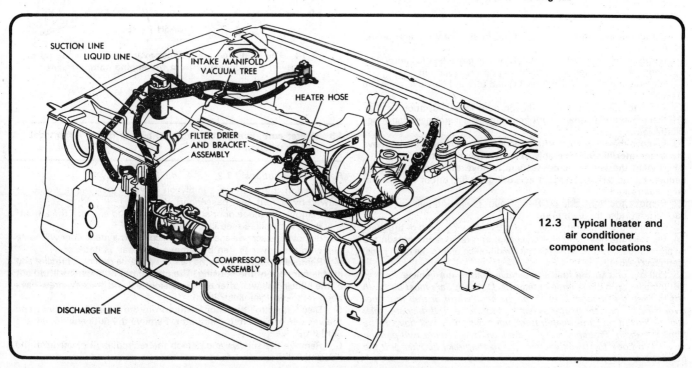

12.3 Typical heater and air conditioner component locations

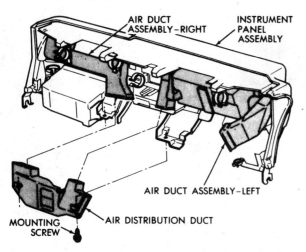

12.8 Center air distribution duct installation details

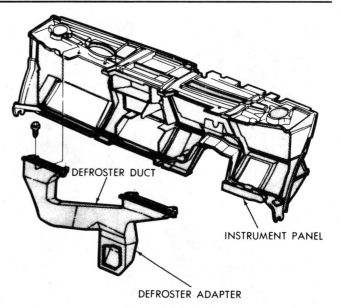

12.9 Defroster mounting details

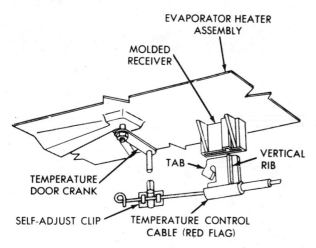

12.11 Heater/air conditioner control cable mount details

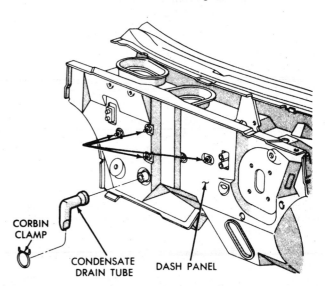

12.14 The heater/air conditioner evaporator retaining nuts (arrows) are accessible from the engine compartment

4 Disconnect the vacuum lines at the intake manifold and heater water valve.
5 Remove the cowl side trim panel and the glove box (Chapter 11).
6 Remove the heater/air conditioning control (Section 11).
7 Remove the console (Chapter 11) and forward console mounting bracket.
8 Remove the center air distribution duct (see illustration).
9 Pull the defroster adapter out from under the instrument panel (see illustration).
10 Remove the clamp, pull off the condensate drain tube, then remove the evaporator/heater assembly-to-dash retaining nuts.
11 Unplug the wiring harnesses and remove the control cable by depressing the tab on the flag, then pulling the flag out of the receiver (see illustration).
12 Remove the right side cowl-to-plenum brace and pull the carpet back from under the evaporator unit.
13 Remove the evaporator/hanger strap and swing it out of the way.
14 Remove the heat shield (if equipped) and remove the nuts in the engine compartment retaining the air conditioning unit to the dash panel (see illustration).
15 Pull the unit to the rear and remove it from the vehicle.
16 Installation is the reverse of removal. **Caution:** *Care must be taken not to hang the vacuum lines up on the accelerator or trap them between the evaporator/heater assembly and the dash. If they are, they will be kinked and the evaporator/heater assembly will have to be removed again. Proper routing of the lines may require two people. Be sure to hook the black vacuum line to the brake booster and the gray one to the water valve.*

13 Heater core, evaporator core and blower motor — removal and installation

Refer to illustrations 13.2, 13.4 and 13.10

1 Position the heater assembly on a workbench (as it would be viewed by a front seat passenger).
2 Remove the mode door actuator arm nut and squeeze the arm off the shaft with pliers (see illustration).
3 Remove the screws and detach the cover. The mode door will come out with the cover (it can be removed if repair is necessary).
4 Remove the screw from the heater core tube retaining bracket and lift the core out of the housing. The evaporator core can be lifted out of the housing as well after the expansion seal and bracket screw have been removed (see illustration).
5 Disconnect the actuator linkage from the recirculation door and the vacuum lines from the actuator. Remove the nuts and detach the actuator.
6 Remove the screws and detach the recirculation cover from the housing.

Chapter 3 Cooling, heating and air conditioning systems

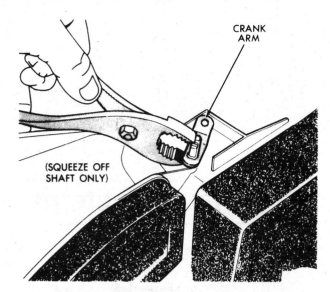

13.2 Remove the actuating arm by squeezing it off the the shaft with pliers

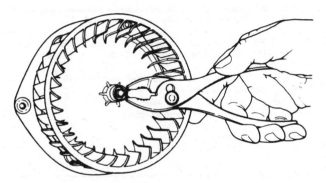

13.10 To remove the blower wheel, use pliers to release the clamp and then slide the clamp and blower off the motor shaft

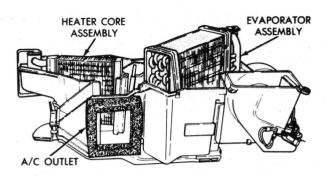

13.4 Heater core/evaporator locations

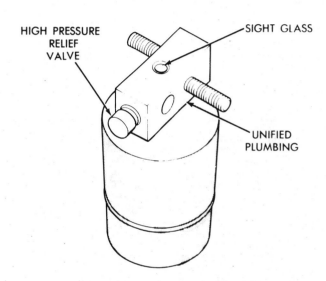

14.4 Air conditioner filter-drier sight glass location

7 The recirculation door can now be removed from the housing as well.
8 Remove the evaporator/heater housing vent tube.
9 Remove the screws and lift out the blower motor and fan assembly.
10 To detach the fan, remove the retaining clamp and slide it off the motor shaft (see illustration).
11 The motor is held in place with three screws.
12 Installation is the reverse of removal.

14 Air conditioning system — description and testing

Refer to illustration 14.4

Warning: *The air conditioning system is pressurized at the factory and requires special equipment for service and repair. Any work should be left to your dealer or a refrigeration shop. Do not, under any circumstances, disconnect the air conditioning hoses while the system is under pressure.*

1 The air conditioning system consists of a condenser mounted in front of the radiator, an evaporator mounted under the dash, a belt-driven compressor incorporating a clutch, a filter-drier which contains a high pressure relief valve and associated hoses.
2 The temperature in the passenger compartment is lowered by transferring the heat in the air to the refrigerant in the evaporator and then passing the refrigerant through the condenser.
3 Maintenance is confined to keeping the system properly charged with refrigerant, the compressor drivebelt adjusted properly and making sure the condenser is free of leaves and other debris.
4 The sight glass located on the top of the filter-drier can give some indication of the refrigerant level (see illustration).
5 With the control on A/C, the fan switch on High and the temperature lever on Cool, run the system for several minutes. The temperature in the vehicle should be approximately 70°F (21 °C).
6 The system has a full refrigerant charge if the sight glass is clear, the air conditioner compressor clutch is engaged, the inlet line to the compressor is cool and the discharge line is warm.
7 If the glass is clear, the clutch is engaged but there is no difference in temperature between the inlet and discharge lines, the refrigerant charge is very low.
8 Continuous foam or bubbles in the sight glass is another symptom of low refrigerant. Occasional foam or bubbles under certain conditions, such as very high or low temperatures in the vehicle interior, is acceptable.

Chapter 4 Part A Fuel and exhaust systems

Contents

Air filter replacement . . . . See Chapter 1	Multi-point EFI fuel injector — removal and installation 18
Electronic Fuel Injection (EFI) — description and checking . . . 6	Multi-point EFI throttle body — removal and installation 14
Exhaust system check . . . . See Chapter 1	Multi-point EFI Throttle Position Sensor (TPS) — removal
Exhaust system — removal and installation 21	and installation 15
Fuel filter replacement . . . . See Chapter 1	Single-point EFI Automatic Idle Speed (AIS) motor — removal
Fuel lines and fittings — replacement 7	and installation 12
Fuel pump — removal and installation 5	Single-point EFI fuel injector — removal and installation 10
Fuel system check . . . . See Chapter 1	Single-point EFI pressure regulator — removal and
Fuel system pressure release 4	installation 9
Fuel tank — removal and installation 19	Single-point EFI throttle body — removal and installation 8
General information 1	Single-point EFI Throttle Position Sensor (TPS) — removal
Heated inlet air system general check . . . . See Chapter 1	and installation 11
Intake and exhaust manifolds (non-turbo	Single-point EFI throttle body temperature sensor (1986
models) — removal, inspection and installation 20	models) — removal and installation 13
Multi-point EFI Automatic Idle Speed (AIS) motor — removal	Throttle cable — removal and installation 2
and installation 16	Throttle pedal — removal and installation 3
Multi-point EFI fuel injector rail assembly — removal	
and installation 17	

Specifications

General

Fuel tank capacity	See Chapter 1
Intake and exhaust manifold gasket surface warpage limit . . .	0.006 in per 12 inches (0.15 mm per 300 mm)

Torque specifications	Ft-lbs (unless otherwise noted)	Nm
EFI fuel line fittings	15	20
EFI hose clamp screw	10 in-lbs	1
Fuel filter screw	6	8
Intake and exhaust manifold (non-turbocharged		
models) bolts or nuts	17	23
Single-point EFI		
Automatic Idle Speed (AIS) motor-to-throttle		
body screws	20 in-lbs	2
throttle body-to-manifold bolts	17	23
pressure regulator screw	4	5
throttle position sensor screw	20 in-lbs	2
fuel injector cap Torx screw	40 in-lbs	5
Multi-point EFI		
fuel rail bolts	19	25
throttle body-to-intake manifold bolt	40	54
throttle position sensor-to-throttle body	17 in-lbs	2

1 General information

The fuel system consists of a rear-mounted fuel tank, an electric fuel pump (located in the fuel tank) which delivers fuel to the Electronic Fuel Injecton (EFI), and associated hoses, lines and filters. The exhaust system is made up of pipes, heat shields, a muffler and a catalytic converter. The catalytic converter requires that only unleaded fuel be used in the vehicle.

2 Throttle cable — removal and installation

Refer to illustrations 2.3a, 2.3b, 2.3c and 2.3d

1 Working inside the vehicle, disconnect the throttle cable from the pedal shaft (Section 3). Working inside the vehicle, remove the retainer clip from the cable assembly housing at the grommet (see illustration 3.1).

2 Working in the engine compartment, pull the cable housing end

Chapter 4 Part A Fuel and exhaust systems

2.3a Pull the throttle linkage retaining clip off with needle nose pliers

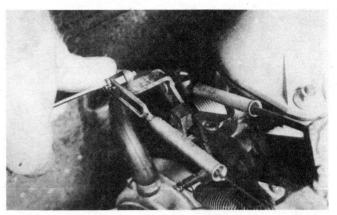

2.3b Use a small screwdriver to pry off the cruise control cable clip

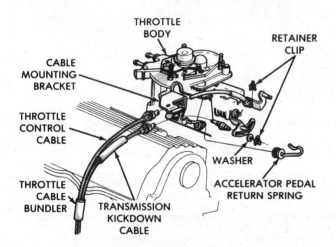

2.3c Single-point EFI throttle cable details

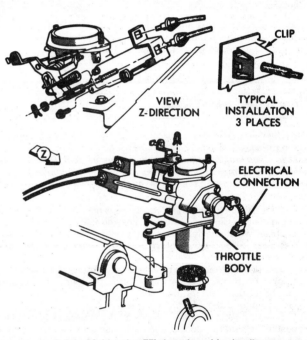

2.3d Multi-point EFI throttle cable details

fitting out of the firewall grommet, making sure the grommet remains in place.
3 Remove the retainer clips and disconnect the throttle cable and cruise control cable (if equipped) from the throttle body (see illustrations). Use pliers to compress the end fitting tabs so the cable can be separated from the mounting bracket (see illustrations).
4 To install, insert the cable housing into the cable mounting bracket on the engine and attach the cable clevis to the throttle body with the retainer clip. Insert the cable through the firewall grommet and connect it to the throttle pedal.

3 Throttle pedal — removal and installation

Refer to illustration 3.1
1 Remove the cable retainer and disengage the throttle cable from the pedal shaft and bracket (see illustration).
2 Working in the engine compartment, remove the pedal assembly retaining nuts.
3 Working inside the vehicle, detach the pedal assembly from the dash panel and remove it.
4 Installation is the reverse of removal.

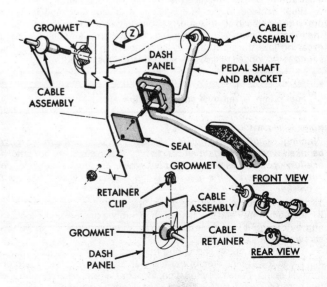

3.1 Throttle pedal details

Chapter 4 Part A Fuel and exhaust systems

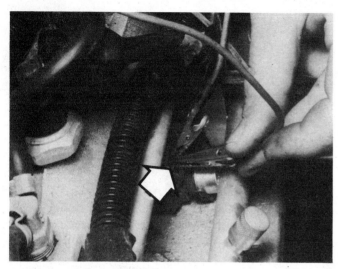

4.5 Bleed the fuel pressure in short bursts by touching the fuel injector terminal (arrow) with the jumper wire clip (multi-point fuel injection shown)

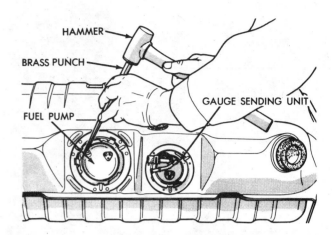

5.2 Use a hammer and a brass drift to unscrew the fuel pump locking ring

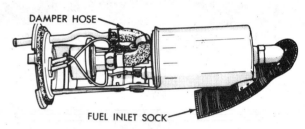

5.4 Inspect the fuel pump damper hose and the fuel inlet filter sock for damage

4 Fuel system pressure release

Refer to illustration 4.5

Warning: *Gasoline is extremely flammable, so extra precautions must be taken when working on any part of the fuel system. Do not smoke or allow open flames or bare light bulbs near the work area. Also, do not work in a garage if a natural gas-type appliance with a pilot light is present.*

1   The fuel system of fuel injected models is pressurized, even when the engine is off. Consequently any time the fuel system is worked on (such as when the fuel filter is replaced) the system must be depressurized to avoid the spraying of fuel when a component is disconnected.
2   Loosen the fuel tank cap to release any pressure in the tank.
3   Unplug the harness connector at the throttle body (single-point) or one of the fuel injectors (multi-point).
4   Ground one of the injector terminals.
5   Connect a jumper wire between the other terminal and the positive (+) post of the battery. Touch the end of the jumper wire to the terminal for no longer than five seconds to depressurize the fuel system (see illustration). Do not energize the injector for more than five seconds to avoid damage to the fuel injector. It is recommended that the pressure be bled in several spurts of one to two seconds to make sure the injector system is not damaged. The fuel pressure can be heard escaping into the throttle body or combustion chamber. When the sound is no longer heard, the system is depressurized.

5 Fuel pump — removal and installation

Refer to illustrations 5.2 and 5.4

Warning: *Gasoline is extremely flammable, so extra precautions must be taken when working on any part of the fuel system. Do not smoke or allow open flames or bare light bulbs near the work area. Also, do not work in a garage if a natural gas-type appliance with a pilot light is present.*

1   Remove the fuel tank (Section 19).
2   Use a hammer and a brass punch to remove the fuel pump locking ring by driving it in a counterclockwise direction until it can be unscrewed (see illustration).
3   Lift the fuel pump and O-ring from the fuel tank.
4   Clean the sealing area of the fuel tank and install a new O-ring on the fuel pump. Prior to installation, inspect the sock-like filter on the fuel pump suction tube for damage or contamination, replacing it with a new one if necessary (see illustration).
5   Place the fuel pump in position in the tank, install the locking ring and use the hammer and brass drift to lock the pump in place.
6   Install the fuel tank.

6 Electronic Fuel Injection (EFI) — description and checking

Two types of Electronic Fuel Injection (EFI) are used on these models; single-point and multi-point. Single-point EFI is used on non-turbocharged models and multi-point on turbocharged models.

Both types are similar in operation and are controlled by the logic module and power module (see Chapter 5) in combination with a variety of sensors and switches. On both the single- and multi-point EFI the conventional carburetor is replaced by a throttle body and injector (single point) or throttle body, fuel rail and injectors (multi-point). On single-point EFI the fuel is mixed with air in the throttle body and sprayed into the intake manifold, which directs it to the intake ports and cylinders. On multi-port EFI the fuel is sprayed directly into the ports by the fuel injectors, with the intake manifold supplying only the air.

The EFI system consists of a throttle body, fuel rail (multi-point), fuel injector(s), pressure regulator, throttle position sensor (TPS), automatic idle speed (IAS) motor and throttle body temperature sensor, all of which interact with the power module and the logic module.

Because of the complexity of the EFI system, the home mechanic can do very little in the way of diagnosis because of the special techniques and equipment required. However, checking of the EFI system components and electrical and vacuum connections to make sure they are secure and not obviously damaged is one thing the home mechanic can do which can often detect a potential or current problem. Since the logic and power modules are completely dependent on the information provided by the many sensors and vacuum connections, a simple visual check and tightening of loose connections can save diagnostic time and a possibly unneccessary trip to the dealer.

Damaged or faulty EFI components can be replaced using the procedures in the following Sections. Additional information on the logic and power modules and the EFI system can be found in Chapter 5.

Chapter 4 Part A Fuel and exhaust systems

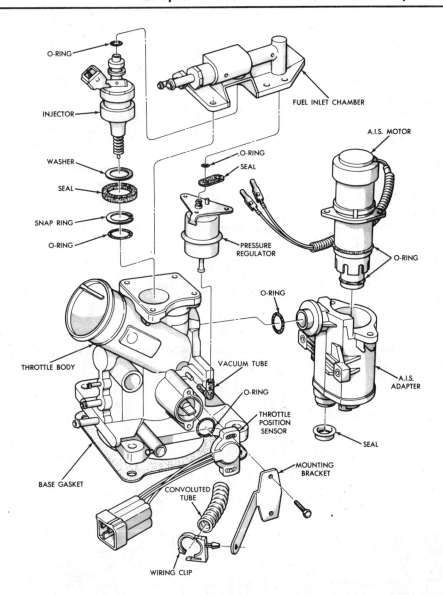

8.4a 1984 and 1985 model single-point EFI throttle body — exploded view

7 Fuel lines and fittings — replacement

Warning: *Gasoline is extremely flammable, so extra precautions must be taken when working on any part of the fuel system. Do not smoke or allow open flames or bare light bulbs near the work area. Also, do not work in a garage if a natural gas-type appliance with a pilot light is present.*

Note: *Because the EFI system is under considerable pressure, always replace any clamp which is released or removed with a new one.*

1 Remove the air cleaner assembly.
2 Release the fuel system pressure (Section 4).
3 Disconnect the negative cable at the battery. Place the cable out of the way so it cannot accidentally come in contact with the negative terminal of the battery, as this would once again allow power into the electrical system of the vehicle.
4 Loosen the hose clamps, wrap a cloth around each end of the hose to catch the residual fuel and twist and pull to remove the hose.
5 Remove the fuel fittings, taking care to note the inlet diameter and remove the copper washers.
6 When installing fuel fittings, make sure the inlet diameters match and always use new copper washers. Tighten to the specified torque.
7 When replacing hoses, always use hoses marked EFI/EFM and new original equipment-type clamps only.
8 Connect the negative cable, start the engine and check for leaks.
9 Install the air cleaner assembly.

8 Single-point EFI throttle body — removal and installation

Refer to illustrations 8.4a, 8.4b and 8.4c

Warning: *Gasoline is extremely flammable, so extra precautions must be taken when working on any part of the fuel system. Do not smoke or allow open flames or bare light bulbs near the work area. Also, do not work in a garage if a natural gas-type appliance with a pilot light is present.*

Removal

1 Remove the air cleaner assembly.
2 Release the fuel system pressure (Section 4).
3 Disconnect the negative cable at the battery. Place the cable out of the way so it cannot accidentally come in contact with the negative terminal of the battery, as this would once again allow power into the electrical system of the vehicle.
4 Disconnect the vacuum hoses and electrical connectors (see illustrations).
5 Disconnect the throttle linkage and (if equipped) the cruise control and transaxle kickdown cable.
6 Remove the throttle return spring.
7 Place rags or newspapers under the fuel hoses to catch the residual fuel. Loosen the clamps, wrap a cloth around each fuel hose and grasp and pull off each hose in turn. Remove the copper washers from the hoses, noting their locations.
8 Remove the mounting bolts or nuts and lift the throttle body from the manifold.

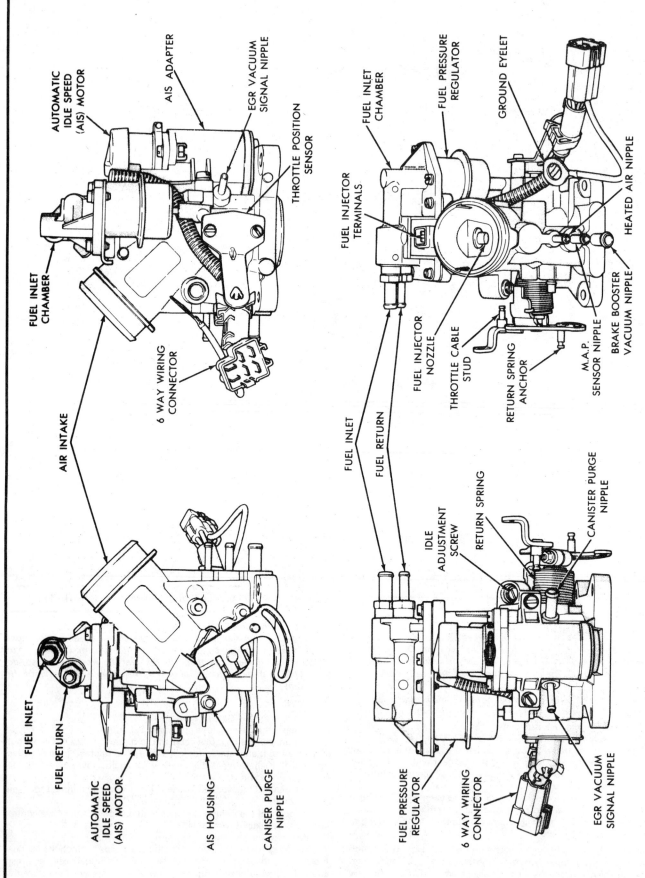

8.4b 1984 and 1985 model single-point EFI throttle body external details

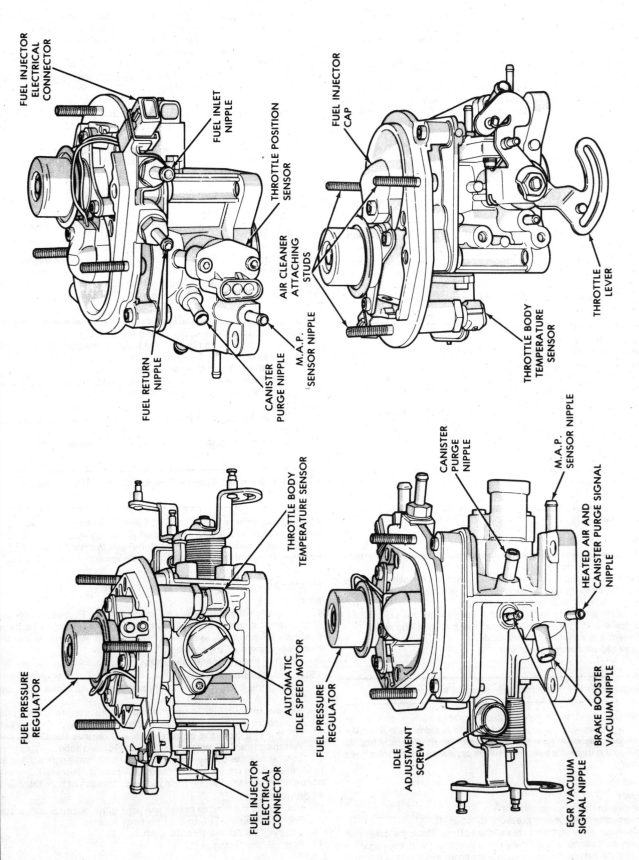

8.4c 1986 model single-point EFI throttle body details

Chapter 4 Part A Fuel and exhaust systems

Installation

9 Inspect the mating surfaces of the throttle body and the manifold for nicks, burrs and debris that could cause air leaks.
10 Using a new gasket, place the throttle body in position and install the mounting bolts or nuts.
11 Tighten the bolts or nuts to the specified torque, following a criss-cross pattern.
12 Check all of the vacuum hoses and electrical connectors for damage, replacing them with new parts if necessary, and install them.
13 Connect the throttle linkage and (if equipped) cruise control and kickdown cable.
14 Connect the throttle return spring.
15 Using new clamps and washers, install the fuel hoses.
16 Check the operation of the throttle linkage.
17 Install the air cleaner assembly.
18 Connect the negative battery cable.
19 Start the engine and check for fuel leaks.

9 Single-point EFI pressure regulator — removal and installation

Warning: *Gasoline is extremely flammable, so extra precautions must be taken when working on any part of the fuel system. Do not smoke or allow open flames or bare light bulbs near the work area. Also, do not work in a garage if a natural gas-type appliance with a pilot light is present.*

Removal

1 Remove the air cleaner assembly.
2 Release the fuel system pressure (Section 4).
3 Disconnect the negative cable at the battery. Place the cable out of the way so it cannot accidentally come in contact with the negative terminal of the battery, as this would once again allow power into the electrical system of the vehicle.
4 On 1984 and 1985 models, remove the fuel inlet chamber (refer to illustrations 8.4a and 8.4b), then remove the vacuum tube from the regulator and throttle body.
5 On 1984 and 1985 models, remove the vacuum tube from the regulator and throttle body.
6 Wrap a cloth around the fuel inlet chamber to catch any residual fuel, which is under pressure.
7 Withdraw the pressure regulator from the throttle body.
8 Carefully remove the O-ring from the pressure regulator, followed by the gasket.

Installation

9 Place a new gasket in position on the pressure regulator and carefully install a new O-ring.
10 On 1986 models, place the pressure regulator in position on the throttle body, press it into position and install the three retaining screws. Tighten the screws to the specified torque.
11 On 1984 and 1985 models, install the vacuum tube.
12 On 1984 and 1985 models, install the fuel inlet chamber.
13 Connect the negative battery cable.
14 Install the air cleaner assembly.

10 Single-point EFI fuel injector — removal and installation

Warning: *Gasoline is extremely flammable, so extra precautions must be taken when working on any part of the fuel system. Do not smoke or allow open flames or bare light bulbs near the work area. Also, do not work in a garage if a natural gas-type appliance with a pilot light is present.*

Removal

All models
1 Remove the air cleaner assembly.
2 Release the fuel system pressure (Section 4).
3 Disconnect the negative cable at the battery. Place the cable out of the way so it cannot accidentally come in contact with the negative terminal of the battery, as this would once again allow power into the electrical system of the vehicle.

1984 and 1985 models
4 Remove the four screws retaining the fuel inlet chamber to the throttle body.
5 Remove the fuel pressure regulator-to-throttle body vacuum tube.
6 Wrap a cloth around the fuel inlet chamber to catch the residual fuel and lift the inlet chamber and injector off the throttle body.
7 Withdraw the injector from the inlet chamber (see illustration 8.4a).
8 Peel the upper and lower O-rings from the fuel injector.
9 Remove the snap-ring retaining the seal and washer on the injector, followed by the seal and washer.

1986 models
10 Remove the bolt retaining the injector cap and lift the cap off the injector using two screwdrivers inserted in the slots provided for this purpose (see illustration 8.4c).
11 Pry the injector from the pod using a screwdriver inserted in the holes in the side of the electrical connector. After removal, make sure the lower injector O-ring is removed from the pod.

Installation

1984 and 1985 models
12 Install a new O-ring, washer and seal on the injector, retaining them with the snap-ring.
13 Insert the injector into the fuel inlet chamber.
14 Place the injector/fuel inlet chamber in position on the throttle body and install the pressure regulator-to-throttle body vacuum tube.
15 Place the assembly into the throttle body and install the retaining screws. Tighten the screws to the specified torque.

1986 models
16 Install new O-rings on the injector and injector cap (a new injector should already have the upper O-ring installed).
17 Place the injector in the pod and position the injector so the cap (which is keyed to the injector) can be installed without interference.
18 Rotate the cap and injector so they line up with the attachment hold, push down to ensure a good seal and install the retaining screws. Tighten the screws to the specified torque.

All models
19 Connect the negative battery cable, start the engine and check for leaks.
20 Turn off the engine and install the air cleaner assembly.

11 Single-point EFI Throttle Position Sensor (TPS) — removal and installation

Warning: *Gasoline is extremely flammable, so extra precautions must be taken when working on any part of the fuel system. Do not smoke or allow open flames or bare light bulbs near the work area. Also, do not work in a garage if a natural gas-type appliance with a pilot light is present.*

Removal

All models
1 Disconnect the negative cable at the battery. Place the cable out of the way so it cannot accidentally come in contact with the negative terminal of the battery, as this would once again allow power into the electrical system of the vehicle.
2 Remove the air cleaner assembly.

1984 and 1985 models
3 Unplug the 6-way wiring connector (see illustrations 8.4a and 8.4b).
4 Remove the two TPS-to-throttle body retaining screws, unclip the wiring clip from the convoluted tube and remove the mounting bracket.
5 Lift the TPS from the throttle shaft and remove the O-ring.
6 Pull the three TPS wires from the convoluted tube.
7 Look inside the 6-way connector and use a small screwdriver to lift the locking tab for each of the TPS wire blade terminals. Disconnect each TPS blade (noting their positions for ease of reinstallation).

1986 models
8 Disconnect the throttle cables and unplug the electrical connector.
9 Remove the two TPS-to-throttle body screws.
10 Lift the TPS off the throttle shaft.
11 Remove the O-ring.

Installation

1984 and 1985 models
12 Making sure they go into the proper locations, insert the wire blade

Chapter 4 Part A Fuel and exhaust systems

terminals into the throttle body connector.
13 Insert the TPS wires into the convoluted tube.
14 Place the TPS and new O-ring in position with the mounting bracket in position on the throttle body and install the retaining screws. Tighten the screws to the specified torque.
15 Install the wiring clips to the convoluted tube and connect the 6-way connector.

1986 models
16 Install the TPS with a new O-ring on the throttle body and install the retaining screws. Tighten the screws to the specified torque.
17 Plug in the electrical connector and connect the throttle cable.

All models
18 Install the air cleaner.
19 Connect the negative battery cable.

12 Single-point EFI Automatic Idle Speed (AIS) motor — removal and installation

Warning: Gasoline is extremely flammable, so extra precautions must be taken when working on any part of the fuel system. Do not smoke or allow open flames or bare light bulbs near the work area. Also, do not work in a garage if a natural gas-type appliance with a pilot light is present.

Removal
All models
1 Disconnect the negative cable at the battery. Place the cable out of the way so it cannot accidentally come in contact with the negative terminal of the battery, as this would once again allow power into the electrical system of the vehicle.
2 Remove the air cleaner assembly.

1984 and 1985 models
3 Remove the two screws retaining the AIS adapter to the throttle body (see illustration 8.4a).
4 Remove the wiring clips followed by the two AIS wires from the throttle body connector. Use a small screwdriver to lift each locking tab (noting the position of each wire for ease of reinstallation) and disconnect each blade terminal.
5 Carefully withdraw the assembly from the rear of the throttle body. Note that the O-ring at the top and the seal at the bottom may fall off the adapter so it is important to keep track of them.
6 Remove the O-ring and seal.
7 To separate the motor from the adapter, remove the two retaining screws (but not the clamp), lift the motor out and remove the O-ring.

1986 models
8 Unplug the 4-pin connector on the AIS.
9 Remove the throttle body temperature sensor (Section 13).
10 Remove the two retaining screws.
11 Withdraw the AIS from the throttle body, making sure that the O-ring does not fall into the throttle body opening.

Installation
1984 and 1985 models
12 If the AIS motor has been removed from the adapter, install new O-rings and carefully work the motor into the adaptor and install the retaining screws.
13 To install the AIS motor/adaptor assembly, first install a new O-ring and seal on the adaptor.
14 Place the assembly carefully in position on the back of the throttle body, making sure the O-ring and seal are in place and install the retaining screws. Tighten the screws to the specified torque.

1986 models
15 Prior to installation, make sure the pintle is in the retracted position. If the retracted pintle measurement is more than 1.26 inches (32 mm), the AIS must be taken to a dealer or properly equipped shop to be retracted.
16 Install a new O-ring and insert the AIS into the housing, making sure the O-ring is not dislodged.
17 Install the two retaining screws. Tighten the screws to the specified torque.
18 Plug the 4-pin connector into the AIS.
19 Install the throttle body temperature sending unit (Section 13).

All models
20 Install the air cleaner assembly and connect the negative battery cable.

13 Single-point EFI throttle body temperature sensor (1986 models) — removal and installation

Warning: Gasoline is extremely flammable, so extra precautions must be taken when working on any part of the fuel system. Do not smoke or allow open flames or bare light bulbs near the work area. Also, do not work in a garage if a natural gas-type appliance with a pilot light is present.

Removal
1 Disconnect the negative cable at the battery. Place the cable out of the way so it cannot accidentally come in contact with the negative terminal of the battery, as this would once again allow power into the electrical system of the vehicle.
2 Remove the air cleaner assembly.
3 Disconnect the throttle cable from the throttle body, remove the two cable bracket screws and lay the bracket aside.
4 Unplug the wiring connector by pulling downward.
5 Remove the sensor by unscrewing it (refer to illustration 8.4c).

Installation
6 Apply a thin coat of heat transfer compound to the tip of the new sensor.
7 Screw the sensor into the throttle body and tighten it securely.
8 Plug in the wiring connector.
9 Connect the throttle linkage.
10 Install the air cleaner and connect the negative battery cable.

14 Multi-point EFI throttle body — removal and installation

Refer to illustrations 14.2, 14.4, 14.5, 14.8 and 14.11
Warning: Gasoline is extremely flammable, so extra precautions must be taken when working on any part of the fuel system. Do not smoke or allow open flames or bare light bulbs near the work area. Also, do not work in a garage if a natural gas-type appliance with a pilot light is present.

Removal
1 Disconnect the negative cable at the battery. Place the cable out of the way so it cannot accidentally come in contact with the negative terminal of the battery, as this would once again allow power into the electrical system of the vehicle.
2 Remove the air cleaner hose and the throttle body adapter (see illustration).

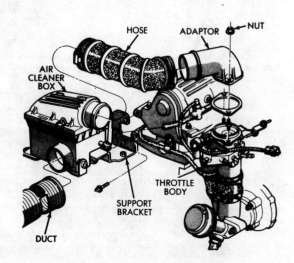

14.2 Multi-point EFI throttle body air cleaner hose and adapter details

Chapter 4 Part A Fuel and exhaust systems

3 Remove the return spring, disconnect the throttle cable and remove (if equipped) the cruise control and automatic transaxle kickdown cable.
4 Unbolt and remove the throttle cable bracket (see illustration).
5 Unplug the electrical connector (see illustration).
6 Disconnect the vacuum hoses from the throttle body.
7 Loosen the turbocharger-to-throttle body hose clamp.
8 Remove retaining bolts and detach the throttle body from the intake manifold (see illustration).

Installation

9 Place the throttle body in position and install the mounting bolts. Tighten them to the specified torque in a criss-cross pattern.
10 Tighten the turbocharger hose clamp.
11 Connect the vacuum hoses (see illustration).
12 Plug in the electrical connector.
13 Install the throttle bracket and connect the throttle cable, return spring and (if equipped) the cruise control and automatic transaxle kickdown cable.
14 Install the air cleaner hose and adapter.
15 Connect the negative battery cable.

15 Multi-point EFI Throttle Position Sensor (TPS) — removal and installation

Refer to illustration 15.2

Removal

1 Disconnect the negative cable at the battery. Place the cable out of the way so it cannot accidentally come in contact with the negative terminal of the battery, as this would once again allow power into the electrical system of the vehicle.
2 Unplug the 6-way electrical connector (see illustration).
3 Remove the TPS-to-throttle body retaining screws.
4 Unclip the convoluted plastic wiring tube and remove the remove the mounting bracket.
5 Withdraw the TPS from the throttle shaft and remove the O-ring.
6 Look inside the 6-way connector and locate the TPS wire blade terminals. Noting their locations for installation to the same positions, use a small screwdriver to lift each locking tab and disconnect the TPS.

14.4 Remove the two throttle cable bracket bolts

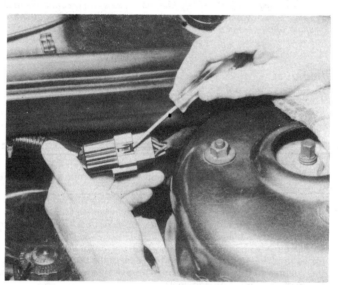

14.5 Bend the clip up with a small screwdriver before separating the electrical connector

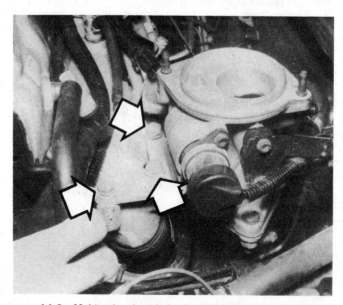

14.8 Multi-point throttle body retaining bolt locations (arrows)

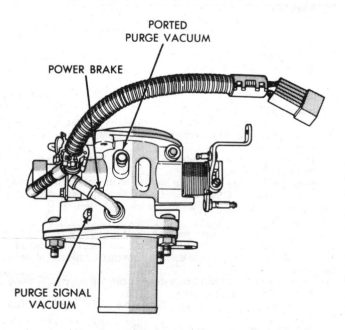

14.11 Multi-point EFI throttle body vacuum connections

Chapter 4 Part A Fuel and exhaust systems

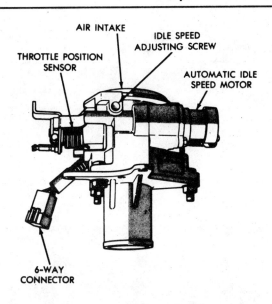

15.2 Multi-point EFI throttle body component layout

Installation

7 Insert the TPS blade terminals into the 6-way connector.
8 Push the TPS wires into the convoluted plastic wiring tube.
9 Install the TPS and new O-ring with the mounting bracket onto the throttle body. Tighten the screws to the specified torque.
10 Install the clips on the convoluted plastic wiring tube.
11 Plug in the 6-way connector and connect the negative battery cable.

16 Multi-point EFI Automatic Idle Speed (AIS) motor — removal and installation

Removal

1 Disconnect the negative cable at the battery. Place the cable out of the way so it cannot accidentally come in contact with the negative terminal of the battery, as this would once again allow power into the electrical system of the vehicle.
2 Unplug the 6-way electrical connector.
3 Remove the wiring clips and then remove the two AIS wires from the 6-way connector. Using a small screwdriver, lift each locking tab and, noting their locations for ease of reassembly, remove each blade terminal.
4 Remove the two AIS motor-to-throttle body retaining screws. Do not remove the clamp (see illustration 15.2).
5 Remove the AIS motor from the throttle body, making sure that both O-rings remain on the motor.

Installation

6 Carefully position the AIS motor (using new O-rings — new motors should be already equipped with O-rings) in place in the throttle body.
7 Install the mounting screws and tighten them securely.
8 Route the AIS wiring to the connector and plug in the AIS blade terminals to the correct positions.
9 Connect the wiring clips and plug in the 6-way connector.
10 Connect the negative battery cable.

17 Multi-point EFI fuel injector rail assembly — removal and installation

Refer to illustration 17.4, 17.7, 17.9 and 17.10
Warning: *Gasoline is extremely flammable, so extra precautions must be taken when working on any part of the fuel system. Do not smoke or allow open flames or bare light bulbs near the work area. Also, do not work in a garage if a natural gas-type appliance with a pilot light is present.*

Removal

1 Release the fuel system pressure (Section 4).
2 Disconnect the negative cable at the battery. Place the cable out of the way so it cannot accidentally come in contact with the negative terminal of the battery, as this would once again allow power into the electrical system of the vehicle.
3 Loosen the supply hose clamp at the fuel rail inlet and pull the hose off.
4 Disconnect the fuel pressure regulator vacuum hose, remove the two bracket bolts, loosen the regulator hose clamp at the end of the rail and then remove the hose and regulator (see illustration).

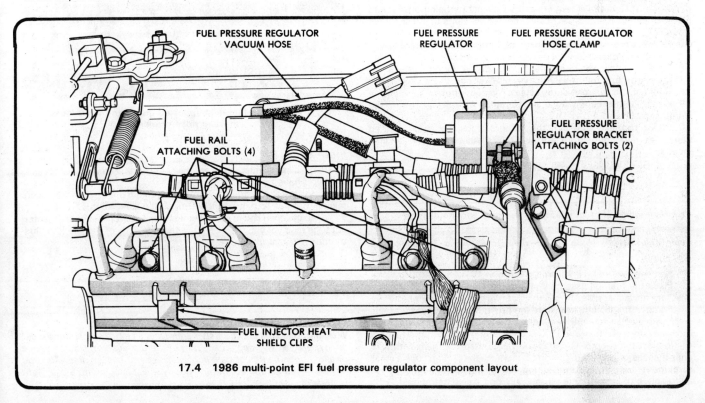

17.4 1986 multi-point EFI fuel pressure regulator component layout

Chapter 4 Part A Fuel and exhaust systems

17.7 A small screwdriver can be used to pry off the heat shield clips

5   Remove the fuel rail-to-camshaft cover bracket bolt.
6   Unplug the fuel injector wiring connector.
7   Remove the fuel injector heat shield clips, using a small screwdriver (see illustration).
8   Remove the four fuel rail attaching bolts.
9   Grasp the fuel rail and injector assembly securely and pull the injectors straight out of the ports. Taking care not to damage the rubber injector O-rings, remove the rail assembly from the vehicle (see illustration). The fuel injectors must not be removed until the fuel rail is detached from the vehicle.
10 Support the fuel rail and disconnect the remaining fuel hoses (see illustration).

Installation

11 Prior to installation, make sure the injectors are securely seated into the receiver cup with the lock rings in place.
12 Inspect the injector holes to make sure they are clean.
13 Lubricate the injector O-rings with clean engine oil.
14 Insert the injector assemblies carefully into their holes and install the bolts and ground straps. Tighten the bolts evenly in a criss cross pattern so that the injectors are seated evenly by being drawn into place. Once the injectors are seated, tighten the bolts to the specified torque.
15 Connect the injector wiring harness to the injectors and fasten it into the wiring clips.
16 Install the heat shield clips and connect the wiring harness.
17 Install the fuel rail-to-valve cover bracket bolt.
18 Connect the fuel pressure regulator vacuum hose.
19 Install the fuel supply hose and clamp on the rail and tighten the clamp securely.
20 Check to make sure that the ground straps, hoses and wiring harnesses and connectors are securely installed in their original locations.
21 Connect the negative battery cable.

18 Multi-point EFI fuel injector — removal and installation

Refer to illustrations 18.3a, 18.3b, 18.4a, 18.4b, 18.4c, 18.4d and 18.6

Removal

1   Remove the fuel rail assembly (Section 17).
2   Place the fuel rail assembly on a clean work surface so the fuel injectors are accessible.
3   Remove the injector clip from the fuel rail and injector by prying it off with a small screwdriver and pull the injector straight out of the receiver cup (see illustrations).
4   Inspect the injector O-rings for damage. Replace them with new ones if necessary (see illustration). To replace the O-rings, use a small screwdriver to remove the old O-ring, roll the new one into the groove and apply a drop of engine oil to the O-ring (see illustrations).

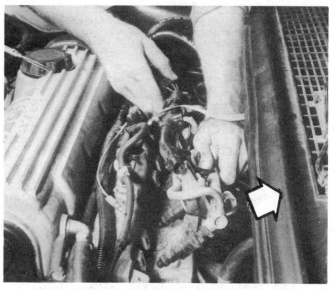

17.9 Grasp the fuel rail securely and pull the injectors out of the intake manifold

17.10 With the fuel rail supported out of the way, loosen the fuel hose clamps

Installation

5   Prior to installation, lubricate the O-ring with a drop of clean engine oil if you have not already done so.
6   Install the top of the injector carefully into the fuel rail receiver cup, taking care not to damage the O-ring (see illustration).
7   Slide the open end of the injector clip into the top slot of the injector, onto the receiver cup ridge and into the side slots of the clip (see illustrations 18.3a and 18.3b).
8   Install the fuel rail (Section 17).

19 Fuel tank — removal and installation

Refer to illustrations 19.5, 19.6 and 19.13
Warning: *Gasoline is extremely flammable, so extra precautions must be taken when working on any part of the fuel system. Do not smoke or allow open flames or bare light bulbs near the work area. Also, do not work in a garage if a natural gas-type appliance with a pilot light is present.*

Chapter 4 Part A Fuel and exhaust systems

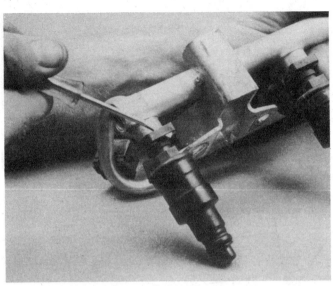

18.3a Insert a small screwdriver under the clip and pry it off

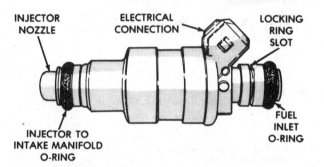

18.4a The fuel injector uses two different size O-rings

18.4c Push the new O-ring over the tip of the injector and into the groove

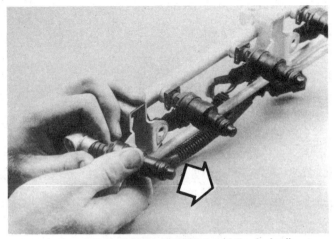

18.3b Pull the injector straight out of the fuel rail receiver cup

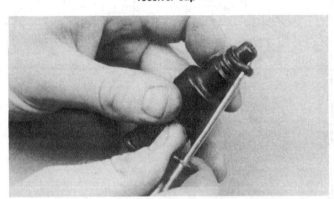

18.4b Insert a small screwdriver under the O-ring and pry it off — be careful not to damage the injector tip

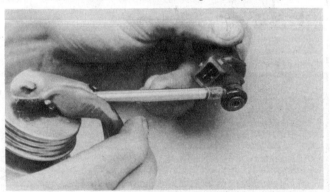

18.4d A drop of clean engine oil will allow the injector to seat more easily

18.6 Insert the injector carefully so that it goes straight into the receiver cup

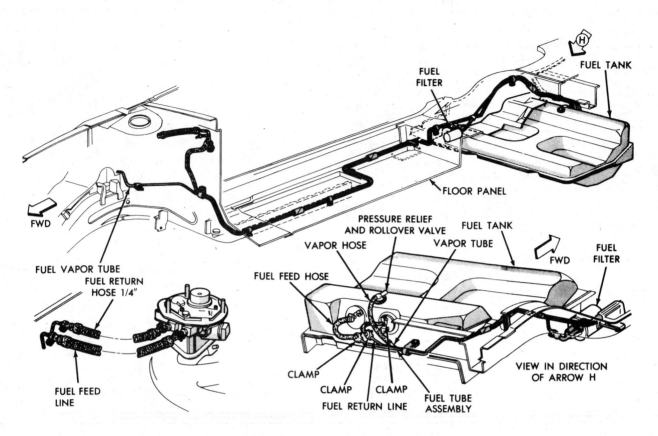

19.5 Fuel lines and tank layout

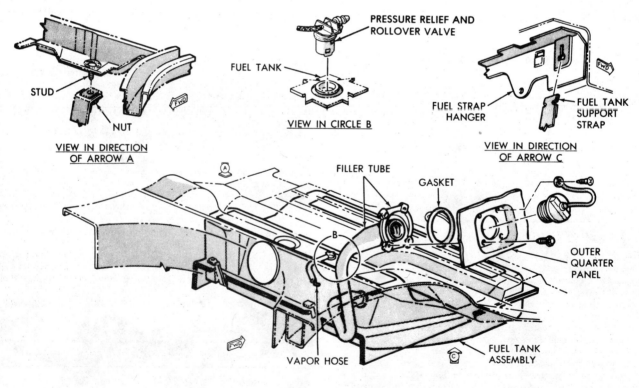

19.6 Fuel filler tube and tank details

Chapter 4 Part A Fuel and exhaust systems

Removal

1 Release the fuel system pressure (Section 4).
2 Disconnect the negative cable at the battery. Place the cable out of the way so it cannot accidentally come in contact with the negative terminal of the battery, as this would once again allow power into the electrical system of the vehicle.
3 Raise the rear of the vehicle and support it securely on jackstands.
4 Remove the gas cap.
5 Disconnect the fuel feed line (the large diameter supply line leading to the fuel injection unit) located in the engine compartment (see illustration), connect a hose and drain or siphon the tank into a metal container. **Warning:** *Do not use your mouth to start the siphoning action.*
6 Remove the screws retaining the filler tube to the body (see illustration). It may be necessary to remove the right rear wheel to provide access to the filler tube.
7 Disconnect all wires and hoses from the tank (label them first to avoid problems during installation).
8 Remove the mounting strap retaining nuts, lower the tank slightly and remove the filler tube.
9 Lower the tank further and support it while disconnecting the rollover/vapor separator valve hose.
10 Remove the tank and insulator pad.

Installation

11 To install the tank, raise it into position with a jack, connect the rollover/vapor separator valve hose and place the insulator pad on the top. Connect the filler tube. **Caution:** *Be sure the vapor vent hose is not pinched between the tank and floor pan.*
12 Raise the tank with the jack, connect the retaining strap and install the retaining nuts. Tighten them securely.
13 Connect the fuel lines and wiring (see illustration) and install the filler tube retaining screws.
14 Fill the fuel tank, install the cap, connect the negative battery cable and check for leaks.

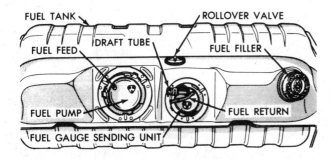

19.13 Fuel tank connection details

20 Intake and exhaust manifolds (non-turbocharged models) — removal, inspection and installation

Refer to illustration 20.15

Removal

1 Disconnect the negative cable at the battery. Place the cable out of the way so it cannot accidentally come in contact with the negative terminal of the battery, as this would once again allow power into the electrical system of the vehicle.
2 Depressurize the fuel system (Section 4).
3 Drain the cooling system (Chapter 1).
4 Remove the air cleaner assembly and disconnect the vacuum, fuel and electrical connections from the EFI throttle body.
5 Disconnect the throttle linkage.
6 Remove the power steering pump drivebelt.
7 Remove the power brake vacuum hose from the intake manifold.
8 On some Canadian models it may be necessary to remove the coupling hose from the diverter valve-to-exhaust manifold air injection tube.
9 Remove the water crossover hoses.
10 Raise the vehicle and support it securely on jackstands.

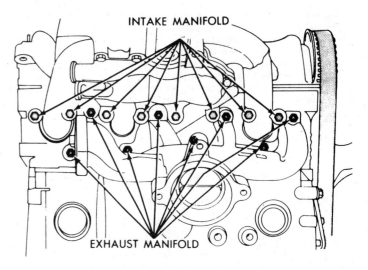

20.15 Intake and exhaust manifold bolt locations

From under the vehicle
11 Remove the power steering pump and secure it out of the way with the hoses still attached.
12 Unbolt and remove the intake manifold support bracket.
13 Remove the EGR tube from the exhaust manifold.
14 On some Canadian models it may be necessary to remove the air injection tube assembly.
15 Remove the intake manifold bolts (see illustration).
16 Lower the vehicle.

In the engine compartment
17 Grasp the intake manifold securely and lift if from the engine.
18 Remove the retaining nuts and detach the exhaust manifold from the engine.

Inspection

19 Carefully scrape all traces of gasket material and sealing compound from the cylinder head and manifold contact surfaces. **Caution:** *Do not gouge the cylinder head, as it is made of aluminum.*
20 Use a straight edge to check the gasket surfaces of the manifolds for flatness (See the Specifications for the allowable warpage limit).
21 Inspect the manifolds for cracks and corrosion.
22 Replace any damaged or warped manifolds with new ones.

Installation

23 Apply a thin coat of gasket sealant to the manifold sides of the new gaskets and place them in position.
24 Place the exhaust manifold in position and install the nuts. Tighten the nuts starting at the center and working outward in both directions in a criss-cross pattern until the specified torque is reached.
25 Place the intake manifold in position.
26 Raise the vehicle.
27 From under the vehicle install the intake manifold retaining bolts finger tight. Tighten the bolts, starting at the center and working outward in both directions, in a criss-cross pattern until the specified torque is reached.
28 The remainder of the installation procedure is the reverse of removal.

21 Exhaust system — removal and installation

Refer to illustration 21.4a, 21.4b, 21.4c and 21.4d

1 The exhaust system should be inspected periodically for leaks, cracks and damaged or worn components (Chapter 1).
2 Allow the exhaust system to cool for at least three hours prior to inspecting or beginning work on it.
3 Raise the vehicle and support it securely on jackstands.

4 Exhaust system components can be removed by removing the heat shields, unbolting and/or disengaging them from the hangers and removing them from the vehicle (see illustrations). Pipes on either side of the muffler must be removed by cutting with a hacksaw. Install the new muffler using new U-bolts. If parts are rusted together, apply a rust dissolving fluid (available at auto supply stores) and allow it to penetrate prior to attempting removal.

5 After replacing any part of the exhaust system, check carefully for leaks before driving the vehicle.

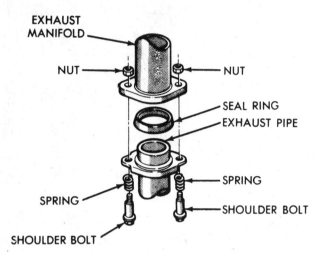

21.4a Exhaust manifold-to-pipe connection details

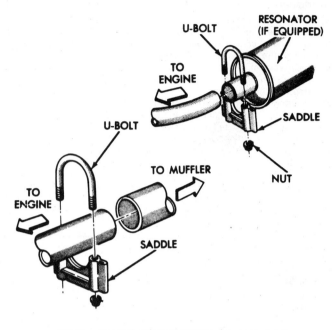

21.4b Typical exhaust system front slip joint connection details

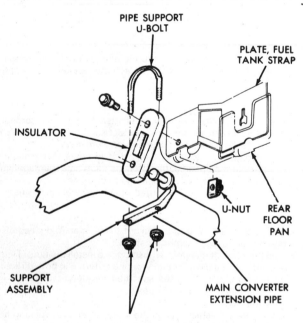

21.4c Typical catalytic converter mount details

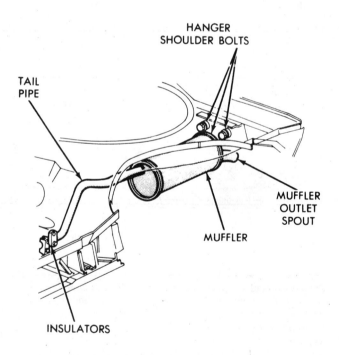

21.4d Typical tail pipe and muffler details

Chapter 4 Part B Turbocharger

Refer to Chapter 13 for information on 1987 and later models

Contents

Air filter replacement .... See Chapter 1	Turbocharger — removal and installation 3
General information 1	Turbocharger — checking 2
Intake and exhaust manifolds — removal and installation 4	

Specifications

Intake and exhaust manifold warpage limit	0.006 in per 12 inches (0.15 mm per 300 mm)	
Torque specifications	**Ft-lbs**	**Nm**
Air cleaner box support bracket bolt	40	54
Coolant tube nut	30	41
Exhaust manifold flange nut	20	28
Exhaust manifold nut	17	23
Exhaust pipe-to-manifold bolt and nut	20	28
Intake manifold bolt	17	23
Oil line tube nut	11	14
Throttle body-to-intake manifold bolt	40	54
Turbocharger discharge hose clamp screw	3	4
Turbocharger-to-exhaust manifold nut	40	54
Turbocharger heat shield bolt	9	12
Turbocharger fuel rail bolt	20	28
Turbocharger support bracket		
to-cylinder block bolt	40	54
to-turbocharger	20	27

1 General information

The turbocharger system increases power by using an exhaust gas-driven turbine to pressurize the fuel/air mixture as it enters the combustion chamber. The amount of boost (intake manifold pressure) is controlled by the wastegate (exhaust bypass valve). This is operated by a spring-loaded actuator assembly which controls the maximum boost level by allowing some of the exhaust gas to bypass the turbine. The wastegate is controlled by the logic module.

2 Turbocharger — checking

1 While a comparatively simple design, the turbocharger is a precision device which can be severely damaged by an interrupted oil or coolant supply or loose or damaged ducting.
2 Due to the special techniques and equipment required, any checking or diagnosis of suspected problems should be left to your dealer. The home mechanic can, however, check the connnections and linkages for security, damage or obvious faults.
3 Because each turbocharger has its own distinctive sound, a change in the noise level can be a sign of potential problems.
4 A high-pitched or whistling sound is a symptom of an inlet air or exhaust gas leak.
5 If an unusual sound issues from the vicinity of the turbine, the turbocharger can be removed and the turbine wheel inspected. **Warning:** *All checks must be made with the engine off and cool to the touch and the turbocharger stopped or injury could result.* Operating the turbocharger without all the ducts and filters installed is dangerous and can result in damage to the turbine wheel blades.
6 Reach inside the housing and turn the turbine wheel to make sure it turns freely. If it does not, this could be a sign that the cooling oil has sludged or coked from overheating. Push inward on the shaft wheels and check for binding. The wheels should rotate freely with no binding or rubbing on the housing.
7 Inspect the exhaust manifold for cracks and loose connections.
8 Because the turbine wheel rotates at speeds up to 140,000 rpm, severe damage can result from the interruption of coolant or contamination of the oil supply to the turbine bearings. Check for leaks in the coolant and oil inlet lines or obstructions in the oil drain back line, as this can cause severe oil loss through the turbocharger seals. Burned oil on the turbine housing is a sign of this. **Caution:** *Any time a major engine bearing such as a main, connecting rod or camshaft bearing is replaced, the turbocharger should be flushed with clean oil.*

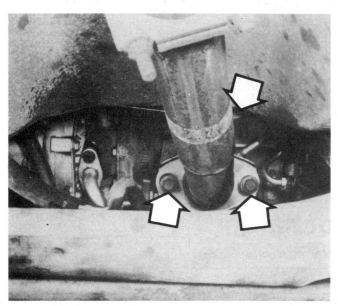

3.4 Exhaust pipe bolts and oxygen sensor connector locations (arrows)

3.6 Disconnect the turbocharger oil drain back hose at the block nipple (arrow) and push it down in the direction shown with a screwdriver

3 Turbocharger — removal and installation

Refer to illustrations 3.4, 3.5, 3.6 and 3.13

Removal

1 Disconnect the negative cable at the battery. Place the cable out of the way so it cannot accidentally come in contact with the negative terminal of the battery, as this would once again allow power into the electrical system of the vehicle.
2 Drain the cooling system (Chapter 1).
3 Raise the front of the vehicle and support it securely on jackstands.
4 Working under the vehicle, disconnect the exhaust pipe and unplug the oxygen sensor connector (see illustration).
5 Remove the turbocharger housing-to-engine block support bracket at the block (see illustration).
6 Loosen the oil drain return hose clamps and push the hose down on the block nipple using a screwdriver (see illustration).
7 Disconnect the turbocharger coolant tube nut at the block outlet (located below the power steering pump bracket) and the tube support bracket at the cylinder block.
8 Lower the vehicle.
9 Working in the engine compartment, remove the air cleaner assembly, along with the throttle body adapter, hose and air cleaner box and support.
10 Disconnect the throttle linkage and electrical connections and vacuum hoses from the throttle body.
11 Loosen the turbocharger-to-throttle body inlet hose clamps. Because the clamps are difficult to reach, a small wrench or socket will make loosening easier.
12 Remove the three retaining bolts and lift the throttle body off.
13 Loosen only the turbocharger-to-intake manifold discharge hose clamps (see illustration).
14 Locate the fuel rail out of the way by removing the hose retainer bracket screw, the four bracket screws from the intake manifold and the two bracket retaining clips and then securing the fuel rail out of the way, complete with injectors, wiring harness and fuel lines.
15 Disconnect the oil line at the turbocharger bearing housing.
16 Remove the three attaching screws and remove the heat shield.
17 Disconnect the coolant return tube and hose assembly from the turbocharger and water box and remove the tube support bracket.
18 Remove the four turbocharger-to-exhaust manifold retaining nuts. To make this job easier, it is a good idea to apply penetrating oil to the exposed threads and allow it to soak in for a few minutes.
19 To remove the turbocharger, lift the assembly off the exhaust manifold studs and push the it down toward the passenger side then up and out of the engine compartment.

3.13 Loosen only the two turbocharger discharge hose clamps (arrows)

Installation

20 Carefully clean the contact surfaces of the turbocharger and manifold.
21 Place the turbocharger in positon on the manifold studs, making sure the discharge tube is in place between the intake manifold and turbocharger.
22 Apply anti-seize compound to the studs and install the nuts. Tighten the nuts to the specified torque.
23 Connect the oil line and tighten the tube nut to the specified torque.
24 Install the bracket to the turbocharger and tighten the bolt to the specified torque.
25 Install the heat shield.
26 Connect the coolant tube and tighten the tube nut to the specified torque.
27 Install the bracket-to-engine bolt. Tighten to the specified torque.
28 Install the fuel rail (refer to Part A of this Chapter).
29 Tighten the turbocharger discharge hose clamp securely.
30 Place the throttle body in position and install the retaining bolts. Tighten the bolts to the specified torque.

3.5 Turbocharger and related components — exploded view

31 Connect the throttle body hose and tighten the clamp screw securely.
32 Connect the throttle linkage, electrical and vacuum connections to the throttle body.
33 Install the throttle body adapter, hose and cleaner box, support and the air cleaner assembly.
34 Raise the front of the vehicle and support it securely on jackstands.
35 From under the vehicle, connect the coolant tube and install the tube nut. Tighten the nut to the specified torque.
36 Connect the oil drain return hose.
37 Connect the exhaust pipe and install the exhaust pipe articulated joint shoulder bolts.
38 Connect the oxygen sensor.
39 Lower the front of the vehicle.
40 Fill the cooling system and connect the negative battery cable.

4 Intake and exhaust manifolds — removal and installation

Refer to illustration 4.20

Warning: *Gasoline is extremely flammable, so extra precautions must be taken when working on any part of the fuel system. Do not smoke or allow open flames or bare light bulbs near the work area. Also, do not work in a garage if a natural gas-type appliance with a pilot light is present.*

1 On these models the cylinder head, complete with turbocharger assembly and intake and exhaust manifolds, must be removed as a unit before the manifolds can be removed.

Removal

2 Disconnect the negative cable at the battery. Place the cable out of the way so it cannot accidentally come in contact with the negative terminal of the battery, as this would once again allow power into the electrical system of the vehicle.
3 Drain the cooling system (Chapter 1).
4 Raise the front of the vehicle and support it securely on jackstands.
5 From under the vehicle, disconnect the exhaust pipe from the manifold and unplug the oxygen sensor electrical connections.
6 Remove turbocharger-to-engine block support bracket (see illustration 3.5).
7 Loosen the oil drain back tube connector hose clamps and push the tube down on block fitting.
8 Disconnect turbocharger coolant inlet tube at cylinder block and remove the bracket support.
9 Lower the vehicle.
10 Working in the engine compartment, remove the air cleaner assembly along with the throttle body adapter, hose and air cleaner box and bracket.
11 Disconnect throttle linkage and throttle body electrical connector and vacuum hoses.
12 Locate the fuel rail out of the way by removing the hose retainer bracket screw, the four bracket screws from the intake manifold and the two retaining clips and then securing the fuel rail out of the way, complete with injectors, wiring harness and fuel lines (refer to Chapter 4, Part A).
13 Disconnect upper radiator hose from thermostat housing.
14 Remove cylinder head with manifolds and turbocharger attached as an assembly (Chapter 2).
15 Place the assembly on a clean working surface. Loosen the upper turbocharger discharge hose end clamp. **Note:** *Do not disturb the center deswirler retaining clamp.*
16 Remove the throttle body-to-intake manifold bolts and remove throttle body assembly.
17 Disconnect turbocharger coolant return tube at the water box along with the retaining bracket on the cylinder head.
18 Remove the heat shield.
19 Remove the turbocharger assembly.
20 Remove the intake manifold retaining bolts and washer assemblies and remove intake manifold. Remove the exhaust manifold retaining

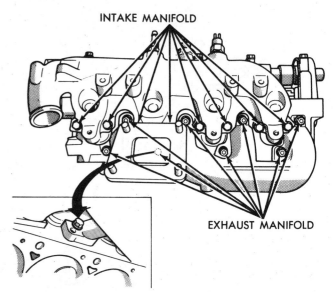

4.20 Intake and exhaust manifold nut/bolt locations

nuts and detach the manifold (see illustration).
21 Discard the gasket and clean the cylinder head and manifold gasket surfaces.
22 Check the gasket mating surfaces for flatness with a straight edge. Refer to the Specifications for the manifold warpage limit.
23 Inspect the manifolds for cracks, corrosion and damage.

Installation

24 Install a new two-sided intake/exhaust Grafoil-type gasket, or equivalent. Do not use sealant on this type of gasket.
25 Place the exhaust manifold in position. Apply anti-seize compound to the threads and install the retaining nuts. Tighten the nuts to the specified torque, working from the center out in both directions, until all the nuts are at the specified torque.
26 Place the intake manifold in position and install the retaining bolts and washers. Tighten the bolts, starting at the center and progressing outward in both directions until all bolts are at specified torque.
27 Connect the turbocharger outlet-to-intake manifold tube and place the turbocharger in position on the exhaust manifold. Apply anti-seize compound to the threads and install the retaining nuts. Tighten the nuts to the specified torque. Tighten connector tube clamps securely.
28 Install the coolant return tube into water box connector, tighten the tube nut to the specified torque and install the tube support bracket to cylinder head.
29 Install the heat shield on the intake manifold and tighten the screws securely.
30 Install the throttle body air horn into the turbocharger inlet tube and install the three throttle body-to-intake manifold bolts. Tighten the bolts to the specified torque.
31 Install the cylinder head (Chapter 2).
32 Connect the turbocharger oil feed line.
33 Install the air cleaner assembly and reconnect the throttle linkage and the electrical and vacuum connections.
34 Install the fuel rail (Chapter 4, Part A).
35 Connect the turbocharger inlet coolant tube to the cylinder block. Tighten the tube nut to the specified torque. Install the support bracket.
36 Install the turbocharger housing-to-engine block support bracket with the bolts finger tight. Tighten first the bolt on the cylinder block, followed by the turbocharger housing bolt, to the specified torque.
37 Reposition the oil drain back tube and tighten the hose clamps.
38 Reconnect the exhaust pipe.
39 Connect the upper radiator hose to the thermostat housing.
40 Fill the cooling system (Chapter 1).
41 Connect the negative battery cable.

Chapter 5 Engine electrical systems

Refer to Chapter 13 for information on 1987 and later models

Contents

Alternator — removal and installation	7
Alternator brushes — removal, inspection and installation	8
Alternator — general information	3
Alternator — maintenance	4
Alternator — special precautions	5
Alternator — troubleshooting and repair	6
Automatic shutdown relay (1984 models) — removal and installation	19
Battery check and maintenance	See Chapter 1
Battery — removal and installation	2
Distributor — removal and installation	21
Engine drivebelt check and adjustment	See Chapter 1
General information	1
Hall Effect assembly — replacement	22
Ignition system — general information	16
Ignition timing check and adjustment	See Chapter 1
Logic module — removal and installation	17
Logic module and power module — description and checking	15
Manifold Absolute Pressure (MAP) sensor — removal and installation	18
Power Loss or Power Limited lamp — general information	14
Power module — removal and installation	20
Spark plug replacement	See Chapter 1
Spark plug wire, distributor cap and rotor check and replacement	See Chapter 1
Starter motor — removal and installation	12
Starter motor — testing on engine	11
Starter solenoid — removal and installation	13
Starting system — general information	10
Voltage regulator — general information	9

Specifications

Distributor direction of rotation	Clockwise
Alternator brush length service limit	0.197 in (5 mm)

Torque specifications	Ft-lbs	Nm
Alternator brush screw	1 to 2	2 to 4
Alternator adjusting bracket bolt	30 to 50	41 to 68
Alternator locking bolt	25	34
Alternator pivot bolt nut	30	41
Battery hold-down nut	8	12

1 General information

The engine electrical system includes the battery, charging system, starter, Electronic Fuel Injection (EFI) system and ignition system. The system is 12-volt with a negative ground.

The charging system consists of the alternator with integral voltage regulator and battery. The starter is operated by the battery's electrical power through the starter relay. The EFI system consists of the power module, logic module, oxygen sensor, the fuel injectors and associated sensors. The ignition system consists of the distributor, ignition coil, spark plugs and associated wires.

Information on the routine maintenance of the ignition, starting and charging systems and battery can be found in Chapter 1.

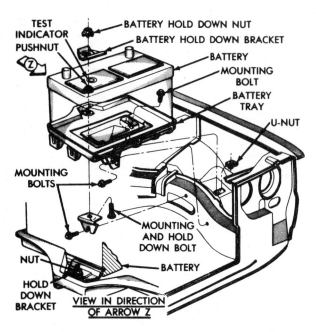

2.1a Battery mount — exploded view

2.1b The hold-down nut (arrow) is located at the base of the battery

2 Battery — removal and installation

Refer to illustrations 2.1a and 2.1b

Warning: *Certain precautions must be followed when checking or servicing the battery. Hydrogen gas, which is very flammable, is produced in the cells, so keep lighted tobacco, open flames and sparks away from the battery. The electrolyte inside is actually dilute sulfuric acid, which can burn your skin and cause serious injury if splashed in the eyes. It will also ruin clothes and painted surfaces.*

1 The battery is located at the left front corner of the engine compartment and is held in place by a hold-down clamp at its base (see illustrations).
2 Always disconnect the negative (–) battery cable first, followed by the positive (+) cable.
3 After the cables are disconnected, remove the nut and hold-down clamp.
4 Remove the battery. **Warning:** *When lifting the battery from the engine compartment, be careful not to twist the case as acid could spurt out of the filler openings.*
5 Installation is the reverse of removal. Be careful not to overtighten the retaining nut as the clamp could damage the battery case.

3 Alternator — general information

The alternator is operated by a drivebelt turned by the crankshaft pulley. The rotor turns inside the stator to produce an alternating current, which is then converted to direct current by diodes. The current is adjusted to battery charging needs by an electronic voltage regulator, which is controlled by the logic and power modules.

4 Alternator — maintenance

1 The alternator requires very little maintenance because the only components subject to wear are the brushes and bearings. The bearings are sealed for life. The brushes should be inspected for wear after about 75,000 miles (120,000 km) and the length compared to the Specifications.
2 Regular maintenance consists of cleaning to remove grease and dirt, checking the electrical connections for tightness and adjusting the drivebelt for proper tension.

5 Alternator — special precautions

Whenever the electrical system is being worked on or a booster battery is used to start the engine, certain precautions must be observed to avoid damaging the alternator.

a) Make sure that the battery cables are never reversed or damage to the alternator diodes will result. The negative (–) cable must always be grounded.
b) The output (+) cable must never be grounded. It should always be connected to the positive battery terminal.
c) Never use a high voltage tester on the alternator.
d) Do not operate the engine with the voltage regulator plug disconnected.
e) When the alternator is to be removed or its wiring disconnected, always disconnect the negative battery cable first.
f) The engine must never be operated with the battery-to-alternator cable disconnected.
g) Disconnect the battery cables before charging the battery from an external source.
h) If a booster battery or charger is used, be sure to observe correct polarity.

6 Alternator — troubleshooting and repair

1 Due to the special training and equipment necessary to test and service the alternator, it is recommended that the vehicle be taken to a dealer or other repair shop with the proper equipment if a problem arises.
2 The most obvious sign of a problem is the alternator warning light on the instrument panel coming on, particularly at low speeds. This indicates that the alternator is not charging. Other symptoms are a low battery state-of-charge, evidenced by dim headlights, and the starter motor turning the engine over slowly.
3 The first check should always be of the drivebelt tension (Chapter 1), followed by making sure that all electrical connections are secure and free of dirt and corrosion.
4 If the drivebelt tension, electrical connections and battery are good, an internal fault in the alternator or voltage regulator is indicated.
5 Due to the special tools and techniques required to work on the alternator, diagnosis and repair should be left to a properly-equipped shop. If the vehicle has considerable miles on it, a good alternative is to replace the alternator with a rebuilt unit.

Chapter 5 Engine electrical systems

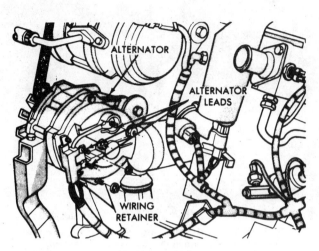

7.2a Typical non-turbocharged engine alternator wiring connections

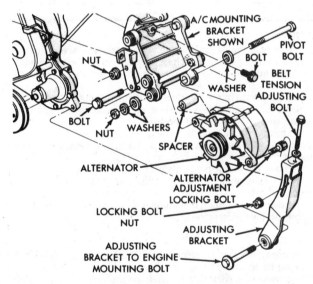

7.3 Alternator mount details

7.2b Typical turbocharged engine alternator connections (arrows)

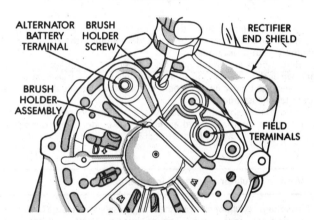

8.2 Remove the screws evenly so the holder won't be damaged ...

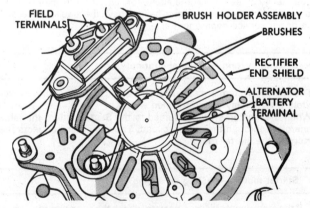

8.3 ... and rotate the brush holder out of the alternator housing

7 Alternator — removal and installation

Refer to illustrations 7.2a, 7.2b and 7.3

1 With the ignition switch in the Off position, disconnect the negative cable at the battery. Place the cable out of the way so it cannot accidentally come in contact with the negative terminal of the battery, as this would once again allow power into the electrical system of the vehicle.
2 Disconnect the alternator wires (see illustrations). Mark each wire and terminal to ensure correct reinstallation.
3 Loosen the adjusting and mounting bolts and remove the drivebelt (see illustration).
4 Remove the mounting/adjusting bolts.
5 Remove the pivot bolt and nut and separate the alternator from the engine.
6 To install the alternator, place it in position and install the pivot bolt and nut finger-tight.
7 Install the drivebelt.
8 Install the locking/adjusting bolts and adjust the drivebelt tension. Tighten the locking bolt.
9 Tighten the pivot bolt and nut.

8 Alternator brushes — removal, inspection and installation

Refer to illustrations 8.2, 8.3, 8.5 and 8.6

1 Disconnect the negative cable at the battery. Place the cable out of the way so it cannot accidentally come in contact with the negative terminal of the battery, as this would once again allow power into the electrical system of the vehicle.
2 Remove the brush holder mounting screws by unscrewing them evenly and alternately a few turns at a time so as not to distort the holder (see illustration).
3 Separate the brush holder from the rear of the alternator and remove it from the housing (see illustration).

Chapter 5 Engine electrical systems

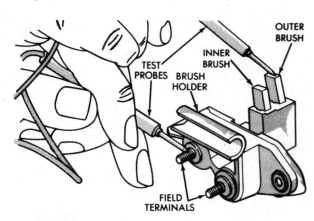

8.5 Be sure continuity exists between each brush and the appropriate field terminal before installing the brush holder assembly (non-turbocharged engine alternator shown)

8.6 Push the brush holder into place making sure the brushes (which are spring loaded) seat securely (turbocharged engine alternator shown)

4 If the brushes are worn beyond the specified limit, or if they do not move smoothly in the brush holder, replace the brush holder assembly with a new one.
5 Before installing the brush holder assembly, check for continuity between each brush and the appropriate field terminal (see illustration).
6 Insert the holder into position, making sure the brushes seat securely (see illustration).
7 Hold the brush holder securely in place and install the screws. Tighten them alternately and evenly so the holder is not distorted.
8 Connect the negative battery cable.

9 Voltage regulator — general information

The voltage regulator controls the charging system voltage by limiting the alternator output voltage. The regulator is a sealed unit and is not adjustable.
If the ammeter fails to register a charge rate or the red warning light on the dash comes on and the alternator, battery, drivebelt tension and electrical connections seem to be fine, have the regulator checked by a dealer service department or a repair shop.

10 Starting system — general information

The starting system is made up of an electric motor, battery, starter switch, starter relay and associated wiring.
When the ignition switch is turned to the Start position, the relay is energized through the control circuit. The relay then connects the battery to the starter motor.

11 Starter motor — testing on engine

1 If the starter motor fails to operate, check the condition of the battery by turning on the headlights. If they glow brightly for several seconds and then gradually dim, the battery is in an uncharged condition.
2 If the headlights continue to glow brightly, check the tightness of the battery cables and the starter wiring. Check the tightness of the connections at the rear of the solenoid.
3 If the battery is fully charged and the wiring is in order, and it still fails to operate, then it will have to be removed from the engine for examination. Before this is done, however, make sure that the pinion gear has not jammed in mesh with the ring gear due either to a broken solenoid spring or dirty pinion gear splines. To release the pinion, engage a low gear (manual transaxle) and, with the ignition switched Off, rock the vehicle backwards and forwards. This should release the pinion from mesh with the ring gear. If the pinion remains jammed, the starter motor must be removed.

12 Starter motor — removal and installation

Refer to illustrations 12.5a, 12.5b, 12.6 and 12.7

Removal

1 Disconnect the negative cable at the battery. Place the cable out of the way so it cannot accidentally come in contact with the negative terminal of the battery, as this would once again allow power into the electrical system of the vehicle.
2 Raise the front of the vehicle and support it securely on jackstands.
3 Remove the right side driveaxle (non-turbocharged models) or intermediate shaft (turbocharged models).
4 On turbocharged models it will be necessary to unbolt the turbocharger brace at the lower starter motor-to-block mount and swivel it out of the way.
5 Remove the lower heat shield (if equipped) (see illustrations).
6 Remove the retaining nuts and disconnect the battery cable and solenoid wire from the starter (see illustration).
7 Remove the mounting bolts/nuts and separate the starter from the bellhousing (see illustration).

Installation

8 To install the starter, place it in position on the studs, install the mounting bolts/nuts and tighten them securely.
9 The remaining installation steps are the reverse of the removal procedures.

13 Starter solenoid — removal and installation

Refer to illustrations 13.2a and 13.2b

1 Remove the starter motor (Section 12).
2 Disconnect the field coil cable, remove the three retaining screws and separate the solenoid from the starter motor (see illustrations).
3 Mark the solenoid housing-to-starter relationship and lift the solenoid from the starter. The plunger and spring may fall out of the solenoid, so make sure the components are properly assembled prior to installation.
4 When installing, insert the solenoid plunger into the starter, hook it over the lever and assemble the spring and solenoid over it. Hold the assembly securely in place while installing the screws. Connect the field coil wire and retaining nut.

14 Power Loss or Power Limited lamp — general information

The Power Loss/Power Limited lamp is located in the instrument panel and ordinarily flashes on briefly and then goes out when the engine is started. The lamp lights and stays on when there is a fault in the EFI system.

Chapter 5 Engine electrical systems

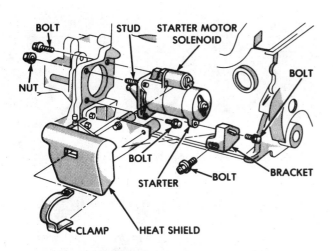

12.5a Typical starter motor mounting details

12.5b Remove the heat shield by squeezing the clamp and pulling it off

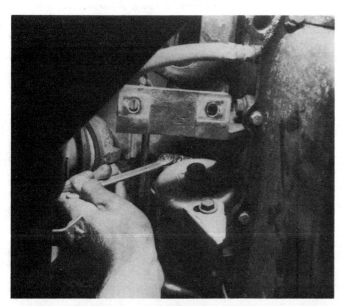

12.6 Removing the battery cable and solenoid wire nuts from the starter motor

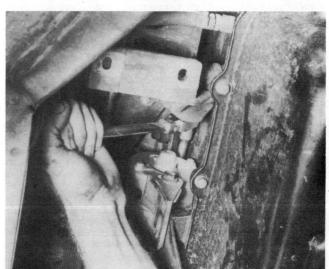

12.7 A socket and long extension are necessary for removing the starter motor-to-bellhousing bolts

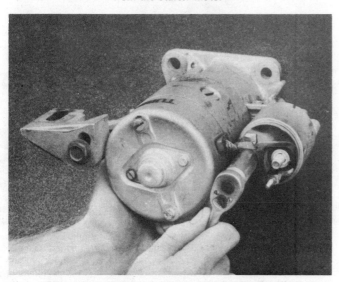

13.2a Use a socket wrench to remove the field coil terminal nut and detach the cable

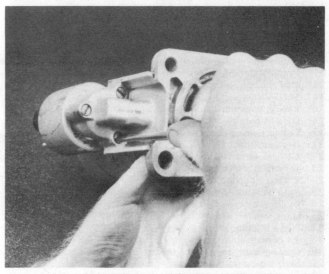

13.2b The solenoid is attached to the starter by three screws

15 Logic module and power module — description and checking

1 The logic module, located in the passenger compartment behind the right kick panel, is a digital computer containing a microprocessor. The power module is located in the left front corner of the engine compartment and works in conjunction with the logic module, insuring that a constant 8 volts is supplied to the various sensors and handling the heavier electric currents including the ground for the ignition coil and power for the fuel injection. It directs the voltage to the logic module and the Hall Effect pickup assembly in the distributor and the Automatic Shutdown relay which energizes the fuel tank mounted electric fuel pump. The logic module receives the input signals from all of the sensors and switches which monitor the engine and then determines the fuel injector operation as well as the spark advance, ignition coil dwell, idle speed, canister purge solenoid operation, cooling fan operation and alternator charge rate.

2 Four components provide basic information so the logic module can operate the EFI system. They are the Manifold Absolute Pressure (MAP) sensor, Throttle Position Sensor (TPS), oxygen sensor and the coolant temperature sensor. The MAP sensor is located on or adjacent to the logic module, is connected to the Throttle Body by a vacuum line and monitors the manifold vacuum. The TPS is located on the throttle body and monitors the position of the throttle lever. The oxygen sensor (located in the exhaust manifold) provides information on the exhaust gas makeup, and the temperature sensor (threaded into the water box) monitors engine operating temperature.

3 Because the EFI system is controlled by the logic module and power module in combination with a variety of sensors and switches, the home mechanic can do very little in the way of diagnosis without factory Diagnostic Read Out Tool number C-4805 or equivalent. Consequently, diagnosis should be confined to inspection and checking of all electrical and vacuum connections to make sure they are secure and not obviously damaged.

4 The logic module is self-testing and a problem in the system will be indicated by the Power Loss/Limited lamp on the dash. The Power Loss/Limited lamp will light when there is electrical system voltage fluctuation or a fault in the MAP, throttle position or coolant temperature sensor circuits. If the fault is severe enough to affect driveability, the logic module will go into a *Limp-In* mode so the vehicle can still be driven.

5 The logic module stores trouble codes, which can be checked using the Power Loss/Limited lamp. Within a five second period, turn the ignition key On-Off-On-Off-On. The Power Loss/Limited lamp will then flash fault codes indicating the area of the fault. The codes are two digit numbers and the Start of Test (88) for example, will be indicated by eight flashes, a pause and eight more flashes. If there is more than one code, the lamp will flash them in order, ending with the End of Message (55) code. Because the fault codes indicate the general location of a fault, simply checking and tightening a vacuum hose or electrical connector can often correct a problem. Any further checking of fault codes should be left to a dealer or properly equipped shop.

Fault codes

88 Start of test
11 Engine not cranked since battery disconnected
12 Memory standby power lost
13* MAP sensor pneumatic circuit
14* MAP sensor electrical circuit
15 Vehicle speed/distance sensor circuit
16* Loss of battery voltage
17 Engine running too cold
21 Oxygen sensor circuit
22* Coolant sensor circuit
23 Throttle body temperature circuit
24* Throttle position circuit
25 Automatic Idle Speed (AIS) motor driver circuit
26 Peak injector circuit has not been reached
27 Logic module fuel circuit internal problem
31 Purge solenoid circuit
33 Air conditioning cutout relay circuit
35 Cooling fan relay circuit
37 Shift indicator light circuit
41 Charging system excess or lack of field current
42 Automatic shutdown relay driver circuit

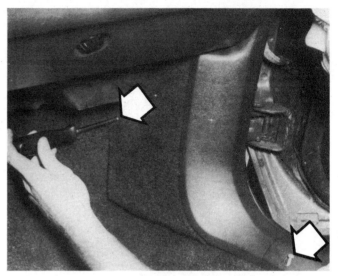

17.2 The cowl trim panel covering the logic module is held in place by two screws (arrows)

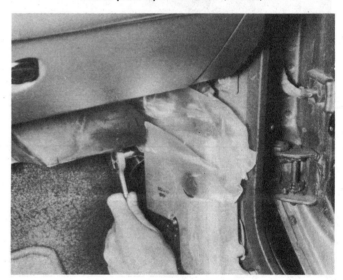

17.3 Use a socket wrench with an extension to remove the logic module mounting bolts

43 Spark interface circuit
44 Battery temperature out of range
46* Battery voltage too high
47 Battery voltage too low
51 Oxygen sensor stuck at lean position
52 Oxygen sensor stuck at rich position
53 Logic module internal problem
55 End of message

* Activates Power Loss/Limited lamp

16 Ignition system — general information

The ignition system is designed to ignite the fuel/air charge entering the cylinders at just the right moment. It does this by producing a high-voltage electrical spark between the electrodes of the spark plugs.

On these vehicles the ignition system and the Electronic Fuel Injection (EFI) system interact to provide the proper fuel flow and ignition curve for all conditions. The ignition system consists of a switch, the ignition coil, the distributor and the logic and power modules. The spark timing is constantly adjusted by the logic and power modules in response to input from the various sensors located on the engine.

Chapter 5 Engine electrical systems

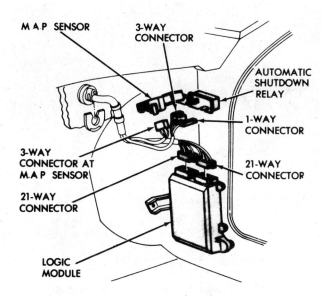

18.3 The MAP sensor and automatic shutdown relay are located adjacent to the logic module on 1984 models

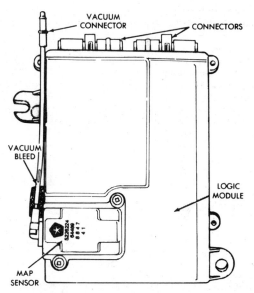

18.6 On 1985 and 1986 models the MAP sensor is located on the logic module

17 Logic module — removal and installation

Refer to illustrations 17.2 and 17.3

Removal
1 Disconnect the negative cable at the battery. Place the cable out of the way so it cannot accidentally come in contact with the negative terminal of the battery, as this would once again allow power into the electrical system of the vehicle.
2 Remove the passenger side cowl trim panel (see illustration).
3 Remove the mounting nuts, lift the logic module off the studs, unplug the connectors and remove the Module from the vehicle (see illustration).

Installation
4 Place the logic module in position, plug in the connectors and install the nuts. Tighten the nuts securely.
5 Install the trim panel and connect the negative battery cable.

18 Manifold Absolute Pressure (MAP) sensor — removal and installation

Refer to illustrations 18.3 and 18.6

1 Disconnect the negative cable at the battery. Place the cable out of the way so it cannot accidentally come in contact with the negative terminal of the battery, as this would once again allow power into the electrical system of the vehicle.

1984 models
2 Remove the glove box (Chapter 11) for access to the MAP sensor.
3 Unplug the vacuum hose and wiring connector, remove the retaining bolts and withdraw the MAP sensor from under the instrument panel (see illustration).
4 To install, place the MAP sensor in position and install the retaining bolts. Tighten the bolts securely and install the glove box.

1985 and 1986 models
5 Remove the logic module (Section 17).
6 Remove the two retaining screws and detach the MAP sensor from the logic module (see illustration).
7 To install, place the MAP sensor in position on the logic module and install the retaining screws.

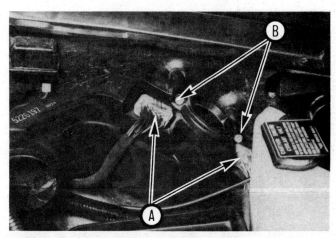

20.4 The power module can be removed after unplugging the electrical connectors (A) and removing the retaining bolts (B)

19 Automatic shutdown relay (1984 models) — removal and installation

1 Disconnect the negative cable at the battery. Place the cable out of the way so it cannot accidentally come in contact with the negative terminal of the battery, as this would once again allow power into the electrical system of the vehicle.
2 Remove the glove box (Chapter 11).
3 Unplug the wiring harness connector, remove the retaining bolt and lower the relay from under the dash (see illustration 18.3).
4 Installation is the reverse of removal.

20 Power module — removal and installation

Refer to illustration 20.4

Removal
1 Disconnect the battery cables (negative first, followed by the positive cable).
2 Disconnect the air cleaner duct from the power module.
3 Remove the battery (Section 2).
4 Remove the retaining bolts, unplug the connectors and lift the power module from the engine compartment (see illustration).

Chapter 5 Engine electrical systems

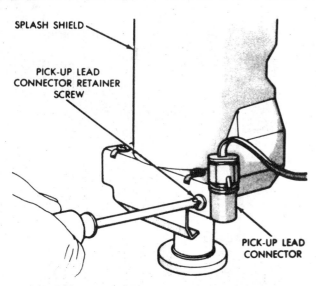

21.3 After unplugging the distributor wire harness connector, remove the screw and detach the connector from the distributor (1984 and 1985 models)

21.6 Mark the relationship of the distributor rotor to the body and the body to the block with white paint

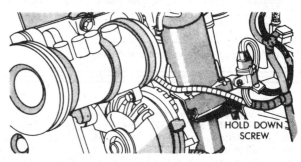

21.7a Distributor hold-down screw or bolt location (1984 and 1985 models)

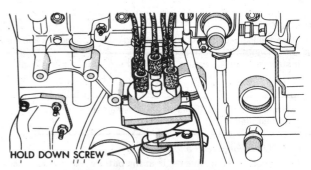

21.7b 1986 model distributor hold-down screw or bolt location

Installation

5 Place the power module in position, install the retaining screws and plug in the connectors.
6 Connect the air cleaner duct.
7 Install the battery and connect the battery cables (positive first, then negative).

21 Distributor — removal and installation

Refer to illustrations 21.3, 21.6, 21.7a and 21.7b

1 Disconnect the negative cable at the battery. Place the cable out of the way so it cannot accidentally come in contact with the negative terminal of the battery, as this would once again allow power into the electrical system of the vehicle.
2 Unplug the distributor pickup coil lead wire at the harness connector.
3 On 1984 and 1985 models, remove the retaining screw and separate the pick-up lead connector from the retainer (see illustration).
4 Remove the distributor splash shield.
5 Loosen the two screws and lift off the distributor cap.
6 Rotate the engine with a large wrench on the crankshaft pulley nut until the distributor rotor is pointed at the engine block. Use white paint to mark the relationship of the rotor to the distributor body and the distributor to the engine block so the distributor can be reinstalled with the rotor in the same exact position (see illustration).
7 Remove the hold-down screw or bolt and carefully lift the distributor out of the engine (see illustrations).
8 To install the distributor, lower it into position, making sure the gasket or O-ring is seated properly. Engage the distributor drive with the auxiliary shaft so that the rotor is aligned with the mark on the engine block made during removal.
9 If the crankshaft was rotated during the time the distributor was removed, it will be necessary to establish the proper relationship between the rotor and the number one piston position. Remove the number one cylinder spark plug, place your finger over the plug hole and rotate the crankshaft until pressure is felt, indicating that the piston is at top dead center on the compression stroke. The pointer on the bellhousing should be aligned with the 0 (TDC) mark on the flywheel. If it is not, continue to turn the crankshaft until the mark and the 0 are lined up.
10 Install the distributor with the rotor pointing at the reference mark previously made and the distributor body and engine block marks aligned.
11 Install the distributor cap.
12 Install the hold-down bolt snugly.
13 Install the splash shield.
14 Install the pickup coil lead and plug the wire into the harness. Connect the negative battery cable.
15 Check the timing (Chapter 1) and tighten the distributor hold-down bolt securely.

22 Hall effect assembly — replacement

Refer to illustrations 22.3, 22.4a, 22.4b, 22.6a and 22.6b

1 Disconnect the negative cable at the battery. Place the cable out of the way so it cannot accidentally come in contact with the negative terminal of the battery, as this would once again allow power into the electrical system of the vehicle.
2 Remove the distributor splash shield and cap.

Chapter 5 Engine electrical systems

22.3 The rotor can be simply lifted off the distributor shaft (1984 and 1985 models)

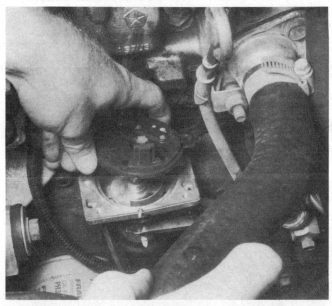

22.4b Lift the Hall Effect pick-up assembly off the distributor shaft

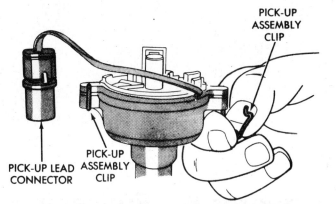

22.4a On some models the retaining clips must be removed before the pick-up assembly can be removed

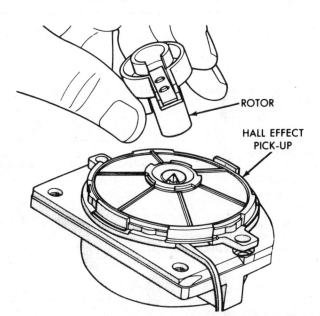

22.6a Remove the 1986 model distributor rotor by lifting it off the Hall Effect pick-up

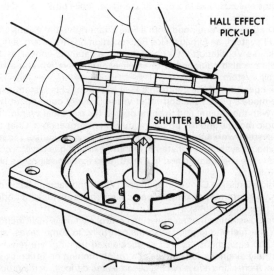

22.6b Lift the Hall Effect pick-up straight up off the distributor shaft (1986 models)

1984 and 1985 models

3 Remove the rotor (see illustration).
4 Remove the clips (if equipped) and lift the pick-up assembly off the distributor shaft (see illustrations).
5 To install, place the assembly carefully into position, making sure the electrical lead retainer is in the locating hole. Install the retaining clips (if equipped) and the rotor, with the ESA marking on top.

1986 models

6 Lift the rotor off the shaft, followed by the pick-up (see illustrations).
7 To install, place the assembly in position carefully, making sure the electrical lead retainer is in the locating hole. Install the rotor with the ESA marking on top.

All models

8 Install the distributor cap and splash shield and connect the negative battery cable.

Chapter 6 Emissions control systems

Refer to Chapter 13 for information on 1987 and later models

Contents

Air aspirator system 8	Heated inlet air system 7
Catalytic converter 4	Oxygen sensor replacement 6
Exhaust Gas Recirculation (EGR) system 5	Positive Crankcase Ventilation (PCV) system 2
Fuel Evaporative Emission Control (EVAP) system 3	Power Loss/Power Limited lamp .... See Chapter 5
General information 1	

Specifications

General

EGR valve test vacuum	10 in-Hg
EGR valve travel	1/8-in (3mm)

Torque specifications	Ft-lbs	Nm
Aspirator tube nut-to-manifold	50	68
Oxygen sensor-to-exhaust manifold..................	20	27
Catalytic converter clamp nut	22	30

1 General information

Refer to illustrations 1.4a and 1.4b

Since these vehicles are equipped with either a 2.2L or a 2.5L engine, many emission control devices are used. Some of these devices or systems are exclusive to a particular engine, while others are applicable to all vehicles.

All engines are equipped with a Fuel Evaporative Emissions Control (EVAP) system, an Exhaust Gas Recirculation (EGR) system, a Positive Crankcase Ventilation (PCV) system and a catalytic converter. Some models are also equipped with an air aspirator system and a heated inlet air system.

The operation of the Electronic Fuel Injection (EFI) system used on these models places the control of virtually every component having to do with the engine, and to some degree the electrical system, under the logic module and the power module microprocessors (Chapter 5). This system works in conjunction with an oxygen sensor located in the exhaust system to constantly monitor exhaust gas oxygen content and vary the spark timing and fuel mixture so that emissions are always within limits.

Vehicle Emission Control Information (VECI) and vacuum hose routing labels with information on your particular vehicle are located under the hood (see illustrations).

Before assuming that an emission control system is malfunctioning, check the fuel and ignition systems carefully. In some cases, special tools and equipment, as well as specialized training, are required to accurately diagnose the causes of a rough running or difficult to start engine. If checking and servicing becomes too difficult, or if a procedure is beyond the scope of the home mechanic, consult a Chrysler dealer service department or a repair shop.

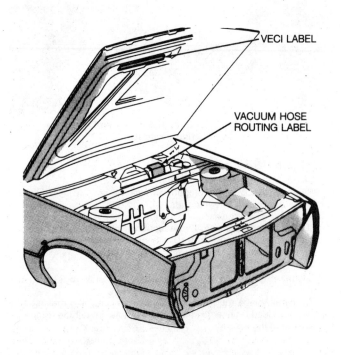

1.4a Vacuum hose routing label and Vehicle Emission Control Information (VECI) label locations

Chapter 6 Emissions control systems

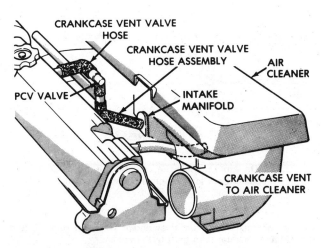

1.4b Typical Vehicle Emission Control Information label

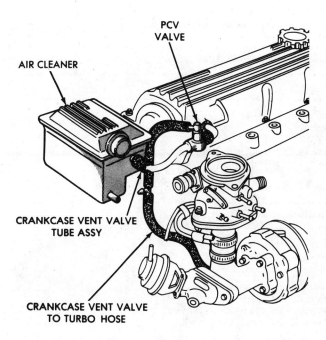

2.2a Typical non-turbocharged engine PCV system

2.2b Turbocharged engine PCV system

2.3 Checking the PCV valve (make sure the hose or grommet [arrow] which seals around the PCV valve is in good condition to prevent air leaks which would affect engine performance)

This does not necessarily mean, however, that the emission control systems are all particularly difficult to maintain and repair. You can quickly and easily perform many checks and do most (if not all) of the regular maintenance at home with common tune-up and hand tools.
Note: *The most frequent cause of emission system problems is simply a loose or broken vacuum hose or wiring connection. Therefore, always check hose and wiring connections first.*

2 Positive Crankcase Ventilation (PCV) system

Refer to illustrations 2.2a, 2.2b and 2.3

General description

1 This system is designed to reduce hydrocarbon emissions (HC) by routing blow-by gases (fuel/air mixture that escapes from the combustion chamber past the piston rings into the crankcase) from the crankcase to the intake manifold and combustion chambers, where they are burned during engine operation.
2 The system is very simple and consists of rubber hoses and a small, replaceable metering valve (PCV valve) (see illustrations).

Checking and component replacement

3 With the engine running at idle, pull the PCV valve out of the mount or hose and place your finger over the valve inlet (see illustration). A strong vacuum will be felt and a hissing noise will be heard if the valve is operating properly. Replace the valve with a new one if it is not functioning as described. *Do not attempt to clean the old valve.*
4 To replace the PCV valve, simply pull it from the hose and install a new one.
5 Inspect the hose prior to installation to ensure that it isn't plugged or damaged. Compare the new valve with the old one to make sure they are the same. Chapter 1 contains additional PCV valve information.

Chapter 6 Emissions control systems

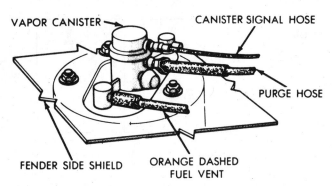

3.6 Check the EVAP canister hoses to make sure they are undamaged and properly installed

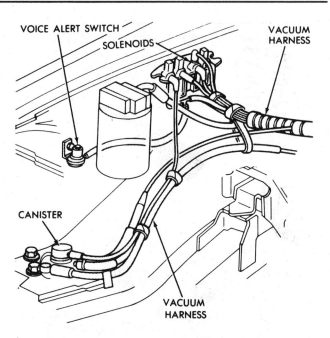

3.11 On some models the canister purge and EGR solenoid are located on the same bracket

3.12 On early models, pull straight out to release the solenoid cover

3 Fuel Evaporative Emission Control (EVAP) system

Refer to illustrations 3.6, 3.11 and 3.12

General description

1 This system is designed to trap and store fuel that evaporates from the fuel injection system and fuel tank which would normally enter the atmosphere in the form of hydrocarbon (HC) emissions.
2 The system consists of a charcoal-filled canister, a purge solenoid, a combination rollover/separator valve and connecting lines and hoses. Later models may also have a valve in the fuel tank vent line which retains vapor until it can be drawn into the canister when the engine is running.
3 When the engine is off and a high pressure begins to build up in the fuel tank (caused by fuel evaporation), the charcoal in the canister absorbs the fuel vapor. When the engine is started cold, the charcoal continues to absorb and store fuel vapor. As the engine warms up, the stored fuel vapors are routed to the intake manifold or air cleaner and combustion chambers where they are burned during normal engine operation.
4 The canister is purged using engine vacuum by the purge solenoid, which is controlled by the logic module. When the engine coolant temperature is below 70°F (145°C) the logic module energizes the solenoid by grounding it so that vacuum will not act on the canister.

Once the coolant temperature rises above 70°F the solenoid is de-energized and vacuum then affects the canister.
5 The relief valve, which is mounted in the fuel tank filler cap, is calibrated to open when the fuel tank vacuum or pressure reaches a certain level. This vents the fuel tank and relieves the high vacuum or pressure.

Checking

6 Check the canister and lines for cracks and other damage (see illustrations).
7 To check the filler cap, look for a damaged or deformed gasket as described in Chapter 1.

Component replacement

Canister

8 The canister is located in the right front corner of the engine compartment, behind the headlight.
9 To replace the canister, disconnect the vacuum hoses, remove the mounting nuts and lower the canister, removing it from beneath the vehicle.
10 Installation is the reverse of removal.

Canister purge/EGR solenoid

11 The canister purge solenoid is located under a cover in the engine compartment on the right inner fender panel. On models equipped with an EGR solenoid, this solenoid is mounted together with the canister purge solenoid (see illustration).
12 To gain access to the solenoid(s), remove the cover. On early models this is accomplished by grasping the cover and pulling it off (see illustration). On later models the cover is retained by a nut.
13 Disconnect the vacuum hose(s), unplug the electrical connector(s), remove the retaining bolt and lift the solenoid(s) and bracket assembly off the fender panel.
14 Installation is the reverse of removal.

4 Catalytic converter

General description

1 The catalytic converter is designed to reduce hydrocarbon (HC) and carbon monoxide (CO) pollutants in the exhaust gases. The converter oxidizes these components and converts them to water and carbon dioxide.

Chapter 6 Emissions control systems

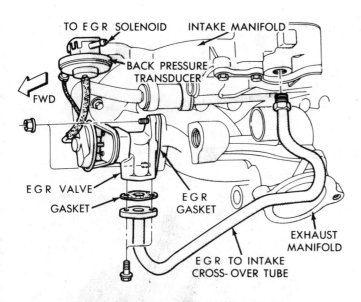

5.5a Non-turbocharged engine EGR system

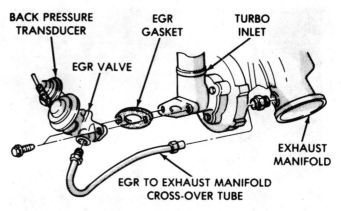

5.5b Turbocharged engine EGR system

2 If large amounts of unburned gasoline enter the catalyst, it may overheat and cause a fire. Always observe the following precautions:
   Use only unleaded gasoline
   Avoid prolonged idling
   Do not run the engine with a nearly empty fuel tank
   Do not prolong engine compression checks
   Avoid coasting with the ignition turned Off

Checking

3 The catalytic converter requires little maintenance and servicing at regular intervals. However, the system should be inspected whenever the vehicle is raised on a lift or if the exhaust system is checked or serviced.
4 Check all connections in the exhaust pipe assembly for looseness and damage. Also check all the clamps for damage, cracks and missing fasteners. Check the rubber hangers for cracks.
5 The converter itself should be checked for damage and dents (maximum 3/4-inch deep) which could affect performance and/or be hazardous to your health. At the same time the converter is inspected, check the heat shields under it, as well as the heat insulator above it, for damage and loose fasteners.

Component replacement

6 Do not attempt to remove the catalytic converter until the complete exhaust system is cool. Raise the vehicle and support it securely on jackstands. Apply penetrating oil to the clamp bolts and allow it to soak in for several minutes.
7 Remove the bolts and the rubber hangers, then separate the converter from the exhaust pipe. Remove the old gaskets if they are stuck to the pipes.
8 Installation of the converter is the reverse of removal. Use new exhaust pipe gaskets and tighten the clamp nuts to the specified torque. Start the engine and check carefully for exhaust leaks.

5 Exhaust Gas Recirculation (EGR) system

Refer to illustrations 5.5a, 5.5b, 5.5c and 5.10

General description

1 This system recirculates a portion of the exhaust gases into the intake manifold in order to reduce the combustion temperatures and decrease the amount of nitrogen oxide (NOx) produced.
2 The main components in the system are the EGR valve, the backpressure transducer and (on some models) the EGR solenoid.
3 The EGR valve is a backpressure type. The amount of exhaust gas admitted is regulated by engine vacuum and the backpressure trans-

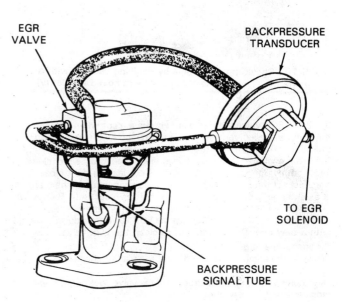

5.5c EGR valve and backpressure transducer component layout

ducer in the signal line. This transducer uses exhaust system backpressure to control the EGR valve vacuum, bleeding the vacuum off to the atmosphere whenever the backpressure at the valve itself drops below the calibrated level. The EGR valve and transducer are controlled by the EGR solenoid (operated by the logic module) which controls the vacuum flow to the EGR valve body in accordance with engine temperature and driving conditions.
4 Symptoms of problems associated with the EGR system are rough idling or stalling when at idle, rough engine performance during light throttle application and stalling during deceleration.

Checking

5 Check all hoses for cracks, kinks, broken sections and proper connections (see illustrations).

Non-turbocharged engine

6 To check the EGR valve operation, bring the engine up to operating temperature with the transaxle in Neutral (tires blocked to prevent movement).
7 Disconnect the hose from the transducer and connect a vacuum pump. Start the engine, raise the engine speed to approximately 2000 rpm, hold it there and apply 10 in-Hg with the vacuum pump. The EGR valve stem should move and stay open for at least thirty seconds if the control system is working properly. Measure the valve travel to make sure it is within the specified limit. If the stem moves but will not stay open, the EGR valve/backpressure transducer assembly is faulty and must be replaced with a new one.

Chapter 6 Emissions control systems

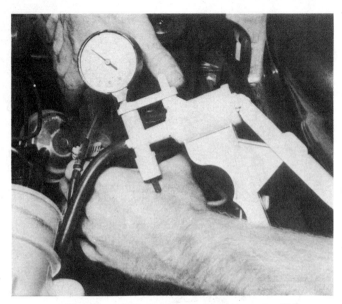

5.10 Check the EGR valve by applying vaccum with a pump (air cleaner assembly partially removed for clarity)

6.1 The oxygen sensor (arrow) is screwed into the exhaust manifold and on some models is accessible only from below

8 If the EGR valve stem does not move except when vacuum from the pump is applied, remove the EFI throttle body (Chapter 4) and clean the EGR ports in the throttle bore and body.

9 If the engine exhibits rough idle, dies when returned to idle or the idle is both rough and slow, the EGR valve is leaking in the closed position. Inspect the EGR tube for leaks at the connection to the manifold. Loosen the tube connection and then tighten it securely. Remove the EGR valve and transducer assembly and inspect the poppet to make sure it is seated. If it is not, replace the EGR valve/transducer assembly with a new one. Do not attempt to clean the EGR valve.

Turbocharged engine

10 With the engine cold, disconnect the vacuum hose from the EGR valve and connect a vacuum pump, apply vacuum and make sure the valve holds vacuum (see illustration). If the valve does not hold vacuum, replace it with a new one.

11 Disconnect the vacuum pump and connect a vacuum gauge to the EGR valve. Start the engine and verify that the gauge reading is zero, indicating that the EGR valve and transducer are holding vacuum.

12 With the engine still cold, increase the engine speed to approximately 2000 rpm and hold it there. If the gauge shows an unsteady vacuum reading above zero, the EGR solenoid is faulty and should be replaced with a new one.

13 Disconnect the vacuum hose between the EGR backpressure solenoid and the EGR solenoid. Connect the vacuum gauge, start the engine and make sure the reading is at least 5 in-Hg of vacuum, indicating that the system is working properly and holding vacuum.

14 Warm the engine to normal operating temperature (the EGR solenoid will now be open), increase the engine speed to approximately 2000 rpm and make sure that the vacuum reading is still at least 5 in-Hg of vacuum. If the reading is less, there is a vacuum leak between the EGR solenoid and the EFI throttle body.

15 Remove the vacuum gauge and reconnect the vacuum line.

16 Disconnect the EGR valve-to-backpressure transducer vacuum hose and connect the vacuum gauge to this line. The gauge should still read approximately 5 in-Hg of vacuum with the engine off. If it does not, replace the EGR valve and backpressure transducer assembly with a new one.

17 Start the engine and raise the engine speed to approximately 2000 rpm. Connect a vacuum pump to the EGR valve vacuum motor and slowly apply vacuum. If the system is operating properly, the engine speed will begin to drop when 2 to 3.5 in-Hg of vacuum is applied.

18 Release the vacuum and allow the engine to slow to idle speed. Apply vacuum with the pump again to make sure the engine speed again drops when 2 to 3.5 in-Hg of vacuum is applied. If the speed does not drop, check the EGR supply tube and passages to make sure they are not blocked, clearing them if necessary. If they are open, then the EGR valve/backpressure transducer and tube are faulty and must be replaced with new units.

19 Apply 10 in-Hg of vacuum to the EGR valve vacuum motor. The system should hold the 10 in-Hg of vacuum for at least ten seconds if it is operating properly. If it does not hold vacuum, the EGR valve/backpressure transducer assembly is faulty and must be replaced with a new unit.

Component replacement

EGR valve/backpressure transducer

20 Disconnect the vacuum hose from the backpressure transducer assembly and then pull the assembly from its mounting clip.

21 Remove the bolts or unscrew the tube nut and disconnect the crossover tube from the EGR valve.

22 Remove the EGR valve retaining bolts and lift the EGR valve/backpressure transducer assembly from the engine.

EGR solenoid

23 The EGR solenoid is located under a cover at the right front corner of the engine compartment adjacent to the EVAP purge solenoid. The cover is held in place by clips on 1984 models and by the solenoid bracket retaining nut on later models.

24 Refer to Section 3 for the solenoid replacement procedure.

6 Oxygen sensor replacement

Refer to illustration 6.1

1 The oxygen sensor must be replaced at the specified interval (Chapter 1). The sensor is threaded into the exhaust manifold and on some models it may be necessary to raise the front of the vehicle and support it securely on jackstands for access from underneath the engine compartment (see illustration).

2 Disconnect the oxygen sensor wire.

3 Use a wrench to unscrew the sensor.

4 Use a tap to clean the threads in the exhaust manifold.

5 If a new sensor is to be installed, apply anti-seize compound to the threads (most new sensors will already have the anti-seize compound on their threads).

6 Install the sensor, tighten it to the specified torque and plug in the connector.

Chapter 6 Emissions control systems

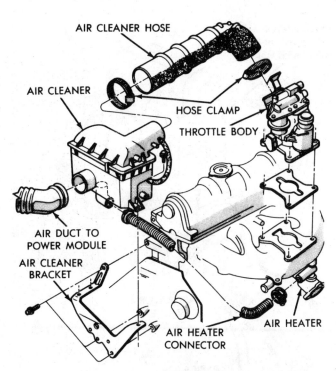

7.1 Heated inlet air system components (non-turbocharged models only) — exploded view

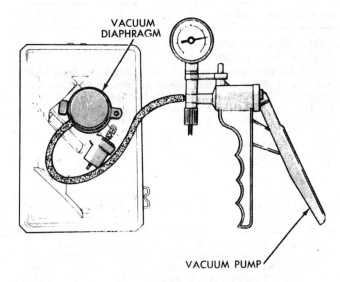

7.8 Checking the inlet air vacuum diaphragm

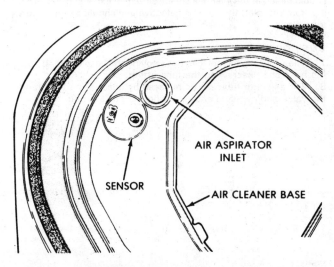

7.16 Heated inlet air sensor location

7 Heated inlet air system

Refer to illustrations 7.1, 7.8 and 7.16

General description

1 This system, used on non-turbocharged engines, is designed to improve driveability and reduce emissions in cold weather by directing hot air from around the exhaust manifold to the air cleaner intake (see illustration).
2 The system is made up of two circuits. When the outside air temperature is below 115°F (46°C), the intake air flows through the flexible connector, up through the air cleaner and into the throttle body.
3 When the air temperature is above 140°F (60°C), air enters the air cleaner through the power module and outside air duct.

Checking

4 Refer to Chapter 1 for the general checking procedure. If the system is not operating properly, check the individual components as follows.
5 Check all vacuum hoses for cracks, kinks, proper routing and broken sections. Make sure the shrouds and ducts are in good condition as well.
6 Remove the air cleaner assembly from the engine and allow it to cool to less than 115°F (46°C). Apply 20 in-Hg of vacuum to the sensor with a hand vacuum pump.
7 The duct door should be in the up (heat on) position with the vacuum applied. If it is not, check the vacuum diaphragm.
8 To check the diaphragm, slowly apply vacuum with the hand pump while observing the door (see illustration).
9 The duct door should not begin to open at less than 2 in-Hg and should be fully open at 4 in-Hg or less. With 20 in-Hg applied, the diaphragm should not bleed down more than 10 in-Hg in five minutes.
10 Replace the sensor with a new unit if it fails any of the tests. Test the new unit as described before reinstalling the air cleaner assembly.

Component replacement

Vacuum diaphragm
11 With the air cleaner removed, disconnect the vacuum hose and drill out the retaining rivet.
12 Disengage the diaphragm by tipping it forward slightly while turning it counterclockwise. Once disengaged, the unit can be removed by moving it to one side, disconnecting the rod from the control door and detaching it from the air cleaner assembly.
13 Check the control door for free travel by raising it to the full up position and allowing it to fall closed. If it does not close easily, free it up. Check the hinge pin for free movement also, using compressed air or spray cleaner to remove any foreign matter.
14 To install the diaphragm, insert the rod end into the control door and position the diaphragm tangs in the slot, turning the diaphragm clockwise until it engages. Rivet the tab in place.
15 Connect the vacuum hose.

Sensor
16 Disconnect the vacuum hoses and use a screwdriver to pry the retaining clips off. Detach the sensor from the housing (see illustration).
17 To install the sensor, place the gasket on the sensor and insert the sensor into the housing.
18 Hold the sensor in place so that the gasket is compressed to form a good seal and install the new retaining clips.
19 Connect the vacuum hoses.

8 Air aspirator system

Refer to illustration 8.5

General description

1 The aspirator system uses exhaust pulsations to draw fresh air from the air cleaner into the exhaust system. This reduces carbon monoxide (CO) and, to a lesser degree, hydrocarbon (HC) emissions.
2 The system is composed of a valve, hoses and tubes between the air cleaner assembly and the exhaust system.
3 The aspirator valve works most efficiently at idle and slightly off idle, where the negative exhaust pulses are strongest. The valve remains closed at higher engine speeds.

Checking

4 Aspirator valve failure results in excessive exhaust system noise from under the hood and hardening of the rubber hose from the valve to the air cleaner.
5 If exhaust noise is excessive, check the aspirator tube-to-exhaust manifold joint and the valve and air cleaner hose connections for leaks (see illustration). If the manifold joint is leaking, retighten the tube fitting to the specified torque. If the hose connections are leaking, install new hose clamps (if the hose has not hardened). If the hose has hardened, replace it with a new one.
6 To determine if the valve has failed, disconnect the hose from the inlet. With the engine idling (transmission in Neutral), the exhaust pulses should be felt at the inlet. If a steady stream of exhaust gases is escaping from the inlet, the valve is defective and should be replaced with a new one.

Component replacement

7 The valve can be replaced by removing the hose clamp, detaching the hose and unscrewing the tube fitting.
8 The aspirator tube can be replaced by unscrewing the fittings at the valve and manifold and removing the bracket bolt.

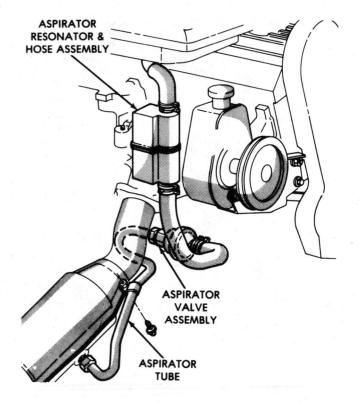

8.5 Air aspirator system components

Chapter 7 Part A Manual transaxle

Refer to Chapter 13 for information on 1987 and later models

Contents

Fluid level check .... See Chapter 1	Manual transaxle speedometer gear assembly — removal and installation 4
Gearshift linkage — check and adjustment 2	
General information 1	Manual transaxle — removal and installation 3
Manual transaxle service and repair 5	

Specifications

General

Transaxle type	5-speed, synchronized in all forward gears
Fluid type and capacity	See Chapter 1

Torque specifications	Ft-lbs	Nm
Gearshift housing-to-case bolt	21	28
Gearshift lever nut*	21	28
Anti-rotational strut (non-turbocharged models)		
upper bolt and nut	40	54
lower bolt and nut	40	54
Anti-rotational damper (turbocharged models)		
upper nut	16	22
lower bolt and nut	40	54
Anti-rotational strut or damper bracket nuts	17	23
Fill plug**	24	33
Transaxle case-to-engine block bolt	70	95
Mount-to-block and case bolt	70	95
Shift linkage adjusting pin	9	12
Speedometer gear bolt	60 in-lbs	7
End cover bolt	21	28
Selector cable adjusting screw	55 in-lbs	6
Crossover cable adjusting screw	55 in-lbs	6

* This nut must be replaced with a new one each time it is removed
** Metal plug only — rubber plug is a press fit

1 General information

The manual transaxle combines the transmission and differential assemblies into one compact unit.

The gearshift features a manual reverse lockout device and synchromesh is used on all forward speeds. A cable-operated shift mechanism is used on all models.

2 Gearshift linkage — check and adjustment

Refer to illustrations 2.3a, 2.3b, 2.4, 2.6, 2.7 and 2.8

1 In the event of hard shifting, disconnect both cables at the transaxle and then operate the shifter from the drivers seat. If the shift lever moves smoothly through all positions with the cables disconnected, then the cable linkage itself should be adjusted as described below.

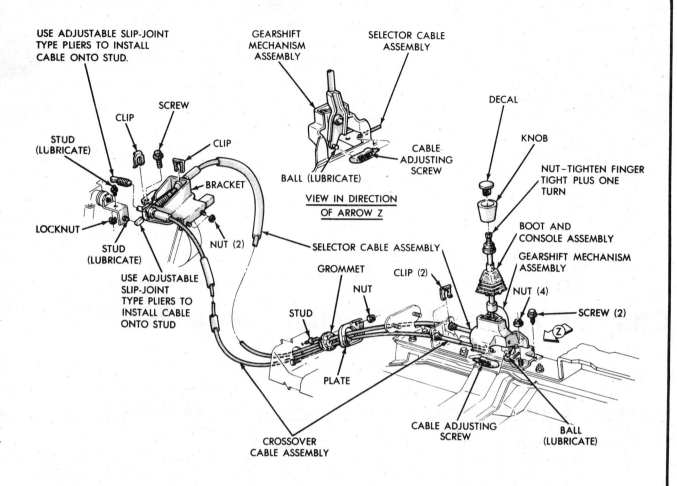

2.3a Shift linkage components — exploded view

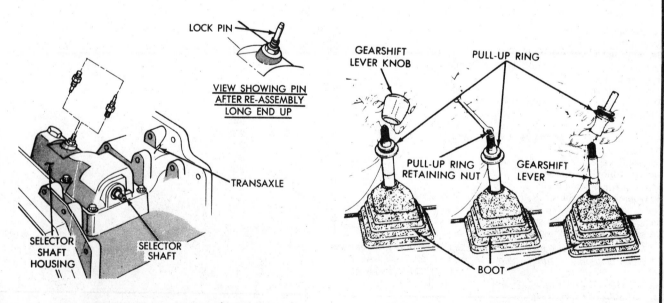

2.3b Use the lock pin to secure the transmission selector shaft prior to cable linkage adjustment

2.4 Gearshift knob and pull-up ring removal details

Chapter 7 Part A Manual transaxle

2 Raise the hood and place a pad or blanket over the left fender to protect it.
3 Remove the lock pin from the transaxle selector shaft housing. Reverse the lock pin so the longer end is down, reinstall it into the hole and move the selector shaft in. When the lock pin aligns with the hole in the selector shaft, thread it into place so the shaft is locked in the 1st/2nd Neutral position (see illustrations).
4 Unscrew the gearshift knob. Remove the reverse lockout pull-up ring retaining nut and lift off the pull-up ring (see illustration). Remove the rubber boot and the console (Chapter 11).
5 Fabricate two 5-inch long adjusting pins from 5/16-inch wire. Bend one end of each pin at right angles so the pins are easy to grasp.
6 Insert one adjusting pin into the crossover cable hole of the shift mechanism and the other into the selector cable hole (see illustration).
7 Use an in-lbs torque wrench to tighten the selector cable adjusting screw to the specified torque (see illustration).
8 Use the in-lbs torque wrench to tighten the crossover cable adjusting screw (see illustration).
9 Remove the adjusting pins from the shift mechanism.
10 Install the console, pull-up ring and nut. Screw the shift knob back on and tighten it securely. If the shift pattern does not line up after the knob has been installed, remove the pattern using a small screwdriver and snap it back in properly aligned.
11 Unscrew the lock pin from the selector housing and reinstall it with the longer end up.
12 Check the shifter operation in 1st and Reverse and make sure the reverse lockout mechanism works properly.

3 Manual transaxle — removal and installation

Refer to illustrations 3.3, 3.6, 3.8, 3.10, 3.12 and 3.17

1 Disconnect the negative cable at the battery. Place the cable out of the way so it cannot accidentally come in contact with the negative terminal of the battery, as this would once again allow power into the electrical system of the vehicle.
2 Remove the hood (Chapter 11), raise the front of the vehicle and support it securely on jackstands.
3 Attach a lifting eye to the number four cylinder exhaust manifold bolt and support the engine with the special support fixture (see illustration). Alternatively, if due care is taken, a jack can be used to support the engine from below.
4 Disconnect the gearshift linkage, clutch cable and speedometer drive gear.
5 Remove the front wheels and the left splash shield.

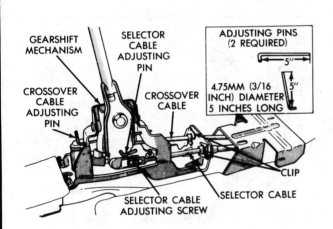

2.6 Fabricate the cable adjusting pins and install them as shown here

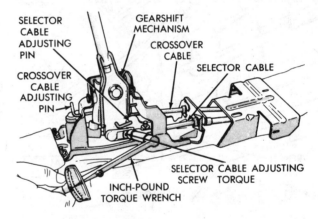

2.7 Adjusting the selector cable with a torque wrench

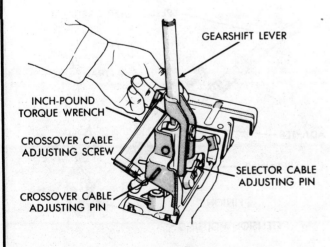

2.8 Adjusting the crossover cable with a torque wrench

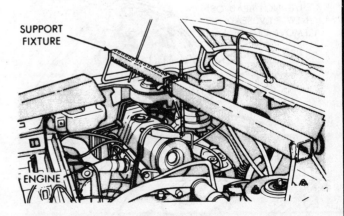

3.3 Support the engine from above with the special fixture as shown here or a with a jack from underneath

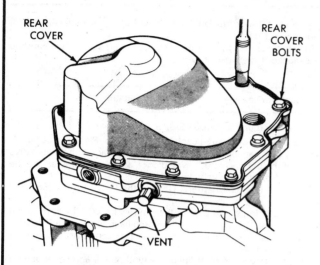

3.6 Use a socket wrench and extension to remove the rear cover bolts

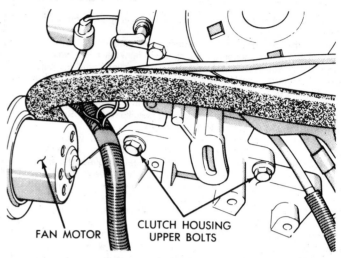

3.8 Upper clutch housing bolt location

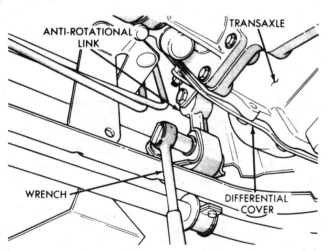

3.10 Removing the anti-rotational link (non-turbocharged models)

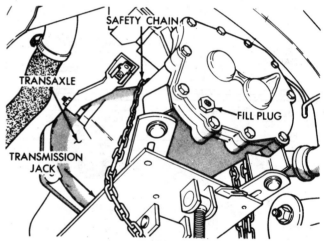

3.12 Support the transaxle with a jack when removing it (note the safety chain)

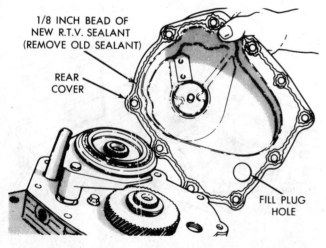

3.17 Apply RTV-type sealant as shown (be sure to go around the bolt holes) when installing the rear cover

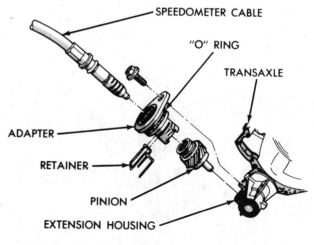

4.1 Speedometer gear assembly — exploded view

Chapter 7 Part A Manual transaxle

6 Place a large drain pan (at least three quarts capacity) under the transaxle and drain the fluid by removing the rear cover (see illustration).
7 Place a jack under the transaxle to support its weight.
8 Remove the clutch housing bolts (see illustration).
9 Remove the left engine mount.
10 Unbolt the anti-rotational link (non-turbocharged engine) or damper (turbocharged engine) between the chassis and transaxle (see illustration).
11 Remove the driveaxles as described in Chapter 8.
12 Carefully pull the transaxle away from the engine and lower it to the floor (see illustration).
13 When installing the transaxle, locating pins can be fabricated and used in place of the top two bolts. Cut the heads off two proper size bolts with a hacksaw and remove any burrs from the ends with a file or grinder. Use a hacksaw to cut slots in the ends of the locating pins so they can be unscrewed with a screwdriver and replaced with bolts.
14 Install the driveaxles (Chapter 8).
15 Install the anti-rotational link or damper and tighten the nut to the specified torque.
16 Install the left engine mount and tighten the bolts to the specified torque.
17 If the transaxle rear cover has not been installed prior to this point, clean the cover and mating surface thoroughly. Apply a 1/8-inch bead of RTV-type sealant to the cover (see illustration) and install the cover and bolts.
18 Fill the transaxle to the bottom of the fill plug hole with the specified fluid (Chapter 1).
19 Install the splash shield and the front wheels.
20 Remove the support from the engine and connect the battery cable.
21 Lower the front of the vehicle.
22 Check the gearshift linkage operation to make sure all gears engage smoothly and easily. If they do not, adjust the linkage as described in Section 2.

4 Manual transaxle speedometer gear assembly — removal and installation

Refer to illustration 4.1

1 The speedometer gear assembly is located in the differential extension housing (see illustration).
2 Remove the retaining bolt and carefully work the speedometer assembly up and out of the extension housing.
3 Remove the retainer and separate the pinion from the adapter.
4 Check the speedometer cable to make sure that transaxle fluid has not leaked into it. If there is fluid in the cable, remove the adapter and replace the small O-ring with a new one. Reconnect the cable to the adapter.
5 Install a new O-ring and connect the adapter to the pinion gear, making sure the retainer is securely seated.
6 Make sure the mating surfaces of the adapter and the extension housing are clean, as any debris could cause misalignment of the gear.
7 Attach the assembly to the transaxle, install the retaining bolt and tighten it to the specified torque.

5 Manual transaxle service and repair

Because of the special tools and expertise required to disassemble, overhaul and reassemble the transaxle, it is recommended that it be left to a dealer service department or a transmission repair shop.

Chapter 7 Part B Automatic transaxle

Contents

Automatic transaxle service and repair	9
Automatic transaxle speedometer gear assembly — removal and installation	8
Automatic transaxle — removal and installation	7
Band adjustment	5
Fluid and filter change	4
Fluid level check	See Chapter 1
Gearshift linkage — adjustment......................	2
General information	1
Neutral start and back-up light switch — check and replacement	6
Throttle cable — adjustment	3

Specifications

General

Transaxle type	3-speed automatic
Fluid type and capacity................................	See Chapter 1

Band adjustment

Kickdown..	Back off 2-1/2 turns from 72 in-lbs (8 Nm)
Low-Reverse (rear)	Back off 3-1/2 turns from 41 in-lbs (5 Nm)
Low-Reverse band end gap	0.080 in (2 mm)

Torque specifications	Ft-lbs	Nm
Differential cover bolts	14	19
Oil pan bolts	14	19
Filter-to-valve body screws	40 in-lbs	5
Kickdown band adjusting screw	72 in-lbs	8
Kickdown band locknut	35	47
Low-Reverse band adjusting screw	41 in-lbs	5
Low-Reverse band locknut	10	14
Neutral start and back-up light switch	25	34
Throttle cable adjustment bracket lock screw	9	12
Speedometer gear bolt	5.2	7
Torque converter-to-driveplate bolts		
1984 and 1985	40	54
1986	55	74

1 General information

The automatic transaxle combines a 3-speed automatic transmission and differential assembly into one unit. Power from the engine passes through the torque converter and the transmission to the differential assembly and then to the driveaxles.

All models feature a transaxle oil cooler with the oil-to-air cooler element mounted in front of the radiator.

2 Gearshift linkage — adjustment

Refer to illustrations 2.2 and 2.4

1 Place the gearshift lever in Park.
2 Working in the engine compartment, loosen the gearshift cable clamp bolt on the transaxle bracket (see illustration).
3 Pull the shift lever on the transmission all the way to the front detent (Park) position.

Chapter 7 Part B Automatic transaxle

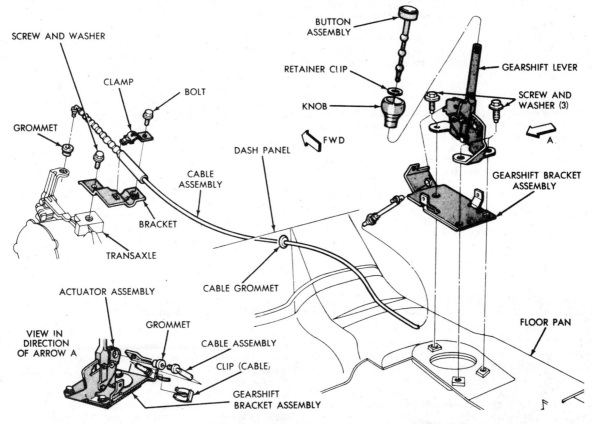

2.2 Automatic transaxle shift linkage components — exploded view

2.4 When tightening the shift cable clamp bolt, maintain pressure with one hand on the shift lever (arrow) as the bolt is tightened

4 Keep pressure on the shift lever and tighten the cable clamp bolt (see illustration).
5 Check the shift lever in the Neutral and Drive positions to make sure it is within the confines of the lever stops. The engine must start only when the lever is in the Park or Neutral positions.

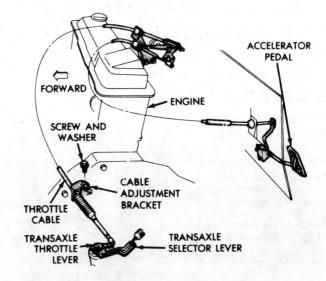

3.3 Throttle cable layout — 1984 and 1985 models

3 Throttle cable — adjustment

Refer to illustrations 3.3 and 3.6

1 The throttle cable controls a valve in the transaxle which governs shift quality and speed. If shifting is harsh or erratic, the throttle cable should be adjusted.
2 The adjustment must be made with the engine at normal operating temperature.

1984 and 1985 models

3 Loosen the cable adjustment bracket lock screw (see illustration).
4 To ensure proper adjustment, the bracket must be free to slide

Chapter 7 Part B Automatic transaxle

back-and-forth. If necessary, remove it and clean the slot and sliding surfaces as well as the screw.
5   Slide the bracket to the left (toward the engine) to the limit of its travel. Release the bracket and move the throttle lever all the way to the right, against the internal stop, then tighten the adjustment bracket lock screw to the specified torque.

1986 models
6   Loosen the cable mounting bracket lock screw and position the bracket so the alignment tabs are in contact with the transaxle casting (see illustration). Tighten the lock screw to the specified torque.
7   Release the cross-lock on the cable assembly by pulling up on it. To ensure proper adjustment, the cable must be free to slide all the way toward the engine, against the stop, after the cross-lock is released.
8   Move the transaxle throttle control lever clockwise as far as possible (against the internal stop) and press the cross-lock down into the locked position.
9   Do not lubricate any of the throttle linkage components on later models.

All models
10 Check the cable action. Move the transaxle throttle cable all the way forward, release it slowly and make sure that it returns completely.

4   Fluid and filter change

Refer to illustrations 4.3, 4.4, 4.6 and 4.8

1   The automatic transaxle fluid and filter should be changed, the magnet cleaned and the bands adjusted (Section 5) at the recommended intervals.
2   Raise the vehicle and support it securely on jackstands.
3   Place a container under the transaxle oil pan. Loosen the pan bolts, completely removing those across the rear of the pan. Tap the corner of the pan (see illustration) to break the seal and allow the fluid to drain into the container (the remaining bolts will prevent the pan from falling at this time). Remove the remaining bolts and lower the pan.
4   Remove the filter screws and detach the filter (a Torx bit may be required for the screws) (see illustration).
5   Refer to Section 5 and adjust the bands before proceeding with the fluid change.
6   Install the new gasket and filter and tighten the screws (see illustration).
7   Carefully remove all traces of old gasket sealant from the oil pan and the transaxle body (don't nick or gouge the sealing surfaces). Clean the magnet in the pan using a clean, lint-free cloth. On 1984 models, the magnet is located in the differential cover. Remove the differential cover, take the magnet out of the recess and clean it.
8   Apply a 1/8-inch bead of RTV-type sealant to the oil pan gasket surface and position it on the transaxle. Install the bolts and tighten them to the specified torque following a criss-cross pattern. Work up to the final torque in three or four steps. On 1984 models, install the magnet in the differential cover, apply a 1/8-inch bead of RTV-type sealant around the contact surface and install it following the same procedure (see illustration).
9   Lower the vehicle and add four quarts of the specified fluid to the transaxle. Start the engine and allow it to idle for at least one minute, then move the shift lever through each of the positions, ending in Park or Neutral. Check for fluid leakage around the oil pan and differential cover.
10 Add more fluid until the level is 1/8-inch below the Add mark on the dipstick.

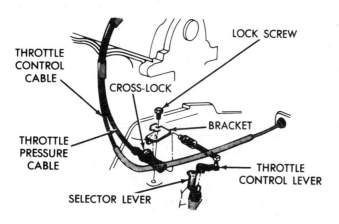

3.6 Throttle cable layout — 1986 models

4.3 Use a soft-face hammer to break the corner of the oil pan loose and drain the fluid

4.4 Removing the filter screws

Chapter 7 Part B Automatic transaxle

11 Drive the vehicle until the fluid is hot, then recheck the level (see Chapter 1).
12 Make sure the dipstick is seated completely or dirt could get into the transaxle.

5 Band adjustment

Refer to illustrations 5.7, 5.9, 5.10, 5.11a, 5.11b, 5.12 and 5.15

1 The transaxle bands should be adjusted when specified in the maintenance schedule or at the time of a fluid and filter change (Section 4).

Kickdown band

2 The kickdown band adjustment screw is located at the top left side of the transaxle case.

3 On some models the throttle cable may interfere with band adjustment. If so, mark its position and then remove the throttle cable adjustment bolt. Move the cable away from the band adjustment screw.
4 Loosen the locknut approximately five turns and make sure the adjusting screw turns freely.
5 Tighten the adjusting screw to the specified torque.
6 Back the adjusting screw off the specified number of turns.
7 Hold the screw in position and tighten the locknut to the specified torque (see illustration).

Low-Reverse band

8 To gain access to the Low-Reverse band, it is necessary to remove the oil pan (Section 4).
9 To determine if the band is worn excessively, remove the Low-Reverse pressure plug from the transaxle case and apply 30 psi of air pressure to the port (see illustration).

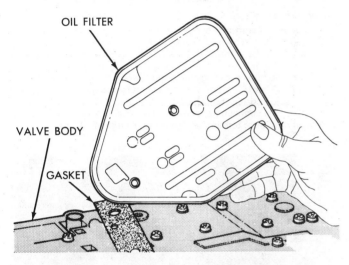

4.6 Be sure to install the new gasket before attaching the new filter to the transaxle

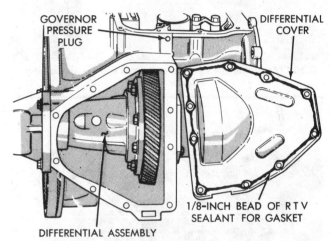

4.8 Apply RTV sealant to the differential cover as shown (1984 models)

5.7 Tighten the kickdown band locknut with a box-end wrench while holding the screw so it doesn't turn

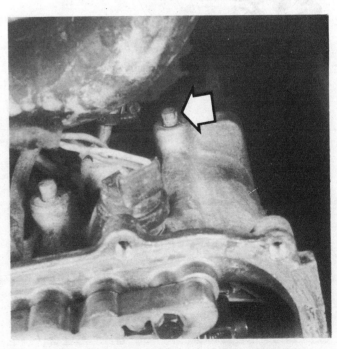

5.9 Low-Reverse pressure plug location (arrow)

Chapter 7 Part B Automatic transaxle

10 Measure the gap between the band ends and compare it to the Specifications (see illustration). If it is less than specified, the band should be replaced with a new one.
11 To proceed with adjustment, pry off the parking rod E-clip and remove the rod (see illustrations).
12 Loosen the locknut approximately five turns (see illustration). Use an in-lbs torque wrench to tighten the adjusting screw to the specified torque.
13 Back the screw off the specified number of turns.
14 Hold the adjusting screw in position and tighten the locknut to the specified torque.
15 Push the shift pawl in the transaxle case to the rear and reinstall the parking rod (see illustration).
16 Install the oil pan and refill the transaxle (Section 4).

6 Neutral start and back-up light switch — check and replacement

Refer to illustration 6.6

1 The neutral start and back-up light switch is located at the lower front edge of the transaxle. The switch controls the back-up light and the starting of the engine in Park and Neutral. The center terminal of the switch grounds the starter solenoid circuit when the transaxle is in Park or Neutral, allowing the engine to start.
2 Prior to checking the switch, make sure the gearshift linkage is properly adjusted (Section 2).
3 Unplug the connector and use an ohmmeter to check for continuity between the center terminal and the case. Continuity should exist only

5.10 Measuring the Low-Reverse band end clearance with a feeler gauge

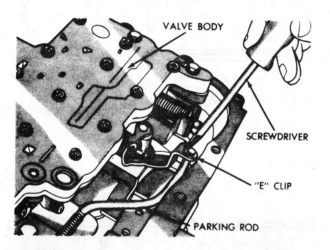
5.11a The parking rod E-clip can be removed with a screwdriver

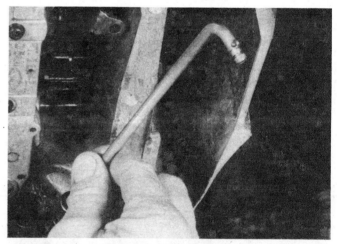
5.11b Removing the parking rod

5.12 Loosening the Low-Reverse band locknut

5.15 The shift pawl must be pushed back before the parking rod can be inserted

Chapter 7 Part B Automatic transaxle

when the transaxle is in Park or Neutral.
4   Check for continuity between the two outer terminals. Continuity should exist only when the transaxle is in Reverse. No continuity should exist between either outer terminal and the case.
5   If the switch fails any of the tests, replace it with a new one.
6   Position a drain pan under the switch to catch the fluid released when the switch is removed. Unscrew the switch and detach it from the transaxle (see illustration).
7   Move the shift lever from Park to Neutral while checking that the switch operating fingers are centered in the opening.
8   Install the new switch, tighten it to the specified torque and plug in the connector. Repeat the checks on the new switch.
9   Check the fluid level and add fluid as required (see Chapter 1).

7 Automatic transaxle — removal and installation

Refer to illustrations 7.6, 7.7, 7.11a, 7.11b, 7.12, 7.15, 7.16, 7.18, 7.19, 7.20 and 7.21

Removal

1   Disconnect the negative cable at the battery. Place the cable out of the way so it cannot accidentally come in contact with the negative terminal of the battery, as this would once again allow power into the electrical system of the vehicle.
2   Drain the cooling system (Chapter 1).
3   Disconnect the heater hoses and move them out of the way.
4   Remove the transaxle shift and throttle position cables and fasten them out of the way.
5   Remove the air cleaner assembly.
6   Support the engine from above with the special fixture (see illustration) or from below with a jack (place a block of wood between the jack and the engine oil pan to prevent damage).
7   Remove the upper bellhousing bolts (see illustration).
8   Raise the vehicle and support it securely. The engine must be supported in the raised position as well. Remove the axle cotter pins and nuts.
9   Remove any under vehicle splash shields.
10 Remove the speedometer drive gear and unplug all electrical connectors.
11 Disconnect the fluid cooler lines at the transaxle and plug them (see illustrations).

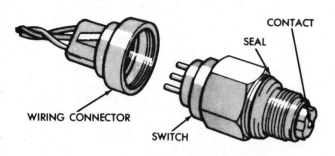

6.6 Neutral start and back-up light switch

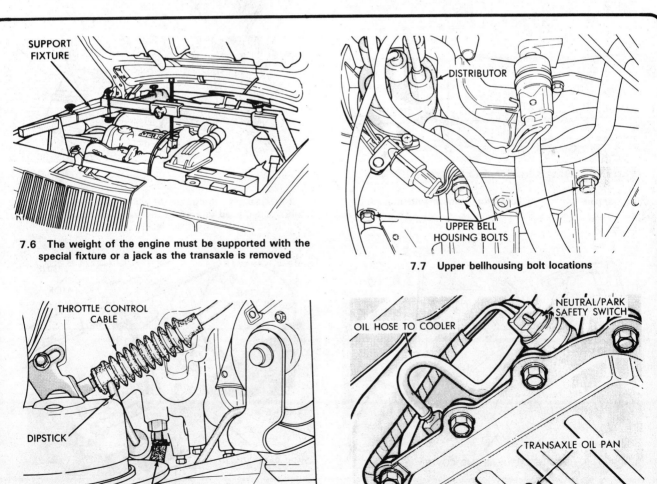

7.6 The weight of the engine must be supported with the special fixture or a jack as the transaxle is removed

7.7 Upper bellhousing bolt locations

7.11a The transaxle oil cooler return hose ....

7.11b ... and supply hose should be detached at the transaxle

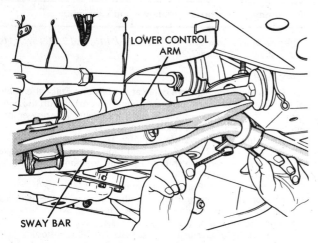

7.12 Removing the sway bar mount bolts

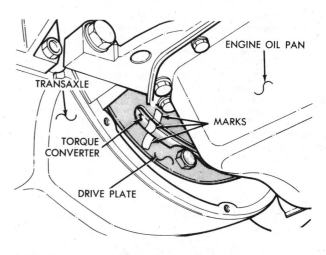

7.15 Be sure to mark the torque converter and driveplate so they can be reinstalled in the same relative position

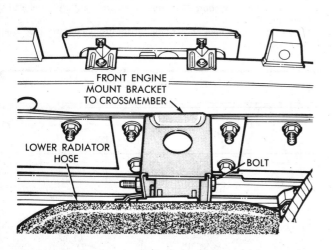

7.16 Engine mount bracket-to-front crossmember location

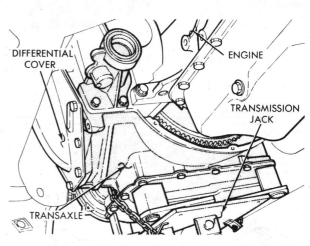

7.18 Support the transaxle with a jack (note the safety chain used to prevent the transaxle from falling off the jack)

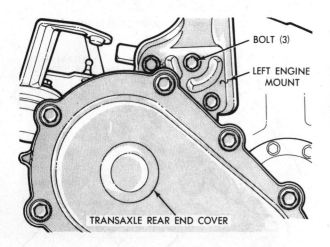

7.19 Left transaxle-to-engine mount bolt locations

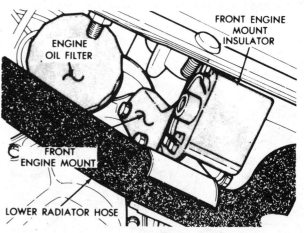

7.20 Front chassis-to-engine mount bolt locations

Chapter 7 Part B Automatic transaxle

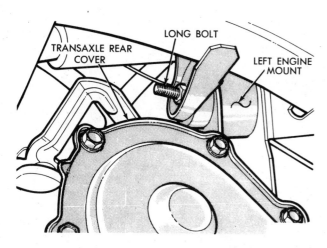

7.21 Left engine mount bolt location

12 Loosen the sway bar bushing bolts, unbolt the ends from the lower control arms and pull the sway bar down out of the way (see illustration).
13 Remove the driveaxles and (if equipped) the intermediate shaft (Chapter 8).
14 Remove the lower bellhousing cover to provide access to the torque converter.
15 Mark the torque converter-to-driveplate relationship so they can be reinstalled in the same position (see illustration). Remove the torque converter-to-driveplate bolts (remove the plug in the right splash shield and turn the crankshaft with a socket and extension on the pulley bolt to gain access to the driveplate bolts).
16 Remove the engine mount bracket from the front crossmember (see illustration).
17 Remove the starter and wiring harness assembly.
18 Support the transaxle with a jack (see illustration).
19 Remove the left transaxle-to-engine mount bolts (see illustration).
20 Remove the front chassis-to-engine mount bolts from the engine and transaxle (see illustration).
21 Remove the left engine mount (see illustration).
22 Carefully pry the transaxle away from the engine.
23 Pull the transaxle away from the engine, making sure the torque converter remains on the input shaft.
24 Lower it from the engine compartment, taking care not to contact the inner end of the lower suspension arm.

Installation

25 To install the transaxle, raise it into position with the torque converter in place on the input shaft.
26 Move the transaxle into place against the engine, align the bolt holes and install the upper bellhousing bolts.
27 Install the left engine mount.
28 Align the torque converter and driveplate marks made during removal, install the bolts and tighten them to the specified torque.
29 Install the bellhousing cover and the front mount.
30 Install the starter and electrical harness.
31 Install the driveaxles.
32 Install the sway bar.
33 Plug in the transaxle electrical connectors.
34 Install the under vehicle splash shields (if equipped).
35 Install the axle nuts, lower the vehicle and tighten the nuts as described in Chapter 8.
36 Install the air cleaner assembly.
37 Connect the transaxle cooler lines and tighten them securely.
38 Connect the transaxle shift and throttle cables.
39 Connect the heater hoses and refill the cooling system.
40 Refill the transaxle (Section 4).
41 Connect the negative battery cable.

8 Automatic transaxle speedometer gear assembly — removal and installation

Refer to Section 4 in Part A. The procedure is the same regardless of the type of transaxle involved.

9 Automatic transaxle service and repair

Because of the special tools, equipment and expertise required to disassemble, overhaul and reassemble the transaxle, it is recommended that it be left to a dealer service department or a transmission repair shop.

Chapter 8 Clutch and driveaxles

Contents

Constant velocity (CV) joints — disassembly, inspection and reassembly 6	Driveaxles, constant velocity (CV) joints and boots — check 3
Constant velocity (CV) joint boot — removal and installation 7	Driveaxles — removal and installation 5
Clutch — removal, inspection and installation 2	General information 1
	Intermediate shaft (turbocharged models) — removal and installation 4

Specifications

Clutch

Flywheel runout limit	0.003 in (0.07 mm)
Clutch lining wear limit	0.015 in (0.38 mm)
Pressure plate warpage limit	0.020 in (0.50 mm)

Driveaxle length

1984 models
  GKN driveaxle
    right (tan tape)... 10.1 to 10.4 in (257 to 265 mm)
    left (silver tape).. 10.0 to 10.6 in (254 to 269 mm)
  Citroen driveaxle
    right (red tape).. 9.5 to 9.9 in (241 to 251 mm)
    left (yellow tape).. 9.4 to 10.0 in (238 to 255 mm)

1985 models
  GKN driveaxle
    right (tan tape)... 10.1 to 10.4 in (257 to 265 mm)
    left (silver tape).. 10.0 to 10.6 in (254 to 269 mm)
    right (blue tape).. 19.9 to 20.3 in (505 to 515 mm)
    left (blue tape).. 10.2 to 10.9 in (259 to 277 mm)
  ACI driveaxle
    right (green tape)... 18.8 to 19.1 in (477 to 585 mm)
    left (green tape)... 9.0 to 9.6 in (229 to 244 mm)
  GKN/ACI driveaxle
    right (orange tape).. 19.4 to 19.7 in (492 to 500 mm)
    left (orange tape).. 9.6 to 10.2 in (243 to 258 mm)
  Citroen driveaxle
    right (red tape).. 9.5 to 9.9 in (241 to 251 mm)
    left (yellow tape).. 9.4 5 to 10.0 in (238 to 255 mm)
    right (white tape).. 18.9 to 19.4 in (480 to 492 mm)
    left (white tape).. 9.4 to 10.0 in (238 to 255 mm)
  SSG driveaxle
    right (gold tape).. 17.8 to 18.5 in (457 to 469 mm)
    left (gold tape).. 8.5 to 9.1 in (216 to 232 mm)

Chapter 8 Clutch and driveaxles

1986 models
- GKN driveaxle
  - right (blue tape) .... 19.9 to 20.3 in (505 to 515 mm)
  - left (blue tape) .... 10.2 to 10.9 in (259 to 277 mm)
  - right (tan tape) .... 10.1 to 10.4 in (257 to 265 mm)
  - left (silver tape) .... 10.0 to 10.6 in (254 to 269 mm)
- ACI driveaxle
  - right (green tape) .... 18.8 to 19.1 in (477 to 485 mm)
  - left (green tape) .... 9.9 to 9.6 in (229 to 244 mm)
- GKN/ACI driveaxle
  - right (orange tape) .... 19.4 to 19.7 in (492 to 500 mm)
  - left (orange tape) .... 9.6 to 10.2 in (243 to 158 mm)
- Citroen driveaxle
  - right (white tape) .... 18.9 to 19.4 in (480 to 492 mm)
  - left (white tape) .... 9.4 to 10.0 in (238 to 255 mm)
  - right (red tape) .... 9.5 to 9.9 in (241 to 251 mm)
  - left (yellow tape) .... 9.4 to 10.0 in (238 to 255 mm)

Torque specifications	Ft-lbs	Nm
Flywheel-to-crankshaft bolts		
1984 and 1985	65	88
1986	70	95
Pressure plate-to-flywheel bolts	21	28
Clutch cable retainer bolt nut	21	28
Steering knuckle-to-balljoint clamp bolt/nut	70	95
Driveaxle nuts	180	245
Intermediate shaft bearing mount-to-engine bolt	40	54
Wheel lug nuts	95	129

1 General information

Refer to illustrations 1.2 and 1.3

The clutch disc is held in place against the flywheel by the pressure plate springs. During disengagement, such as during gear shifting, the clutch pedal is depressed, which operates a cable, pulling on the release lever so the release bearing pushes on the pressure plate springs, disengaging the clutch.

The clutch pedal incorporates a self-adjusting device which compensates for clutch disc wear (see illustration). A spring in the clutch pedal arm maintains tension on the cable and the adjuster pivot grabs the positioner adjuster when the pedal is depressed and the clutch is released. Consequently the slack is always taken up in the cable, making adjustment unnecessary.

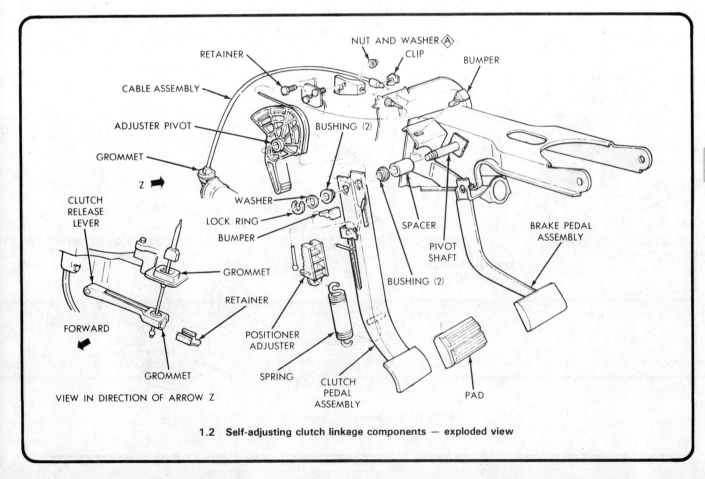

1.2 Self-adjusting clutch linkage components — exploded view

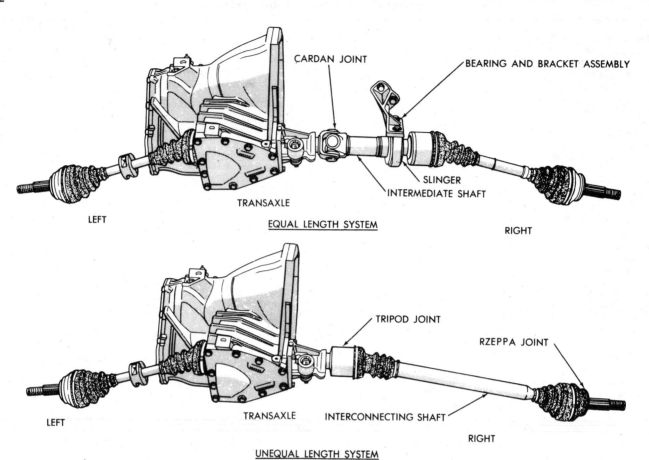

1.3 The two types of driveaxle systems

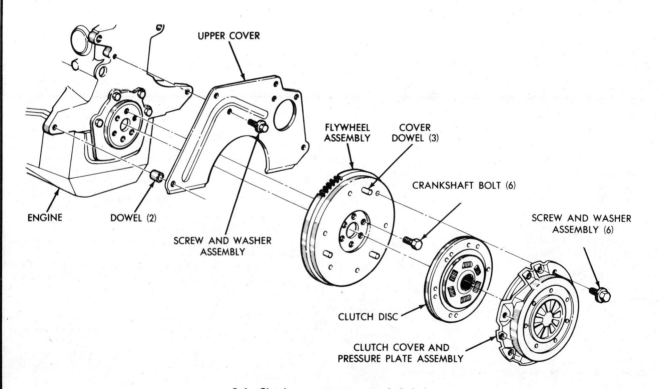

2.4 Clutch components — exploded view

Chapter 8 Clutch and driveaxles

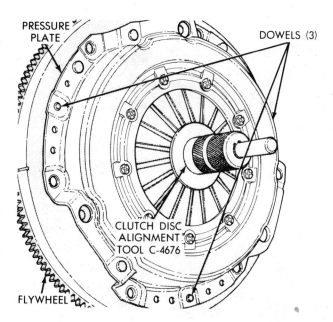

2.16 When installing the clutch disc and pressure plate, a clutch alignment tool must be used to center the disc

Power from the engine passes though the clutch and transaxle to the front wheels by two driveaxles. On non-turbocharged models the driveaxles are unequal length, while turbocharged models are equipped with equal length driveaxles and feature an intermediate shaft incorporating a Cardan joint (see illustration). The driveaxles consist of three sections: the inner splined ends which are held in the differential by clips or springs, two constant velocity (CV) joints and outer splined ends which are held in the hub by a nut. The CV joints are internally splined and contain ball bearings which allow them to operate at various lengths and angles as the driveaxles move through their full range of travel. The CV joints are lubricated with special grease and are protected by rubber boots which must be inspected periodically for cracks, tears and signs of leakage, which could lead to damage of the joints and failure of the driveaxle.

The driveaxles are identified as "GKN", "ACI", "SSG" or "Citroen" assemblies (depending on the manufacturer). Vehicles may be equipped with any of these types. However, they should not be interchanged. The driveaxles on your vehicle can be identified by referring to illustration 6.2.

2 Clutch — removal, inspection and installation

Refer to illustrations 2.4 and 2.16

1 Remove the transaxle (Chapter 7).
2 Mark the position of the pressure plate assembly on the flywheel so it can be installed in the same position.
3 Loosen the pressure plate bolts 1/4-turn at a time, in a criss-cross pattern, to avoid warping the cover.
4 Remove the pressure plate and clutch disc (see illustration).
5 Handle the disc carefully, taking care not to touch the lining surface, and set it aside.
6 Remove the clutch release shaft.
7 Slide the clutch release bearing and fork assembly off the input shaft. Remove the fork from the thrust plate. Inspect the bearing for damage, wear and cracks. Hold the center of the bearing and spin the outer race. If the bearing doesn't turn smoothly or if it is noisy, replace it with a new one.
8 Clean the dust out of the clutch housing with a vacuum cleaner or clean cloth. *Do not use compressed air, as the dust probably contains asbestos, which can endanger your health if inhaled.*
9 Inspect the friction surfaces of the clutch disc and flywheel for signs of uneven contact, indicating improper mounting or damaged clutch springs. Check the surfaces for burned areas, grooves, cracks and other signs of wear. It may be necessary to remove a badly grooved flywheel and have it machined to restore the surface. Light glazing of the flywheel surface can be removed with fine sandpaper. Attach a dial indicator to the engine and, with the contact plunger within the wear circle of the flywheel, rotate the crankshaft 180 degrees. The flywheel runout should be within the specified limit. Be sure to push the crankshaft forward so the end play won't be included in the runout measurement.
10 To determine clutch disc lining wear, measure the distance from the rivet head to the lining surface and compare it to the Specifications. Check the lining for contamination by oil or grease and replace the disc with a new one if any is present. Check the center of the disc to make sure it is clean and dry, shows no signs of overheating and that the springs are not broken. Slide the disc onto the input shaft temporarily to make sure the fit is snug and the splines are not burred or worn.
11 Check the flatness of the pressure plate with a straightedge. Look for signs of overheating, cracks, grooves and ridges. The inner end of the release levers should not show any signs of uneven wear. Replace the pressure plate with a new one if its condition is in doubt.
12 Make sure the cover fits snugly on the flywheel dowels. Replace the cover with a new one if it fits loosely on the dowels.
13 Clean the old grease from the release bearing. Fill the cavities and coat the inner liner surfaces with multi-purpose grease.
14 Lubricate the rounded thrust pads and spring clip cavities of the fork with multi-purpose grease. Make sure the spring clips on the bearing are not distorted and then attach the fork to the bearing by sliding the thrust pads under the spring clips.
15 Position the clutch disc on the flywheel, centering it with an alignment tool.
16 With the disc held in place by the alignment tool, place the pressure plate assembly in position on the flywheel dowels, aligning it with the marks made at the time of removal (see illustration).
17 Install the bolts and tighten them in a criss-cross pattern, one or two turns at a time, until they are at the specified torque.
18 Slide the fork and bearing assembly into position on the bearing pilot.
19 Install the release shaft bushings in the housing and slide the shaft into position. Retain the shaft with the clip which fits into the groove near the large bushing.
20 Install the release lever, retaining it to the shaft with the clip.
21 Install the transaxle.

3 Driveaxles, constant velocity (CV) joints and boots — check

Refer to illustration 3.1

1 The driveaxles, CV joints and boots (see illustration) should be inspected periodically and whenever the vehicle is raised, such as during chassis lubrication. The most common symptom of driveaxle or CV joint failure is knocking or clicking noises when turning.
2 Raise the vehicle and support it securely on jackstands.
3 Inspect the CV joint boots for cracks, leaks and broken retaining bands. If lubricant leaks out through a hole or crack in a boot, the CV joint will wear prematurely and require replacement. Replace any damaged boots immediately (Section 6).
4 Inspect the entire length of each axle for cracks, dents and signs of twisting and bending.
5 Grasp each axle, rotate it in both directions and move it in and out to check for excessive movement, indicating worn splines or loose CV joints.

4 Intermediate shaft (turbocharged models) — removal and installation

Refer to illustrations 4.3 and 4.6

Removal

1 Remove the right driveaxle (Section 5).
2 Remove the retaining bolt and push up on the speedometer cable to disengage the gear from the transaxle.

3.1 Driveaxle and CV joint components

1 Right outer CV joint boot
2 Right inner CV joint boot
3 Intermediate shaft bearing mount
4 Intermediate shaft cardan joint
5 Transaxle extension
6 Transaxle
7 Left inner CV joint boot
8 Left outer CV joint boot

Chapter 8 Clutch and driveaxles

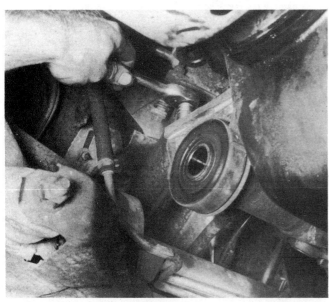

4.3 Use a socket to remove the intermediate shaft bearing mount bolts

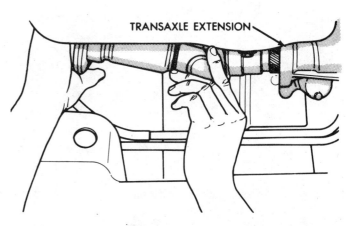

4.6 Support the intermediate shaft while inserting the splines into the transaxle extension

5.4 Remove the suspension arm balljoint-to-steering knuckle clamp bolt nut (arrow)

5.6 Grasp the outer CV joint and pull the steering knuckle out to separate it from the driveaxle

3   Remove the bearing mount assembly bolts (see illustration).
4   Grasp the intermediate shaft/bearing assembly securely with both hands and withdraw it from the transaxle.
5   Have an assistant with a drain pan ready to catch the fluid which will pour from the transaxle with considerable force when the shaft is withdrawn.

Installation

6   Place the intermediate shaft and bearing assembly in position and carefully insert the splined stub yoke into the transaxle extension housing (see illustration).
7   Place the bearing mount in position and install the retaining bolts. Tighten the mount-to-engine bolts to the specified torque.
8   Lubricate the inside splines and the pilot bore of the intermediate shaft with a liberal coat of multi-purpose grease.
9   Install the right driveaxle.

5 Driveaxles — removal and installation

Refer to illustrations 5.4, 5.6, 5.7 and 5.19
Removal

1   Remove the front hub bearing cap, cotter pin, nut lock and wave washer. With the weight of the vehicle on the wheels and an assistant applying the brakes, loosen the axle nut.
2   Raise the vehicle, support it securely on jackstands and remove the front wheel, axle nut and washer.
3   Remove the speedometer gear prior to removing the right axle.
4   Remove the steering knuckle-to-balljoint clamp bolt (see illustration).
5   Pry the lower balljoint stud out of the steering knuckle. **Note:** *The sway bar must be disconnected from the suspension arm to allow enough movement to separate the balljoint.*
6   Grasp the outer CV joint and the steering knuckle and pull the steering knuckle out to separate the driveaxle from the hub (see illustration). Be careful not to damage the CV joint boot. **Caution:** *Do not pry on or damage the wear sleeve on the CV joint when separating it from the hub.*

Chapter 8 Clutch and driveaxles

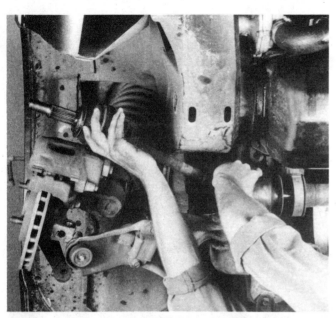

5.7 Support both CV joints as the driveaxle is removed and *do not pull on the shaft*

7 Grasp the CV joints so they will be supported during removal and withdraw the driveaxle from the differential. **Caution:** *Do not pull on the shaft — pull only on the inner CV joint (see illustration).*

8 The driveaxles, when in place, secure the hub bearing assemblies. If the vehicle is to be supported or moved on wheels while the driveaxles are removed, install a bolt through each hub and thread nuts onto them to keep the bearings from loosening.

Installation

9 Prior to installation, clean the wear sleeve on the driveaxle outer CV joint and the seal in the hub. Lubricate the entire circumference of the seal lip and fill the seal cavity with grease. Apply a 1/4-inch bead of grease to the wear sleeve seal contact area as well.

10 Apply a small amount of multi-purpose grease to the splines at each end of the driveaxle. Place the driveaxle in position and carefully insert the inner end of the shaft into the transaxle.

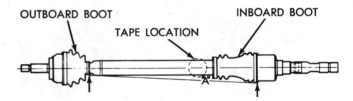

5.19 Measure each driveaxle between the points indicated by the arrows to verify correct length (A)

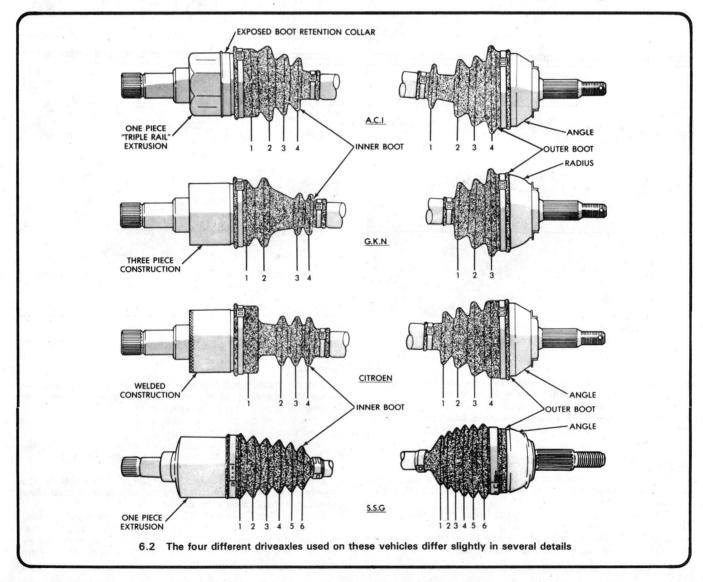

6.2 The four different driveaxles used on these vehicles differ slightly in several details

Chapter 8 Clutch and driveaxles

11 Push the steering knuckle out and insert the outer splined shaft of the CV joint into the hub.
12 Rejoin the balljoint stud to the steering knuckle, install the clamp bolt and tighten it to the specified torque.
13 Install the sway bar ends, if removed (see Chapter 10).
14 Install the speedometer gear.
15 Install the wheels, the washers and the axle nuts.
16 Tighten the driveaxle nuts to the specified torque and install the wave washers, the nut locks and new cotter pins.

Driveaxle position check

17 These vehicles have engine mounts with slotted holes which allow for side-to-side positioning of the engine. If the vertical bolts on the right or left upper engine mounts have been loosened for any reason, or if the vehicle has been damaged structurally at the front end, driveaxle length must be checked and if necessary corrected. A driveaxle that is shorter than required will result in objectionable noise, while a driveaxle that is longer than necessary may result in damage.
18 The vehicle must be completely assembled, the front wheels must be properly aligned and pointing straight ahead and the weight of the vehicle must be on all four wheels.
19 Using a tape measure, check the distance from the inner edge of the outboard boot to the inner edge of the inboard boot on both driveaxles. Take the measurement at the lower edge of the driveaxles (six o'clock position) (see illustration). Note that the required dimension varies with engine type, transaxle type and driveaxle manufacturer.
20 If the dimensions are not as specified, the engine mount bolts can be loosened and the engine repositioned to obtain the specified driveaxle lengths. If the engine cannot be moved enough within the range of the slotted engine mounts, check for damaged or distorted support brackets and side rails.
21 If the engine is moved, refer to Chapter 7 and adjust the shift linkage.

6 Constant velocity (CV) joints — disassembly, inspection and reassembly

Refer to illustrations 6.2, 6.4, 6.6, 6.7, 6.8, 6.9a, 6.9b, 6.11a, 6.11b, 6.15, 6.16a, 6.16b, 6.16c, 6.18, 6.19, 6.23, 6.25a, 6.25b, 6.25c, 6.26a, 6.26b, 6.31a, 6.31b, 6.31c, 6.35, 6.36, 6.37, 6.38, 6.41, 6.43, 6.44, 6.45a, 6.45b, 6.45c, 6.46 and 6.49

1 Obtain a CV joint rebuild or replacement kit.
2 Remove the driveaxles (Section 5) and identify which type they are (see Section 1 and the accompanying illustration).
3 Place one of the driveaxles in a vise, using wood blocks to protect it from the vise jaws, so the CV joint can be easily worked on. If the CV joint has been operating properly with no noise or vibration, replace the boot as described in Section 7. If the CV joint is badly worn or has run for some time with no lubricant due to a damaged boot, it should be disassembled and inspected.

Inner CV joint

4 Remove the clamps and slide the boot back to gain access to the tripod retention system (see illustration).
5 Depending on the type of CV joint assembly, separate the tripod from the housing as follows.

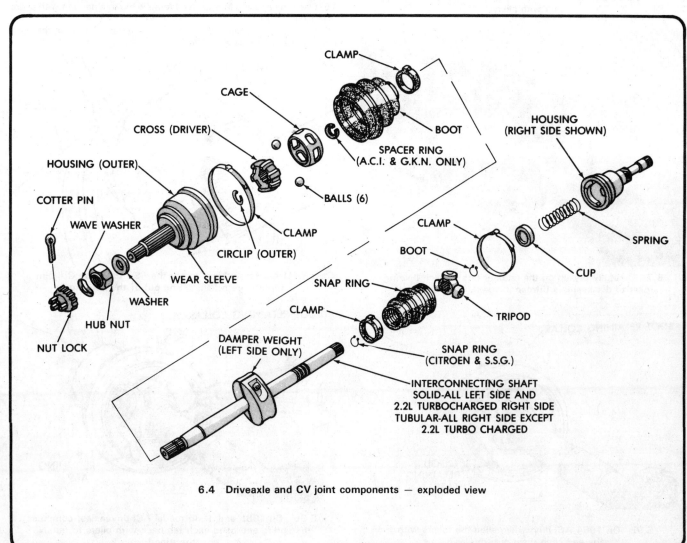

6.4 Driveaxle and CV joint components — exploded view

6   On GKN driveaxles, the retaining tabs are an integral part of the housing cover. Hold the housing and lightly compress the retention spring while bending the tabs back with pliers (see illustration). Support the housing as the retention spring pushes it from the tripod. This will prevent the housing from reaching an unacceptable angle and keep the tripod rollers from being pulled from the tripod studs.
7   Citroen driveaxles utilize a tripod retainer ring which is rolled into a groove in the housing. Slightly deform the retainer ring at each roller with a screwdriver (see illustration). The retention spring will push the housing from the tripod. The retainer ring can also be cut carefully from the housing. New rings are included in the rebuild kit and can be installed by rolling the edge into the machined groove in the housing with a hammer and punch.
8   On SSG driveaxles the tripod is retained in the housing by a wire ring. Pry the ring out of the groove with a screwdriver and slide the tripod from the housing (see illustration).
9   On ACI driveaxles, the tripod retaining pads are part of the boot retaining collar which is staked in place. On 1984 models, place the housing in position so that the three rollers are flush with the retaining tabs. Pull the housing while angling the joint slightly so that the rollers can be snapped through the tabs, one at a time. Be careful not to angle the joint too far as this could cause the tripod rollers to be pulled from the studs. On 1985 and 1986 models, compress the retaining spring lightly while bending the tabs back with pliers (see illustrations). Be sure to support the housing as the spring pushes it from the tripod.
10   When removing the housing from the tripod, hold the rollers in place on the studs to prevent the rollers and needle bearings from falling. After the tripod is out of the housing, secure the rollers in place with tape.
11   Remove the snap-ring (see illustration) and use a brass drift to drive the bearing and tripod assembly from the splined shaft (see illustration).
12   Clean the grease from the tripod assembly. Check for scoring, wear, corrosion and excessive play and replace any damaged or worn components with new ones.
13   Inspect the inner splined area of the bearing tripod for wear and damage, replacing parts as necessary.
14   Remove all of the old grease from the housing. Inspect the housing splines, ball raceways, spring, spring cup and the spherical end of the shaft for wear, damage, nicks and corrosion, replacing parts as necessary.
15   Place the housing in a vise and remove the retainer ring with pliers (see illustration).
16   On 1984 turbocharged models, install a new slinger on the right inner housing (see illustrations).

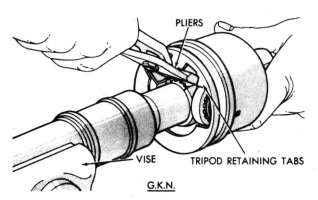

6.6   To separate the inner joint tripod from the housing on GKN driveaxles, the retaining tabs must be bent up with pliers

6.7   Carefully pry up on the retainer ring at each bearing roller to disassemble Citroen driveaxle inner CV joints

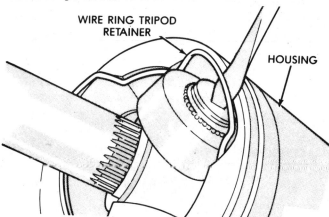

6.8   Use a screwdriver to pry the SSG driveaxle CV joint tripod wire retaining ring out of the housing

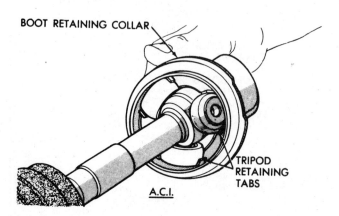

6.9a   On 1984 ACI driveaxles, align the rollers with the tabs and snap them out, one at a time

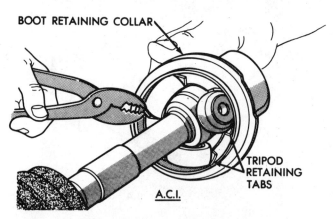

6.9b   On 1985 and 1986 model ACI driveaxles, compress the spring and bend each tab back with pliers to remove the tripod

Chapter 8 Clutch and driveaxles

6.11a The tripod is held on the shaft with a snap-ring

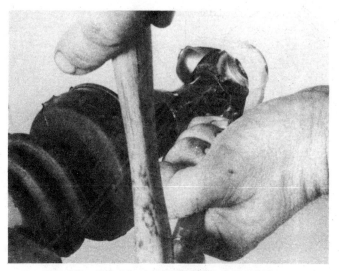

6.11b Drive the bearing tripod off the shaft (note the tape retaining the bearings)

6.15 Detach the Citroen driveaxle retainer ring with pliers

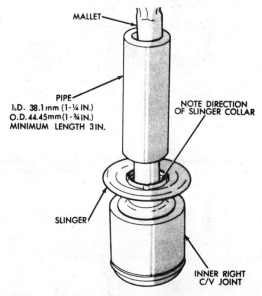

6.16a 1984 turbocharged model inner right CV joint slinger collar installation

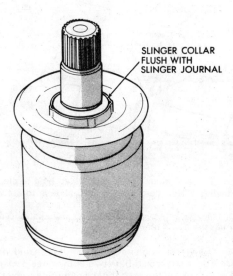

6.16b Proper installation of the 1984 turbocharged model right inner slinger

6.16c Install the rubber seal washer over the splines and into the retaining groove (1984 turbocharged models)

Chapter 8 Clutch and driveaxles

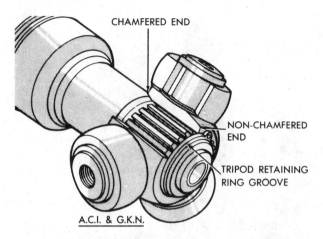

6.18 On GKN and ACI driveaxles, the non-chamfered end of the tripod must face out when it is installed on the shaft

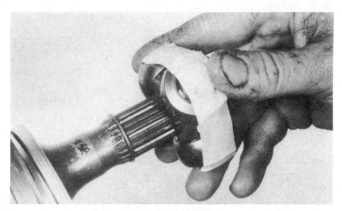

6.19 On GKN driveaxles, make sure the tripod assembly is installed correctly (Citroen and SSG driveaxles are equipped with tripods that can be installed either way)

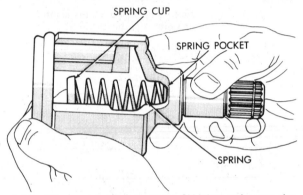

6.23 When assembling the inner CV joint, make sure the spring is seated in the housing pocket and position the cup with the concave side out

6.25a Make sure the bearing grooves in the housing have been greased, then slide the housing over the tripod until it bottoms

6.25b The new retainer ring should be staked in place with a hammer and punch

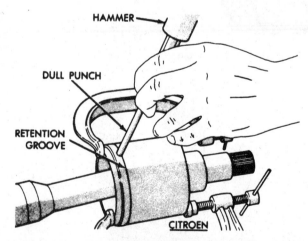

6.25c If the retainer ring is difficult to hold in place while staking it into the groove, use C-clamps to steady it

17 Install the new boot on the axle.
18 On GKN and ACI driveaxles, slide the tripod onto the shaft with the non-chamfered end facing out (next to the snap-ring groove) (see illustration).
19 Citroen and SSG driveaxles are equipped with tripods that can be installed with either end out (both ends are the same) (see illustration). Be sure to first install the wire ring tripod retainer on the interconnecting shaft before sliding the tripod onto the shaft.

20 If necessary, use a section of pipe or a socket and a hammer to carefully tap the tripod onto the shaft until it just clears the snap-ring groove.
21 Install a new snap-ring and make sure it is seated in the groove.
22 On GKN driveaxles, distribute two of the three packets of grease supplied with the kit in the boot and the remaining packet in the housing. On ACI driveaxles, distribute one of the two supplied packs of grease in the boot and the remaining pack in the housing. On SSG

Chapter 8 Clutch and driveaxles

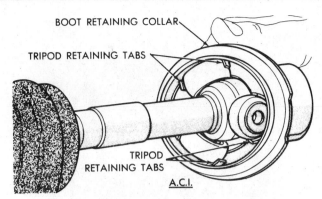

6.26a On 1984 model ACI driveaxles, snap each roller into place, one at time

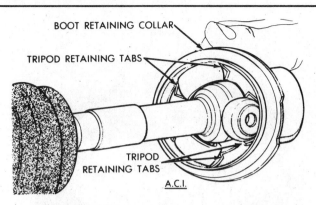

6.26b On 1985 and 1986 ACI driveaxles, press the housing onto the tripod

6.31a On ACI, GKN and Citroen driveaxles, the outer joint housing can be dislodged from the shaft circlip with a soft-face hammer . . .

6.31b . . . and removed by hand

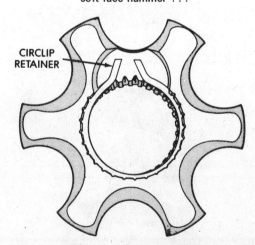

6.31c The SSG driveaxle outer joint is retained by a circlip

driveaxles, distribute one half pack in the housing and the rest in the boot. On Citroen driveaxles, distribute two-thirds of the grease in the packet in the boot and the remaining amount in the housing. Make sure the grease is applied to the bearing grooves in the housing.

23 Position the spring in the housing spring pocket with the cup attached to the exposed end of the spring (see illustration). Apply a small amount of grease to the concave surface of the spring cup.

24 On GKN driveaxles, slip the tripod into the housing and bend the retaining ring tabs down to their original positions. Make sure the tabs can hold the tripod in the housing.

25 On Citroen driveaxles slide the housing over the tripod until it bottoms (see illustration). Install a new retainer ring by rolling the edge into the machined groove in the housing with a hammer and punch (see illustration). If the retainer ring will not stay in place during this operation, hold it with two C-clamps (see illustration). Check the ability of the retainer ring to hold the tripod in the housing.

26 On 1984 model ACI driveaxles, align the tripod rollers with the retaining tabs and housing tracks and snap one roller at a time through the retaining tabs (see illustrations). Make sure the retaining tab holds the tripod securely in the housing. On 1985 and 1986 model ACI driveaxles, slip the tripod into the housing but do not bend the retaining tabs back to their original positions. Reattach the boot instead, which will hold the housing on the shaft. When the driveaxle is reinstalled on the vehicle, make sure the tripod is re-engaged in the housing.

27 On SSG driveaxles, slip the tripod into the housing and install the wire retaining ring. Check to make sure the ring holds the tripod securely in the housing.

28 Make sure the retention spring is centered in the housing spring pocket when the tripod is installed and seated in the spring cup.

29 Install the boot and retaining clamp (Section 7).

Outer CV joint

30 Mount the axleshaft in a vise with wood blocks to protect it, remove the boot clamps and push the boot back.

31 Wipe the grease from the joint. On GKN, ACI and Citroen driveaxles, use a soft-face hammer to drive the housing from the axle (see illustrations). Support the CV joint as this is done and rap the housing sharply on the outer edge to dislodge it from the internal circlip installed on the shaft. On SSG driveaxles a circlip located in a groove in the cross locks the outer joint to the shaft (see illustration). Loosen the damper weight bolts, mark its position and slide the weight and boot toward the inner joint. Remove the circlip with snap-ring pliers and slide the inner joint off the axle.

6.35 Mark the bearing cage, cross and housing relationship after removing the grease

6.36 With the cage and cross tilted, the balls can be removed one at a time

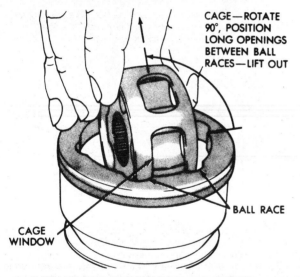

6.37 Bearing cage and cross removal details (outer CV joint)

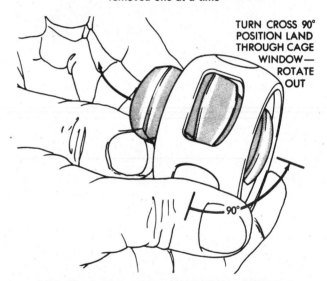

6.38 Bearing cross removal details (outer CV joint)

6.41 The wear sleeve can be pryed off the housing with a large screwdriver if replacement is necessary

6.43 The bearing cross will slide into the cage by aligning one of the lands with the elongated window in the cage

Chapter 8 Clutch and driveaxles

6.44 Lower the cage and cross assembly into the housing with the elongated window aligned with the race

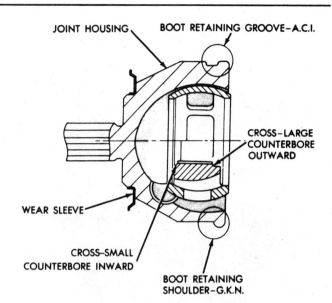

6.45a On GKN and ACI driveaxles, make sure the large cross counterbore faces out when the CV joint is reassembled

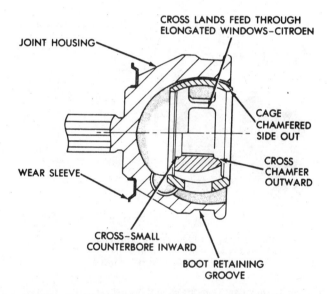

6.45b On Citroen driveaxles, make sure the cross and cage chamfers face out when the CV joint is reassembled

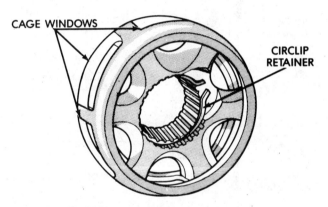

6.45c On SSG driveaxles, the circlip in the cross must face out

32 Slide the boot off the driveaxle. If the CV joint was operating properly and the grease does not appear to be contaminated, just replace the boot (Section 7). Bypass the following disassembly procedure. If the CV joint was noisy, proceed with the disassembly procedure to determine if it should be replaced with a new one.
33 Remove the circlip from the driveaxle groove and discard it (the rebuild kit will include a new circlip). GKN and ACI driveaxles are equipped with a large spacer ring, which must not be removed unless the driveaxle is being replaced with a new one.
34 Clean the axle spline area and inspect for wear, damage, corrosion and broken splines.
35 Clean the outer CV joint bearing assembly with a clean cloth to remove excess grease. Mark the relative position of the bearing cage, cross and housing (see illustration).
36 Grip the housing shaft securely in the wood blocks in the vise. Push down one side of the cage and remove the ball bearing from the opposite side. Repeat the procedure in a criss-cross pattern until all of the balls are removed (see illustration). If the joint is tight, tap on the cross (not the cage) with a hammer and brass drift.
37 Remove the bearing assembly from the housing by tilting it vertically and aligning two opposing elongated cage windows in the area between the ball grooves (see illustration).
38 Turn the cross 90 degrees to the cage and align one of the spherical lands with an elongated cage window. Raise the land into the window and swivel the cross out of the cage (see illustration).
39 Clean all of the parts with solvent and dry them with compressed air (if available).
40 Inspect the housing, splines, balls and races for damage, corrosion, wear and cracks. Check the bearing cross for wear and scoring in the races. If any of the components are not serviceable, the entire CV joint assembly must be replaced with a new one.
41 Check the outer housing wear sleeve for damage and distortion. If it is damaged or worn, pry the sleeve from the housing (see illustration) and replace it with a new one. A special tool is made for this purpose, but a large section of pipe will work if care is exercised (do not nick or gouge the seal mating surface).
42 Apply a thin coat of oil to all CV joint components before beginning reassembly.
43 Align the marks and install the cross in the cage so one of the cross lands fits into the elongated window (see illustration).
44 Rotate the cross into position in the cage and install the assembly in the CV joint housing, again using the elongated window for clearance (see illustration).
45 Rotate the cage into position in the housing. On GKN and ACI driveaxles, the large cross counterbore must face out (see illustration). On Citroen driveaxles, the cage and cross chamfers must face out (see illustration). On SSG driveaxles, the circlip in the cross must face out (see illustration). The marks made during disassembly should face out and be aligned.

Chapter 8 Clutch and driveaxles

6.46 Make sure the marks are aligned properly and that the bearing cross is installed with the correct side out (see text for details)

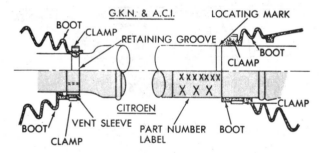

7.9 CV joint rubber boot installation details

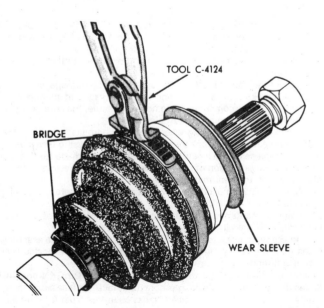

7.12 Squeezing the ladder-type boot clamp bridge with the special tool

6.49 Strike the end of the housing shaft with a soft-face hammer to engage it with the shaft circlip

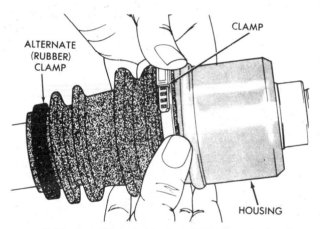

7.11 Installing a ladder-type boot clamp by hand (GKN driveaxles) (note the rubber clamp installed on the small end of the boot)

46 Pack the lubricant from the kit into the ball races and grooves (see illustration).
47 Install the balls into the elongated holes, one at a time, until they are all in position.
48 Place the driveaxle in the vise and slide the boot over it. On ACI, GKN and Citroen driveaxles, install a new circlip in the axle groove, taking care not to twist it. SSG driveaxles have a reuseable circlip which is part of the driver assembly.
49 Place the CV joint housing in position on the axle, align the splines and rap it sharply with a soft-face hammer (see illustration). Make sure the housing is seated on the circlip by attempting to pull it from the shaft.
50 Install the boot (Section 7).
51 Install the driveaxle (Section 5).

7 Constant velocity (CV) joint boot — removal and installation

Refer to illustrations 7.9, 7.11, 7.12, 7.18, 7.19, 7.21a, 7.21b, 7.22, 7.23a, 7.23b and 7.23c

Note: *If the instructions supplied with the replacement boot kit differ from the instructions here, follow the ones with the new boots. A special tool is required to install the factory-supplied boot clamps, so it may be a good idea to leave the entire procedure to a dealer service department. Do-it-yourself kits which offer greatly simplified installation may be available for your vehicle. Consult your auto parts counterman or your local dealer for more information on these kits.*

Chapter 8 Clutch and driveaxles

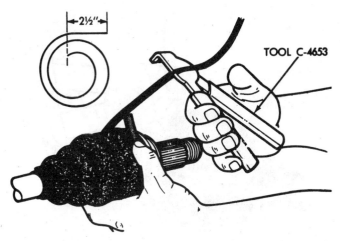

7.18 Wrap the clamp around the boot twice, leaving 2-1/2 inches of extra material, then cut it off

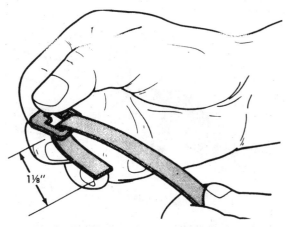

7.19 Pass the strap through the buckle and fold it back about 1-1/8 inch on the inside of the buckle

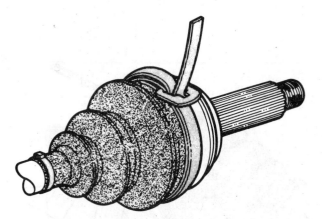

7.21a After installing it on the boot, bend the strap back so it doesn't unwind

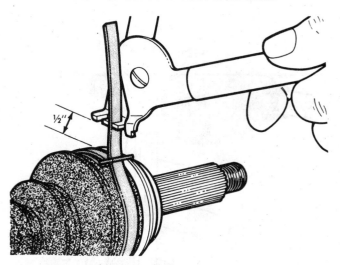

7.21b Attach the special tool about 1/2-inch from the buckle, . . .

1 If the boot is cut, torn or leaking, it must be replaced and the CV joint inspected as soon as possible. Even a small amount of dirt in the joint can cause premature wear and failure. Obtain a replacement boot kit before beginning this procedure.
2 Remove the driveaxle (Section 4).
3 Disassemble the CV joint and remove the boot as described in Section 6.
4 Inspect the CV joint to determine if it has been damaged by contamination or running with too little lubricant. If you have any doubts about the condition of the joint components, perform the inspection procedures described in Section 5.
5 Clean the old lubricant from the CV joint and repack it with the lubricant supplied with the kit.
6 Pack the interior of the new boot with the remaining lubricant.
7 Install the boot and clamps as follows.

GKN, SSG and ACI driveaxles

8 GKN units generally are equipped with metal ladder-type clamps. However, two alternate clamps are also used. They include a small rubber clamp at the shaft end of the inner CV joint and a large spring-type clamp on the housing.
9 If so equipped, slide the small rubber clamp over the shaft. Slide the small end of the boot over the shaft and position it as follows: On right inner joints, the small end of the boot lip must be aligned with the mark on the shaft. On left inner and all outer joints, position the small end of the boot in the groove in the shaft (see illustration).
10 Place the rubber clamp in the boot groove (if so equipped) or install the metal clamp.
11 Make sure the boot is properly located on the shaft, then locate the metal clamp tangs in the slots, making the clamp as tight as possible by hand (see illustration).
12 Squeeze the clamp bridge with tool number C-4124 to complete the tightening procedure (see illustration). Do not cut through the clamp bridge or damage the rubber boot.
13 Reassemble the CV joints and driveaxle components (Section 6).
14 Locate the large end of the boot over the shoulder or in the groove in the housing (make sure the boot is not twisted).
15 Install the spring-type clamp or ladder-type clamp. If a ladder-type clamp is used, repeat the tightening procedure described in Steps 11 and 12.

Citroen driveaxles

16 Slide the boot over the shaft. If installing an outer CV joint boot, position the vent sleeve under the boot clamp groove.
17 On right inner joints, align the boot lip face with the inboard edge of the part number label. If the label is missing, use the mark left by the original boot. On left inner and all outer joints, position the boot between the locating shoulders and align the edge of the lip with the mark made by the original boot. **Note:** *Clamping procedures are identical for attaching the boot to the shaft and the CV joint housing.*
18 Wrap the clamping strap around the boot twice, plus 2-1/2 inches, and cut it off (see illustration).
19 Pass the end of the strap through the buckle opening and fold it back about 1-1/8 inch on the inside of the buckle.
20 Position the clamping strap around the boot, on the clamping surface, with the eye of the buckle toward you. Wrap the strap around the boot once and pass it through the buckle, then wrap it around a second time and pass it through the buckle again.
21 Fold the strap back slightly to prevent it from unwinding itself (see illustration), then open the special tool (C-4653) and place the strap in the narrow slot, about 1/2-inch from the buckle (see illustration).

Chapter 8 Clutch and driveaxles

7.22 ... then push the tool forward and up and engage the hook in the buckle eye

7.23a Close the tool handles slowly to tighten the clamp strap, ...

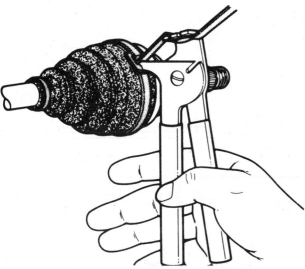

7.23b ... then rotate the tool down while releasing the pressure on the handles (allow the handles to open)

7.23c DO NOT rotate the tool down while applying pressure to the handles

22 Hold the strap with one hand and push the tool forward and up slightly, then fit the tool hook into the buckle eye (see illustration).
23 Tighten the strap by closing the tool handles (see illustration), then rotate the tool down slowly while releasing the pressure on the handles (see illustration). Allow the handles to open progressively, then open the tool all the way and slide it sideways off the strap. **Caution:** *Never fold the strap back or rotate the tool down while squeezing the handles together (if this is done, the strap will be broken) (see illustration).*

24 If the strap is not tight enough, repeat the procedure. Always engage the tool about 1/2-inch from the buckle. Make sure the strap moves smoothly as tightening force is applied and do not allow the buckle to fold over as the strap passes through it.
25 When the strap is tight, cut it off 1/8-inch above the buckle and fold it back neatly. It must not overlap the edge of the buckle.
26 Repeat the procedure for the remaining boot clamps.

Chapter 9 Brakes

Refer to Chapter 13 for information on 1989 and later models

Contents

Brake disc — removal, inspection and installation 3	Master cylinder — removal and installation 8
Brake light switch — removal, installation and adjustment ... 15	Parking brake — adjustment 11
Brake pedal — removal and installation 16	Parking brake cables — removal and installation 12
Brake hydraulic system — bleeding 10	Power brake booster — removal and installation 13
Disc brake caliper — removal, overhaul and installation 4	Power brake booster — testing and overhaul 14
Disc brake pads — inspection .... See Chapter 1	Rear brake shoes — inspection .... See Chapter 1
Disc brake pads — replacement 2	Rear brake shoes — replacement 5
General information 1	Rear wheel cylinder — overhaul 7
Master cylinder — overhaul 9	Rear wheel cylinder — removal and installation 6

Specifications

General
Brake fluid type .... DOT 3 brake fluid

Drum brakes
Standard drum diameter	8.66 in (220 mm)
Minimum drum diameter	Refer to marks on drum
Out-of-round limit	0.002 in (0.05 mm)
Runout limit	0.006 in (0.15 mm)
Brake lining thickness limit	1/8 in (3.2 mm)
Wheel cylinder bore diameter	0.563 in (14.3 mm)

Disc brakes
Pad thickness limit	5/16 in (7.9 mm)
Brake disc thickness	
standard	0.930 to 0.940 in (23.87 to 24.13 mm)
minimum*	0.882 in (22.4 mm)
Brake disc thickness variation limit (1-inch from edge)	0.0005 in (0.013 mm)
Brake disc runout limit (1-inch from edge)	0.005 in (0.13 mm)
Hub runout limit	0.003 in (0.08 mm)
Caliper bore diameter	2.1276 in (54 mm)
Caliper bore honing limit	0.001 in (0.0254 mm)

Refer to marks on disc (they supersede information printed here).

Torque specifications	Ft-lbs	Nm
Master cylinder-to-booster nuts	17 to 25	23 to 34
Power brake booster nuts	17 to 25	23 to 34
Wheel cylinder bolts	75 in-lbs	8
Backing plate-to-rear axle bolts	35 to 55	47 to 75
Disc brake caliper adapter bolts	130 to 190	176 to 258
Disc brake caliper guide pins	18 to 22	25 to 30
Bearing retainer mounting bolt	17 to 25	23 to 34
Parking brake pedal assembly-to-cowl bolts	17 to 25	23 to 34

1 General information

All models are equipped with disc-type front and drum-type rear brakes which are hydraulically operated and vacuum assisted.

The front brakes feature a single piston, floating caliper design. The rear drum brakes are leading/trailing shoe type with a single pivot.

The front disc brakes automatically compensate for pad wear during usage. The rear drum brakes also feature automatic adjustment.

Front drive vehicles tend to wear the front brake pads at a faster rate that rear drive vehicles. Consequently, it is important to inspect the brake pads frequently to make sure they have not worn to the point where the disc itself is scored or damaged. Note that the pad thickness limit on these models includes the metal portion of the brake pad, not just the lining material.

All models are equipped with a cable actuated parking brake which operates the rear brakes.

The hydraulic system is a dual line type with a dual master cylinder and separate systems for the front and rear brakes. In the event of brake line or seal failure, half the brake system will still operate.

2 Disc brake pads — replacement

Refer to illustrations 2.2a, 2.2b, 2.3, 2.5, 2.6 and 2.7

Caution: *Disc brake pads must be replaced on both front wheels at the same time — never replace the pads on only one wheel. Disassemble one brake at a time so the remaining brake can be used as a guide if difficulties are encountered during reassembly.*

1 Raise the front of the vehicle and support it securely. Block the rear wheels and set the parking brake, then remove the front wheels.
2 Remove the hold down spring retainer by pushing down at the center and pushing it outward (see illustrations).
3 Loosen the caliper guide pins sufficiently to allow the caliper to be removed (see illustration).
4 Lift the caliper forward and off the adapter and brake disc. The inner pad will remain with the caliper.
5 Remove the inner pad by grasping it securely and unsnapping the retainer from the piston (see illustration).
6 Remove the outer pad from the caliper. If the noise suppression gasket causes the pad to adhere to the caliper, pry it loose with a screwdriver (see illustration).
7 Support the caliper out of the way with a wire hanger (see illustration). Do not allow the caliper to hang by the brake hose.
8 Measure the brake pad thickness (including the metal backing material) and compare it to the Specifications.
9 Inspect the caliper and adapter for wear, damage, rust and fluid leaks. If the caliper-to-adapter mating surfaces are rusty, clean them thoroughly with a wire brush (the caliper must be free to move as the brakes are applied).
10 Apply Mopar lubricant (No. 2932524) or high temperature brake grease to the adapter-to-brake pad and caliper mating surfaces. Remove the protective paper from the noise suppression gasket on both pads. Install the inner brake pad by pressing the retainer into the piston recess.

Caution: *Do not get any grease on the pad lining material, gasket surface or brake disc.*

11 Place the outer pad in position in the caliper.

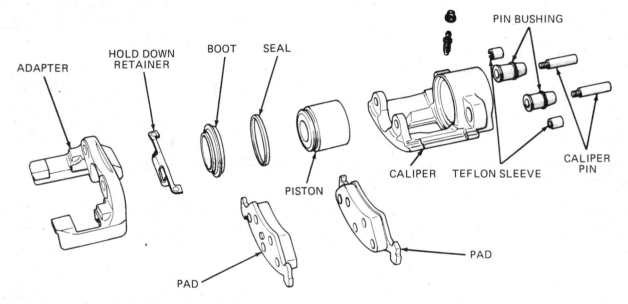

2.2a Disc brake caliper components — exploded view

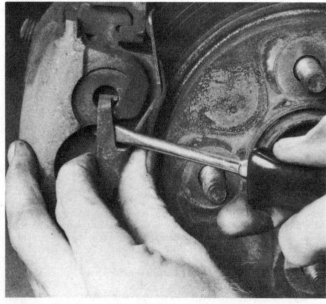

2.2b Push in on the hold down spring retainer and push it out of the caliper

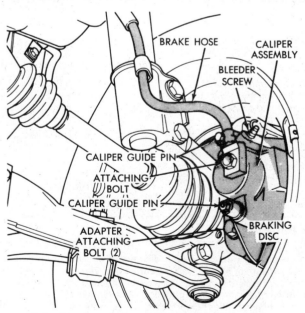

2.3 Loosen only the caliper guide pins, not the attaching bolts

12 Install the caliper over the brake disc and adapter.
13 Install the guide pins finger tight, taking care not to cross-thread them. Tighten the guide pins to the specified torque.
14 Install the pad retainer.
15 Repeat the procedure for the remaining caliper.
16 The remaining steps are the reverse of disassembly. When installing the wheel and tire assemblies, tighten the lug nuts in a criss-cross pattern and work up to the specified torque in two steps (see Chapter 1 for torque recommendations). The first step should be equal to one-half the specified torque.
17 Drive the vehicle and make several stops to wear off any foreign material on the pads and seat the linings on the disc.

3 Brake disc — removal, inspection and installation

Refer to illustrations 3.3a, 3.3b, 3.4, 3.6, 3.8a and 3.8b

1 Raise the front of the vehicle and support it securely. Block the rear wheels and set the parking brake, then remove the front wheels.
2 Remove the caliper and pads (Section 2).
3 Unscrew the disc retainer (see illustration). Mark one wheel stud and the disc so that it can be reinstalled in the same relative position (see illustration) and slide the disc off the hub.

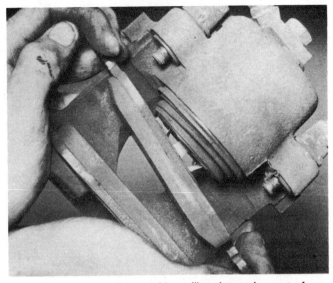

2.5 Unsnap the inner pad by pulling the retainer out of the piston

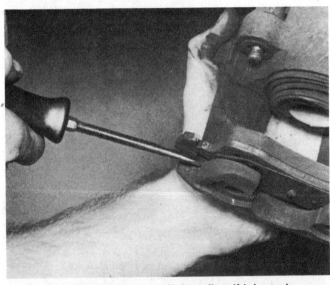

2.6 Pry the outer pad off the caliper if it is stuck

2.7 Hang the caliper out of the way with a piece of wire so the hose is not damaged

3.3a The disc retainer must be unscrewed to remove the disc

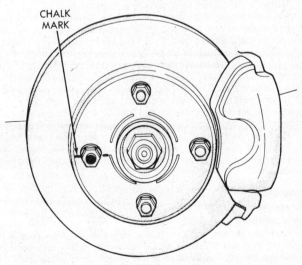

3.3b Mark the disc and the hub so they can be rejoined in the same relative position

3.4 Checking hub runout with a dial indicator

3.6 Checking disc runout with a dial indicator (make sure the lug nuts are in place and tightened evenly)

4 Use a dial indicator to check the hub runout (see illustration). Replace the hub (Chapter 10) if the runout is beyond the specified limit.
5 Inspect the disc for cracks, score marks, rust, deep grooves, burned areas and distortion. If damage goes so deep into the friction surface that the disc cannot be refinished and still maintain the minimum thickness, it must be replaced with a new one.
6 Reinstall the disc in the marked position and install the wheel lug nuts. Check the disc for runout with a dial indicator (see illustration).
7 If the runout is beyond the specified limit, remove the disc and reinstall it 180 degrees from the original position. Recheck the runout and if it is still excessive, either have the disc resurfaced or replace it with a new one.
8 With the disc in its original position, check the thickness at 12 places around its circumference (see illustration). Replace the disc with a new one if the thickness varies more than the specified limit (or have it resurfaced if it can be done without reducing the thickness below the minimum) (see illustration).
9 Install the caliper and pads.
10 Install the front wheels and lower the vehicle.

4 Disc brake caliper — removal, overhaul and installation

Refer to illustrations 4.7, 4.8, 4.9, 4.13 and 4.15

Removal

1 Raise the front of the vehicle and support it securely. Block the rear wheels and set the parking brake, then remove the front wheels.
2 Remove the caliper as described in Section 2.
3 Have an assistant push on the brake pedal very slowly until the piston moves out of the caliper. Do not remove the piston at this time as it will be followed by a gush of brake fluid. The piston should extend from the caliper enough to be easily removed but be retained by the boot. Prop the brake pedal in any position below the first inch of travel to prevent loss of brake fluid. **Warning:** *Do not allow your fingers to come between the piston and the caliper, as serious injury could result.*
4 If both calipers are being serviced, disconnect the flexible brake hose at the frame bracket and plug the metal line so the remaining caliper piston can be forced out of the caliper by hydraulic pressure.
5 Remove the bolt and detach the caliper from the brake hose.
6 Place the caliper on a workbench which has been covered with several layers of newspaper to absorb the brake fluid and remove the piston.

Overhaul

7 Mount the caliper in a vise, pry the dust boot from the bore and discard it (see illustration).
8 Carefully remove the piston seal from the bore using a wooden or plastic tool (see illustration). *Do not use a metal tool of any kind to remove the seal as damage to the caliper bore will result.*
9 Remove the guide pin bushing by pushing it out of its bore with a wooden dowel, then discard it (see illustration).

3.8a Measure the disc thickness with a micrometer at several points around the circumference (1-inch from the edge)

10 Clean the components with brake system cleaner or denatured alcohol and blow them dry with compressed air, if possible. **Warning:** *Do not, under any circumstances, use petroleum-based solvents or gasoline to clean brake parts.*
11 Inspect the cylinder bore for scratches and corrosion. Light scratches or imperfections in the bore can be removed with fine crocus cloth. If deeper scratches are present, they can be removed with a honing tool if the bore is not enlarged more than specified. Replace the caliper with a new one if the bore is badly scored. Inspect the piston for damage and wear. Replace it with a new one if it is badly worn or scratched or if the cylinder bore has been honed.
12 Insert the new guide pin bushing. Compress the bushing flanges with your fingers and work them into position by pushing in on the bushing with your fingertips or a plastic tool until they are seated.
13 Lubricate the new piston seal with brake assembly lubricant or clean brake fluid and install it in the bore groove. Start the seal into the groove and carefully press it into place by working around the circumference with your fingers (see illustration).
14 Coat the inside of the new dust boot with clean brake fluid and attach it to the piston.
15 Insert the piston into the bore and push it in until it bottoms (see illustration).
16 Press the new dust boot into the counterbore with a large socket or section of pipe.
17 Repeat the procedure for the remaining caliper.

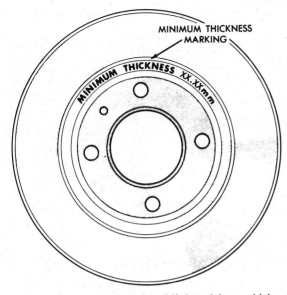

3.8b The disc can be resurfaced if the minimum thickness will not be exceeded

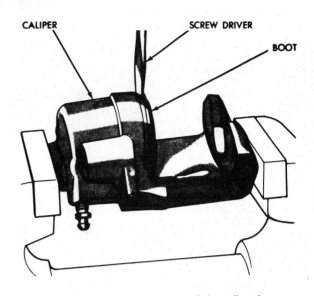

4.7 Prying the dust boot out of the caliper bore

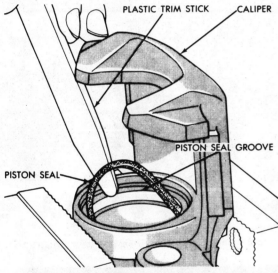

4.8 The piston seal must be removed with a plastic or wooden tool to avoid damage to the bore and seal groove

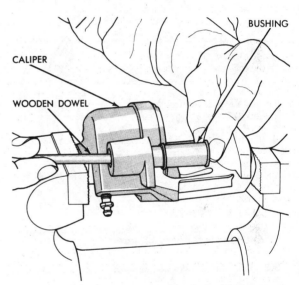

4.9 Removing the guide pin bushing

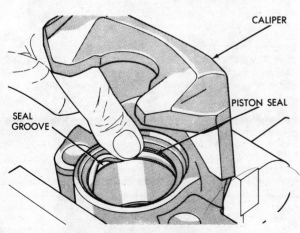

4.13 Start the new seal in the groove and push it in with your fingers (make sure it is not twisted)

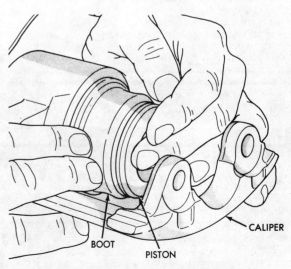

4.15 Install the piston and boot by hand only

Chapter 9 Brakes

Installation

18 Install the caliper(s) as described in Section 2.
19 Attach the brake hose to the caliper, using new sealing washers, and tighten the bolt securely.
20 Bleed the brakes as described in Section 10.
21 Test the brake operation carefully before putting the vehicle into normal service.

5 Rear brake shoes — replacement

Refer to illustrations 5.3, 5.4a, 5.4b, 5.5, 5.6, 5.7, 5.8a, 5.8b, 5.9a, 5.9b, 5.10, 5.11, 5.13, 5.15, 5.16, 5.18, 5.19, 5.20, 5.27 and 5.28

Note: Disassemble one brake at a time so the remaining brake can be used as a guide if difficulties are encountered during reassembly. Always replace the brake shoes on both wheels at the same time — never replace just one set.

1 Raise the rear of the vehicle, support it securely and block the front wheels. Remove the rear wheels.
2 Remove the hub/brake drum assembly as described in Chapter 10.
3 Use brake system cleaner to remove dust and brake fluid from the shoe assembly components (see illustration). **Warning:** *Brake dust contains asbestos, which is harmful to your health. Do not blow it out of the brake shoe assembly with compressed air and do not inhale any of it.*
4 Remove the adjuster lever spring and arm (see illustrations).
5 Back off the adjuster screw star wheel (see illustration)
6 Disconnect the parking brake cable from the trailing shoe with a pliers (see illustration).
7 Remove the hold down springs by depressing them with pliers or the special tool and turning the retainer until the slot aligns with the flattened end of the pin, allowing removal (see illustration).
8 Disengage the brake shoe assembly from the wheel cylinder at the top and the anchor plate at the bottom and remove it from the backing plate (see illustrations).
9 Place the assembly on a work surface and remove the springs (see illustrations).

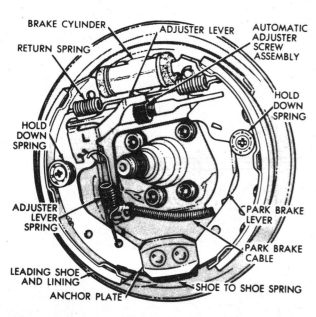

5.3 Rear drum brake components — left side shown

5.4a Use pliers to remove the adjuster spring . . .

5.4b . . . and lift the arm off

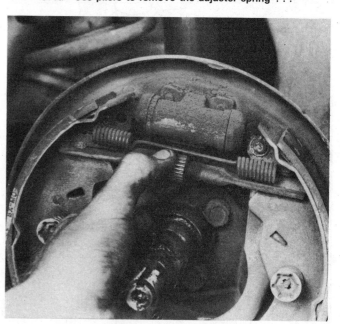

5.5 Back off the adjuster star wheel

Chapter 9 Brakes

5.6 Use pliers to pull the cable from the parking brake lever

5.7 Use a brake tool or pliers to depress the hold down spring and turn the retainer

5.8a Pull the brake shoe assembly out of the wheel cylinder . . .

5.8b . . . and then disengage the bottoms of the shoes from the anchor plate

5.9a Unhook the lower spring . . .

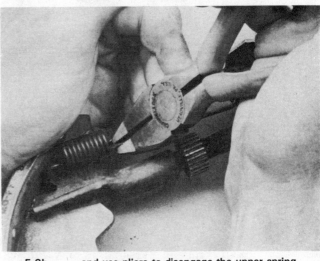
5.9b . . . and use pliers to disengage the upper spring from the shoes

Chapter 9 Brakes

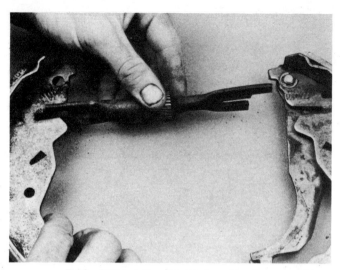

5.10 Remove the adjuster from the shoes

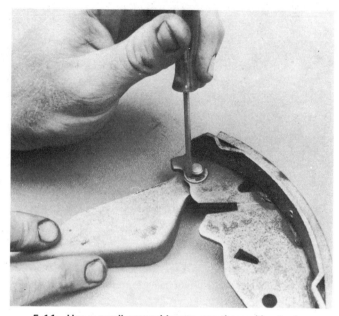

5.11 Use a small screwdriver to pry the parking brake lever retainer clip off

5.13 The maximum allowable diameter of the drum is stamped on it

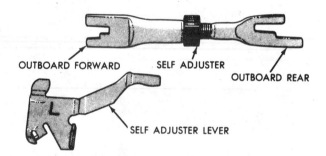

5.15 Adjuster components — exploded view

5.16 Carefully peel off the wheel cylinder boots to check for leakage

5.18 The shoe contact areas on the backing plate must be smooth and lubricated with high-temperature grease

Chapter 9 Brakes

10 Separate the shoes from the adjuster (see illustration).
11 Remove the parking brake lever retainer clip with a small screwdriver (see illustration) and transfer the lever to the new shoe.
12 Inspect the shoe linings to make sure they show full contact with the drum.
13 Check the drum for cracks, score marks and signs of overheating of the contact surface. Measure the inside diameter of the drum and compare it to the size stamped on the drum (see illustration). Minor imperfections in the drum surface can be removed with fine sandpaper. Deeper scoring can be removed by having the drum turned by a repair shop as long as the maximum diameter is not exceeded. Check the drum for runout. Replace the brake drum with a new one if it is not usable.
14 Check the brake springs for signs of discolored paint, indicating overheating, and distorted end coils. Replace them with new ones if necessary.
15 Check the adjuster screw assembly and threads for bent, corroded and damaged components. Replace the assembly if the screw threads are damaged or rusted. Clean the threads and lubricate them with white lithium-based grease (see illustration).
16 Inspect the wheel cylinder boots for damage and signs of leakage (see illustration).
17 Rebuild or replace the wheel cylinder if there is any sign of leakage around the boots.
18 Check for rough or rusted shoe contact areas on the backing plate, then lubricate the contact points with high temperature grease (see illustration).
19 Assemble the shoes over the adjuster and connect the return springs (see illustration).
20 Place the assembly in position, and insert the parking brake cable into the parking brake lever (see illustration)
21 Spread the bottom spring enough to allow the bottom of the shoes to be seated in the anchor plate.
22 Spread the top spring and seat the shoes in the pistons.
23 Insert the pins through the backing plate from the rear and hold them in place while installing the hold down springs and retainers.
24 Install the adjuster lever and connect the return spring.
25 Turn the adjuster until the shoes are retracted sufficiently to allow the drum to be reinstalled.
26 Install the hub/drum assembly and the wheel. Repeat the procedure for the other wheel.
27 Remove the rubber adjusting hole plug in the backing plate (see illustration).
28 Insert a narrow screwdriver through the hole in the backing plate and turn the star wheel until the brake drags slightly as the tire is turned (see illustration).
29 Back off the star wheel until the tire turns freely.
30 Repeat the adjustment on the opposite wheel.
31 Install the plugs in the backing plate access holes.
32 Adjust the parking brake.
33 Lower the vehicle and test for proper operation before placing the vehicle into normal service.

5.19 Assemble the shoes, adjuster and springs prior to installation

5.20 Use pliers to insert the parking brake cable into the lever

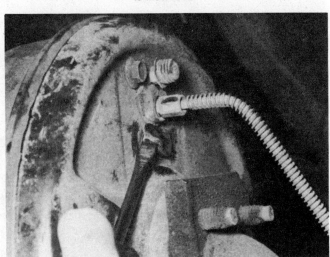

5.27 Pry the rubber adjusting hole plug out with a screwdriver

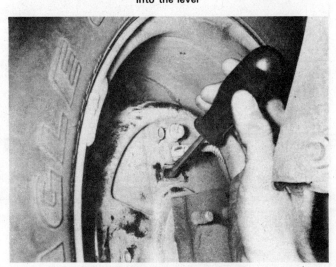

5.28 Insert the screwdriver into the adjuster hole and turn the star wheel

Chapter 9 Brakes

6 Rear wheel cylinder — removal and installation

Refer to illustration 6.4

1 Raise the rear of the vehicle and support it securely, then block the front wheels. Remove the rear wheels.
2 Remove the rear hub/drum (Chapter 10) and brake shoes (Section 5).
3 Disconnect the brake line from the back of the wheel cylinder and plug it.
4 Unbolt the wheel cylinder and remove it from the backing plate (see illustration). Clean the backing plate and wheel cylinder mating surfaces.
5 Apply RTV-type sealant to the wheel cylinder mating surface of the backing plate.
6 To install the wheel cylinder, hold it in position, install the mounting bolts and tighten them to the specified torque.
7 Unplug the brake line, insert it into the wheel cylinder fitting and carefully thread the tube flare nut into place. Once the nut is properly started, tighten it securely with a wrench.
8 Install the brake shoes and the hub/drum.
9 Bleed the brakes.
10 Install the wheels and lower the vehicle. Test for proper brake operation before placing the vehicle into normal service.

7 Rear wheel cylinder — overhaul

Refer to illustration 7.3

1 You must have a clean place to work, clean rags, some newspapers, a wheel cylinder rebuild kit, a container of brake fluid and some denatured alcohol to perform a wheel cylinder overhaul.
2 Remove the wheel cylinder as described in Section 6.
3 Remove the bleeder screw (see illustration) and check to make sure it is not obstructed.
4 Carefully pry the boots from the wheel cylinder and remove them.
5 Push in on one piston and force out the opposite piston, cups and spring with the cup expanders.
6 Clean the wheel cylinder, pistons and spring with clean brake fluid, denatured alcohol or brake system cleaner and dry them with compressed air. **Warning:** *Do not, under any circumstances, use petroleum-based solvents or gasoline to clean brake parts.*
7 Inspect the cylinder bore and piston for score marks and corrosion (pitting). Black stains on the cylinder walls are caused by the cups and will not impair brake operation. If the pistons or wheel cylinder bore are badly scored or pitted, replace them with new parts.
8 Lubricate the components with clean brake fluid or brake assembly

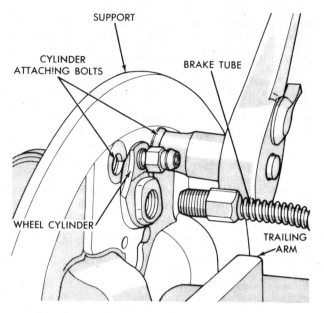

6.4 Removing the wheel cylinder mounting bolts

lubricant prior to installation.
9 With the cylinder bore coated with clean brake fluid or brake assembly lube, install the expansion spring and cup expanders. Install the cups in each end of the cylinder, making sure the open ends of the cups are facing each other.
10 Engage the boot on the piston and slide the assembly into the bore. Carefully press the boot over the cylinder end until it is seated. Repeat the procedure for the remaining boot and piston.
11 Install the bleeder screw.
12 Attach the wheel cylinder to the backing plate by referring to Section 6.

8 Master cylinder — removal and installation

Refer to illustration 8.2

1 Place several layers of newspaper under the master cylinder to catch any spilled brake fluid.

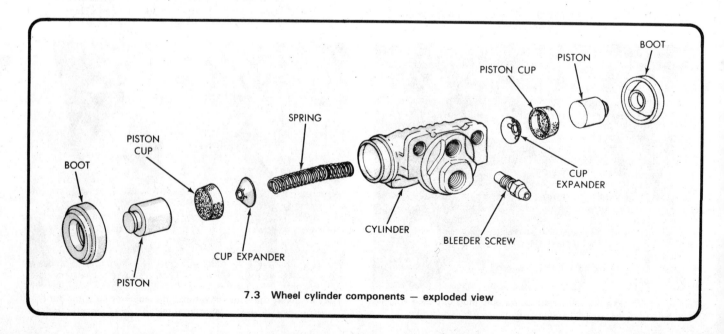

7.3 Wheel cylinder components — exploded view

Chapter 9 Brakes

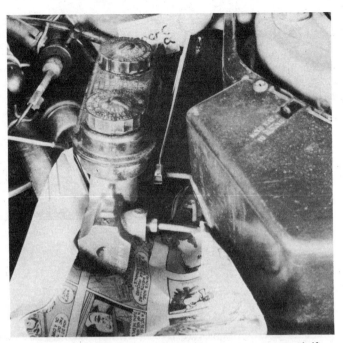

8.2 Loosen the brake line fittings with a flare nut wrench if one is available (otherwise, be careful not to round off the fitting hex)

2 Unscrew the steel line flare nuts (see illustration), remove the lines and cap them. Allow the fluid in the master cylinder to drain into a container.
3 Remove the mounting nuts and detach the master cylinder from the booster.
4 To install the master cylinder, hold it in position, align the pushrod and master cylinder piston and install the mounting nuts. Tighten the nuts to the specified torque.
5 Install the lines and carefully start the flare nuts, taking care not to cross-thread them. After they have been started by hand, tighten them securely with a wrench.
6 Fill the master cylinder reservoir and bleed the brakes.
7 **Note:** *Every time the master cylinder is removed, the complete hydraulic system must be bled.* The time required to bleed the system can be reduced if the master cylinder is filled with fluid and *bench bled* before the master cylinder is installed on the vehicle.
8 Insert threaded plugs of the correct size into the cylinder outlet holes and fill both reservoirs with brake fluid. The master cylinder should be supported in a level manner so that brake fluid will not spill during the bench bleeding procedure.
9 Loosen one plug at a time and push the piston assembly into the bore to force air from the master cylinder. To prevent air from being drawn back into the cylinder, the appropriate plug must be tightened before allowing the piston to return to its original position.
10 Since high pressure is not involved in the bench bleeding process, an alternative to the removal and replacement of the plug with each stroke of the piston assembly is available. Before pushing in on the piston assembly, remove the plug, then depress the piston as described above. Before releasing the piston, however, instead of replacing the plug simply put your finger tightly over the hole to keep air from being drawn back into the master cylinder. Wait several seconds for brake fluid to be drawn from the reservoir into the piston bore, then depress the piston again, removing your finger as the brake fluid is expelled. Be sure to put your finger back over the hole each time before releasing the piston, and when the bleeding procedure is complete for that outlet, replace the plug and snug it before going on to the other port to repeat the procedure.
11 Stroke the piston three or four times for each outlet to ensure that all air has been expelled.
12 Refill the master cylinder reservoirs and install the cover assembly.
Note: *The reservoirs should only be filled to the top of the reservoir divider to prevent overflowing when the cover is installed.*

9 Master cylinder — overhaul

The master cylinder on these models cannot be rebuilt. If problems are encountered, replace it with a new unit.

10 Brake hydraulic system — bleeding

Refer to illustration 10.7

1 If the brake system has air in it, operation of the brake pedal will be spongy and imprecise. Air can enter the brake system whenever any part of the system is dismantled or if the fluid level in the master cylinder reservoir runs low. Air can also leak into the system though a hole too small to allow fluid to leak out. In this case, it indicates that a general overhaul of the brake system is required.
2 To bleed the brakes, you will need an assistant to pump the brake pedal, a supply of new brake fluid, a plastic container, a clear plastic or vinyl tube which will fit over the bleeder screw fitting, and a wrench for the bleeder screw.
3 There are four locations at which the brake system is bled: both front brake caliper assemblies and the rear brake wheel cylinders.
4 Check the fluid level in the master cylinder reservoir. Add fluid if necessary to bring it up to the correct level (see Chapter 1). Use only the recommended brake fluid and do not mix different types. Never use fluid from a container that has been standing uncapped. You will have to check the fluid level in the master cylinder reservoir often during the bleeding procedure. If the level drops too far, air will enter the system though the master cylinder.
5 Raise the vehicle and support it securely on jackstands.
6 Remove the bleeder screw cap from the wheel cylinder or caliper assembly that is being bled. If more than one wheel must be bled, start with the one farthest from the master cylinder.
7 Attach one end of the clear plastic or vinyl tube to the bleeder screw fitting and place the other end in the plastic container, submerged in a small amount of clean brake fluid (see illustration).

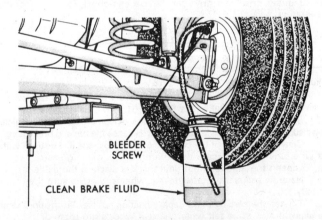

10.7 Brake bleeding equipment setup

8 Loosen the bleeder screw slightly, then tighten it to the point where it is snug yet easily loosened.
9 Have the assistant pump the brake pedal several times and hold it in the fully depressed position.
10 With pressure on the brake pedal, open the bleeder screw approximately one-quarter turn. As the brake fluid is flowing through the tube and into the container, tighten the bleeder screw. Again, pump the brake pedal, hold it in the fully depressed position and loosen the bleeder screw momentarily. Do not allow the brake pedal to be released with the bleeder screw in the open position.
11 Repeat the procedure until no air bubbles are visible in the brake fluid flowing through the tube. Be sure to check the brake fluid level in the master cylinder reservoir while performing the bleeding operation.
12 Completely tighten the bleeder screw, remove the plastic or vinyl tube and install the cap.
13 Follow the same procedure to bleed the other wheel cylinder or

Chapter 9 Brakes

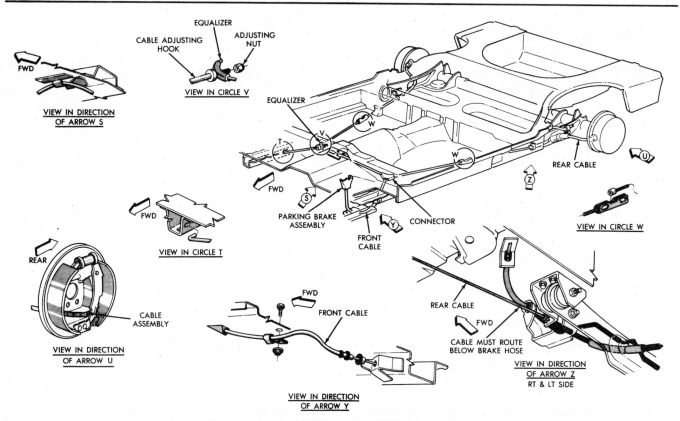

11.3 Parking brake cable layout

caliper assemblies.
14 Check the brake fluid level in the master cylinder to make sure it is adequate, then test drive the vehicle and check for proper brake operation.

11 Parking brake — adjustment

Refer to illustrations 11.3 and 11.6

1 The rear drum brakes must be in proper working order before adjusting the parking brake (see Section 5).
2 Block the front wheels to prevent any movement, raise the vehicle and support it securely with jackstands. Release the parking brake lever.
3 Clean the threads of the cable adjusting hook (see illustration) with a wire brush and lubricate them with multi-purpose grease.
4 Loosen the adjusting nut until there is slack in the cable.
5 Have an assistant rotate the rear wheels to make sure they turn easily.
6 Tighten the parking brake cable adjusting nut (see illustration) until a slight drag can be felt when the rear wheels are rotated.
7 Loosen the nut until the rear wheels turn freely, then back it off two full turns.
8 Apply and release the parking brake several times to make sure it operates properly. It must lock the rear wheels when applied and the wheels must turn easily, without dragging, when it is released.
9 Lower the vehicle.

12 Parking brake cables — removal and installation

Refer to illustration 12.15

1 Raise the vehicle and support it securely.

Front cable

2 Working underneath the vehicle, loosen the adjusting nut and disengage the cable from the connector (see illustration 11.3).
3 Inside the vehicle, lift the floor mat or carpeting and remove the floor pan seal panel.

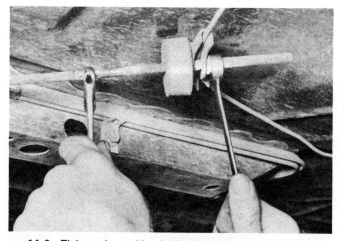

11.6 Tighten the parking brake cable adjusting nut until the brakes drag slightly as the tire is rotated

4 Pull the cable forward to disconnect it from the clevis.
5 Remove the cable assembly through the hole.
6 To install the cable, insert the cable housing retainers into the rail bracket hole and the parking brake pedal assembly. Feed the end of the cable through the holes in the pan and the rail bracket.
7 Install the floor seal.
8 Engage the end of the cable in the lever clevis and seat the cable ends in the parking brake assembly and the rail bracket.
9 Install the floor mat.
10 Engage the rear cable in the equalizer bracket, then adjust it as described in Section 11.

Rear cables

11 Remove the rear wheels and the hub/drum assemblies.
12 Back off the adjusting nut until the cable is slack.
13 Disconnect the rear brake cable from the connector.
14 Disconnect the cable at the brake shoe lever.

Chapter 9 Brakes

179

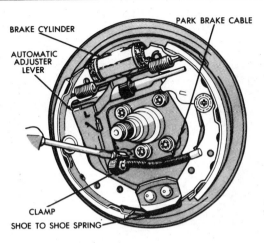

12.15 Removing the rear parking brake cable housing from the backing plate

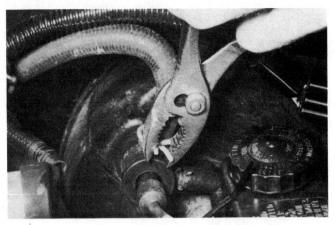

13.3 Use pliers to remove the vacuum hose clamp

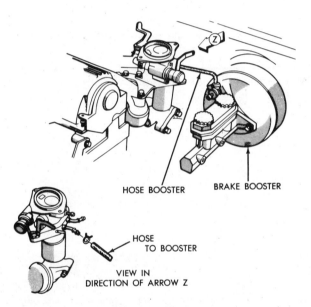

14.6a Brake booster vacuum hose layout — non-turbocharged engine

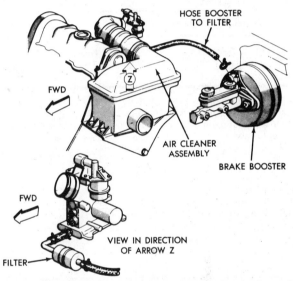

14.6b Brake booster vacuum hose layout — turbocharged engine

15 Use a screw-type hose clamp to compress the retainers so the cable can be removed from the brake backing plate (see illustration). Remove the clamp when the cable is free.
16 Pull the cable from the backing plate and rear axle.
17 To install the cable, insert the cable and the inner housing end through the trailing arm holes and the crossmember cable bracket.
18 Connect the chassis fitting to the bracket.
19 Insert the outer cable end through the support hole and snap the retainer into place.
20 Attach the cable to the lever in the brake shoe assembly, making sure the washer is between the cable spring and parking brake lever.
21 Insert the cable inner end into the connector.
22 Install the hub/drum assemblies.
23 Attach the cable to the connector, then adjust the parking brake (Section 11).

13 Power brake booster — removal and installation

Refer to illustration 13.3

1 Remove the master cylinder (Section 8).
2 On manual transaxle-equipped vehicles, unfasten the clutch cable bracket and pull the wiring harness up away from the strut tower.
3 Disconnect the vacuum hose at the booster (see illustration).
4 Working underneath the dash inside the vehicle, use a small screwdriver to remove the brake pedal-to-booster retainer clip and pin.
5 Remove the brake light switch and plate assembly.
6 Remove the four retaining nuts and lift the booster from the engine compartment.
7 Installation is the reverse of removal. Lubricate the surface of the brake pedal pin with white lithium-based grease and connect the pedal pin to the pushrod with a new retainer clip. Check the operation of the brake light switch and adjust if necessary (Section 15).
8 Check the booster for proper operation (Section 14).

14 Power brake booster — testing and overhaul

Refer to illustrations 14.6a and 14.6b

1 Symptoms of brake booster problems include low pedal and excessive braking effort and the brake pedal dropping after the initial application.
2 With the engine off, depress and release the brake pedal several times to bleed any vacuum from the booster.
3 Depress the pedal and hold it with light (between 15 and 25 lbs) pressure. Start the engine.
4 If the system is operating properly, the pedal should drop slightly and then stop. Subsequent applications will require less pressure.
5 If the booster fails this test, or if the pedal drops after the initial application, check the vacuum supply.
6 Inspect the vacuum hose for cracks and trace it to the manifold or carburetor. Check the connections, check valve and filter (if

equipped) for cracks and leaks, replacing any faulty components (see illustrations).

7 If no defects are found, have the booster operation checked by a dealer service department.

8 The brake booster must not be disassembled for any reason. If the booster is not operating properly, obtain a new or rebuilt unit and install it in the vehicle by referring to Section 13.

15 Brake light switch — removal, installation and adjustment

Refer to illustration 15.1

1 Unplug the wire connectors, grasp the switch and pull it out of the retainer (see illustration).

2 Press the new switch into the retainer and push the switch forward as far as possible. The brake pedal will move forward slightly.

3 Gently pull the pedal back as far as it will go, causing the switch to ratchet back to the correct position. Very little movement is required and no further adjustment is necessary.

16 Brake pedal — removal and installation

Note: *Refer to Chapter 8 for an exploded view illustration of the brake pedal components mentioned in this procedure.*

1 Disconnect the power brake pushrod from the brake pedal.

2 On manual transaxle-equipped models, remove the lock ring from the pivot shaft and carefully withdraw the shaft. Remove the clutch pedal assembly, followed by the brake pedal.

3 On automatic transaxle-equipped vehicles, remove the pivot shaft nut, withdraw the shaft and remove the brake pedal.

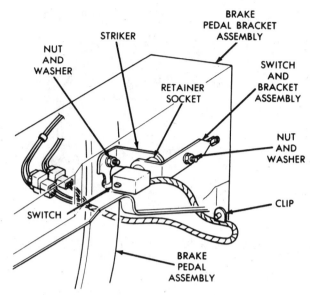

15.1 Brake light switch mounting details

4 To install, place the brake pedal in position and insert the pivot shaft.

5 On manual transaxle models, install the clutch pedal assembly and lock ring.

6 On automatic transaxle models, install the pivot shaft nut and tighten it securely.

Chapter 10 Steering and suspension systems

Refer to Chapter 13 for information on 1987 and later models

Contents

Balljoint check 4	Steering column (fixed) (1986 models) — disassembly
Chassis lubrication .... See Chapter 1	and reassembly 15
Front shock absorber strut and spring assembly — removal,	Steering column (tilt-wheel) — disassembly, inspection
inspection and installation 6	and reassembly 16
General information 1	Steering gear — removal and installation 17
Lower control arm — removal, inspection and installation ... 3	Steering knuckle and hub — removal, inspection
Power steering pump — removal and installation 18	and installation 5
Power steering system — bleeding 19	Steering shaft seal lubrication .... See Chapter 1
Rear hub and bearings — inspection and lubrication 9	Steering system — general information 11
Rear hub/drum assembly — removal and installation 8	Steering wheel — removal and installation 12
Rear shock absorber — removal, inspection	Suspension and steering check .... See Chapter 1
and installation 7	Sway bar — removal and installation 2
Rear spindle — inspection, removal and installation 10	Tie-rod ends — removal and installation 20
Steering angles and wheel alignment — general	Tire and tire pressure checks .... See Chapter 1
information 21	Tire rotation .... See Chapter 1
Steering column — removal and installation 13	Wheels and tires — general information 22
Steering column (fixed) (1984 and 1985	
models) — disassembly, inspection and reassembly 14	

Specifications

Front suspension

Torque specifications	Ft-lbs	Nm
Strut-to-steering knuckle bolts	75*	100*
Strut-to-tower nuts	20	27
Lower control arm balljoint clamp bolt nut	70	95
Lower control arm pivot bolt		
1984 and 1985	105	142
1986	95	129
Lower control arm stub strut nut	70	94
Sway bar bushing retainer nuts	25	34
Hub nut	180	245

* Plus 1/4-turn

Rear suspension

Torque specifications	Ft-lbs	Nm
Hub nut	20 to 25	27 to 34
Trailing arm-to-hanger bracket mounting nuts	40	55
Track bar attaching bolts	65	88
Track bar mounting bracket-to-frame bolt	40	55
Track bar brace/body stud	40	55
Track bar brace-to-stud nut	55	75
Shock absorber mounting bolts	40	55
Brake backing plate-to-stub axle bolts		
1984	45	60
1985 and 1986	40	55

Chapter 10 Steering and suspension systems

Steering gear

Torque specifications	Ft-lbs	Nm
Mounting bolts	21	28
Tie-rod end nut	35	47
Tie-rod jam nut	55	75
Crossmember mounting bolts	90	122

Steering column

Torque specifications	Ft-lbs	Nm
Steering wheel nut	45	61
Column clamp stud	20 in-lbs	2
Column clamp stud nut and bolt	105 in-lbs	12

Power steering pump

Torque specifications	Ft-lbs	Nm
Bracket mounting bolts/nuts		
M-10 stud	35	48
M-10 bolt/nut	30	40
M-8 bolt	21	28

1 General information

Refer to illustrations 1.1 and 1.2

The front suspension is by MacPherson struts. The steering knuckle is located by a lower control arm and both front control arms are connected by a sway bar (see illustration).

The rear suspension features a beam-type axle with coil springs, located by trailing arms and a track bar (see illustration). Damping is handled by vertically mounted shock absorbers located between the axle and the chassis.

The rack and pinion steering gear is located behind the engine and actuates the steering arms which are integral with the steering knuckles. Power assist is optional and the steering column is designed to collapse in the event of an accident. **Note:** *These vehicles use a combination of standard and metric fasteners on the various suspension and steering components, so it would be a good idea to have both types of tools available when beginning work.*

2 Sway bar — removal and installation

Refer to illustrations 2.2a and 2.2b

1 Raise the front of the vehicle, support it securely and remove the front wheels.
2 Remove the sway bar nuts, bolts, and retainers at the control arms (see illustrations).

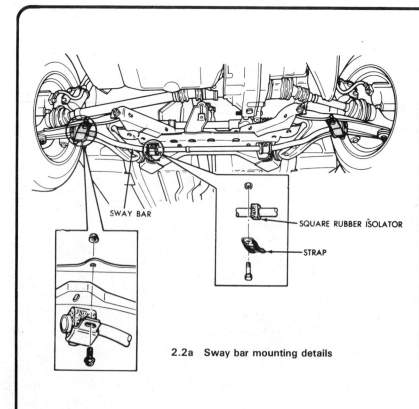

2.2a Sway bar mounting details

2.2b Use a wrench to hold the nut when removing the sway bar bolts

1.1 Front suspension components

1 Tie-rods
2 Sway bar bolts
3 Lower control arm
4 MacPherson struts
5 Sway bar
6 Power steering fluid lines

1.2 Rear suspension components

1 Coil spring
2 Track bar
3 Axle assembly
4 Trailing arm

Chapter 10 Steering and suspension systems

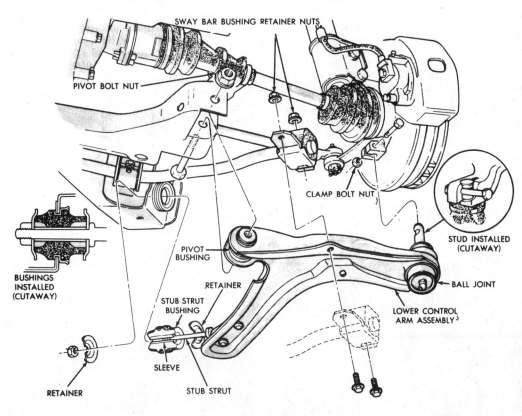

3.2a Lower control arm components — exploded view

3 Unbolt the clamps at the crossmember and remove the sway bar from the vehicle.
4 Check the bar for damage, corrosion and signs of twisting.
5 Check the clamps, bushings and retainers for distortion, damage and wear. Replace the inner bushings by prying them open at the split and removing them. Install the new bushings with the curved surface up and the split facing toward the front of the vehicle. The outer bushing can be removed by cutting it off or hammering it from the bar. Force the new bushing onto the end of the bar so that 1/2-inch of the bar is protruding.
6 Place the upper bushing retainers in position on the crossmember bushings, attach the bar to the crossmember and then install the lower clamps, bolts and nuts.
7 Install the bushing retainers, nuts and bolts at the lower control arm.
8 Raise the lower control arms to normal ride height and tighten the nuts to the specified torque.
9 Install the wheels and lower the vehicle.

3 Lower control arm — removal, inspection and installation

Refer to illustrations 3.2a, 3.2b, 3.4 and 3.5

1 Raise the front of the vehicle, support it securely and remove the front wheels. Detach the sway bar.
2 Remove the through-bolt and nut from the control arm pivot (see illustrations).
3 Remove the rear stub strut nut retainer and bushing.
4 Remove the balljoint clamp bolt and nut from the steering knuckle (see illustration).

3.2b Removing the lower control arm through-bolt (pivot bolt) nut

3.4 Removing the balljoint clamp bolt and nut to release the control arm

Chapter 10 Steering and suspension systems

5 Disconnect the balljoint stud from the steering knuckle, taking care not to separate the inner CV joint (see illustration).
6 Remove the sway bar nuts, separate the control arm and remove it from the vehicle.
7 Remove the rear stub strut bushing and sleeve assembly.
8 Inspect the lower control arm for distortion and the bushings for wear, damage and deterioration. Replace a damaged or bent control arm with a new one. If the inner pivot bushing or the balljoint (Section 4) are worn, take the control arm assembly to a dealer service department or a repair shop, as special tools are required to replace them. The strut bushings can be replaced by sliding them off the strut.
9 Assemble the retainer, bushing and sleeve on the stub strut.
10 Place the control arm in position over the sway bar and attach the stub strut and front pivot to the crossmember.
11 Install the front pivot bolt and stub strut assembly in the crossmember, with the nuts finger tight.
12 Attach the balljoint stud to the steering knuckle and tighten the clamp bolt to the specified torque.
13 Attach the sway bar end to the control arm and tighten the clamp to the specified torque.
14 Install the wheels and lower the vehicle. With the vehicle weight lowered onto the suspension, tighten the front pivot bolt and stub strut nuts to the specified torque.

4 Balljoint check

Refer to illustration 4.2

1 The suspension balljoints are designed to operate without free play.
2 To check for wear, grasp the grease fitting and attempt to move it with the vehicle weight resting on the front suspension (see illustration).
3 If there is any movement, the balljoint is worn and must be replaced with a new one (if the balljoint is worn the grease fitting will move easily). Since special equipment is required to press the balljoint from the control arm, it is recommended that the job be left to a dealer service department or a repair shop.

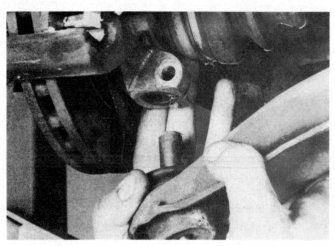

3.5 Pull down sharply to disengage the balljoint stud from the steering knuckle (a pry bar may be required)

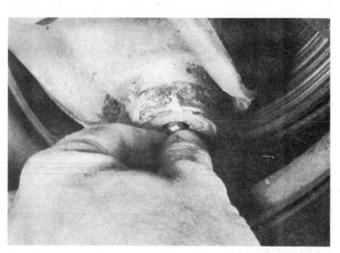

4.2 The balljoint can be checked for wear by attempting to move the grease fitting

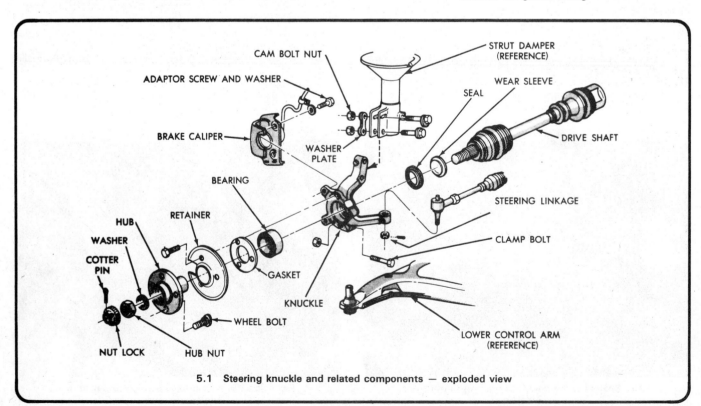

5.1 Steering knuckle and related components — exploded view

Chapter 10 Steering and suspension systems

5 Steering knuckle and hub — removal, inspection and installation

Refer to illustrations 5.1, 5.4, 5.9, 5.11, 5.13, 5.14 and 5.18

Removal

1 With the vehicle weight resting on the front suspension, remove the hub cap, cotter pin, nut lock and spring washer. Loosen, but do not remove, the front hub (axle) nut and wheel nuts (see illustration).
2 Raise the front of the vehicle, support it securely and remove the front wheels.
3 Remove the hub nut and washer.
4 Push the driveaxle in until it is free of the hub (see illustration). It may be necessary to tap on the axle end with a brass drift punch and hammer to dislodge the driveaxle from the hub.
5 Remove the cotter pin and nut and use a puller to disconnect the steering tie-rod from the hub (Section 20).
6 Move the tie-rod out of the way and secure it with a piece of wire.
7 Disconnect the brake hose from the shock strut by removing the bolt and retainer.
8 Remove the caliper and brake pads (Chapter 9), then remove the adapter from the steering knuckle. Taking care not to twist the brake hose, hang the caliper out of the way in the wheel well with a piece of wire.
9 Loosen the sway bar bushing bolts, unbolt the ends from the control arm and pull the sway bar down and out of the way (see illustration).
10 Remove the retainer from the wheel stud and pull the brake disc off.
11 Mark the cam bolt and washer location prior to removal (see illustration).
12 Remove the balljoint pinch bolt and nut and disengage the balljoint from the hub.
13 Remove the steering knuckle-to-strut bolts and nuts (see illustration).
14 With the knuckle and hub assembly in the straight-ahead position, grasp it securely and pull it directly out and off the driveaxle splines (see illustration).

5.4 Push in on the driveaxle while supporting the outer CV joint

5.9 With the bushing mounting bolts loose, the swaybar can be pulled down and out of the way

5.11 Mark the location of the cam bolt head before removing the bolts and separating the strut from the steering knuckle

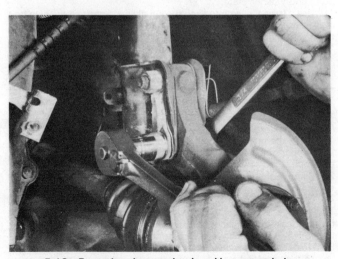

5.13 Removing the steering knuckle-to-strut bolts and nuts

5.14 Pull the steering knuckle out and off the driveaxle splines

188   Chapter 10 Steering and suspension systems

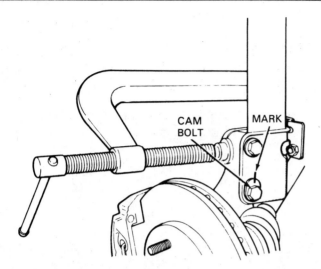

5.18 Use a C-clamp to pull the strut and steering knuckle into alignment

Inspection

15 Place the assembly on a clean working surface and wipe it off with a lint-free cloth. Inspect the knuckle for rust, damage and cracks. Check the bearings by rotating them to make sure they move freely. The bearings should be packed with an adequate supply of clean grease. If there is too little grease, or if the grease is contaminated with dirt, clean the bearings and inspect them for wear, scoring and looseness. Repack the bearings with the specified lubricant. Inspect the grease seals to make sure they are not torn or leaking. Further disassembly will have to be left to your dealer service department or a repair shop because of the special tools required.

Installation

16 Prior to installation, clean the CV joint seal and the hub grease seal with solvent (don't get any solvent on the CV joint boot). Lubricate the entire circumference of the CV joint wear sleeve and seal contact surface with multi-purpose grease (refer to Chapter 8 if necessary).
17 Carefully place the knuckle and hub assembly in position. Align the splines of the axle and the hub and slide the hub into place.
18 Install the knuckle-to-strut bolts and nuts, followed by the balljoint pinch bolt and nut. Adjust the knuckle so that the mark made during removal is aligned with the cam bolt and washer. It may be necessary to use a large C-clamp to pull the steering knuckle and strut together and line up the marks (see illustration).
19 Install the tie-rod end, tighten the nut and install the cotter pin.
20 Install the brake disc, pads and caliper/adapter assembly.
21 Connect the brake hose to the shock strut.
22 Attach the sway bar ends to the control arm and tighten the fasteners to the specified torque.
23 Push the CV joint completely into the hub to make sure it is seated and install the washer and hub nut finger tight.
24 Install the wheels and lower the vehicle.
25 With an assistant applying the brakes, tighten the hub nut to the specified torque. Install the spring washer, nut lock and a new cotter pin.
26 With the weight of the vehicle on the suspension, check the steering knuckle and balljoint nuts to make sure they are tightened properly.
27 Have the vehicle front end alignment checked.

6   Front shock absorber strut and spring assembly — removal, inspection and installation

Refer to illustration 6.6
1 Loosen the front wheel nuts.
2 Raise the vehicle and support it securely. Remove the front wheels.
3 Mark the location of the cam bolt and washer as described in Step 11 of the previous Section.
4 Remove the strut-to-steering knuckle nuts, bolts and washer plate.

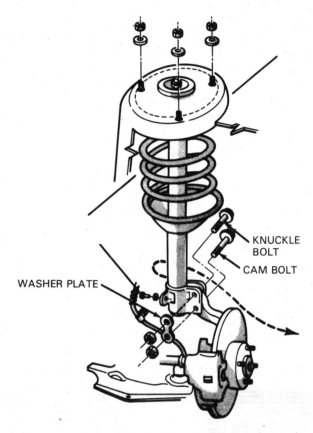

6.6 Shock absorber strut and spring assembly removal details

5 Disconnect the brake hose from the strut.
6 Remove the upper mounting nuts, disengage the strut from the steering knuckle and detach it from the vehicle (see illustration).
7 Checking of the strut and spring assembly is limited to inspection for leaking fluid, dents, damage and corrosion. Further disassembly should be left to your dealer service department or a repair shop because of the special tools and expertise required.
8 To install the strut, place it in position with the studs extending up through the shock tower. Install the nuts and tighten them to the specified torque.
9 Attach the strut to the steering knuckle, then insert the mounting bolts and washer plate.
10 Use a large C-clamp to align the knuckle and strut so the cam bolt marks line up. Install the nuts on the cam and knuckle bolts and tighten them to the specified torque, plus 1/4-turn. Remove the clamp.
11 Attach the brake hose to the strut.
12 Install the wheels and lower the vehicle.

7   Rear shock absorber — removal, inspection and installation

Refer to illustrations 7.2a, 7.2b, 7.3a and 7.3b
1 Raise the rear of the vehicle and support it securely.
2 Support the axle with a jack and remove the rear wheels (see illustrations).
3 Remove the upper and lower shock mounting bolts (see illustrations) and detach the shock absorber.
4 Grasp the shock at each end and pump it in and out several times. The action should be smooth, with no binding or dead spots. Check for fluid leakage. Replace the shock with a new one if it is leaking or the action is rough. Always replace the shocks in pairs.
5 Hold the shock in position and install the bolts. Tighten the upper bolt to the specified torque. Lower the vehicle and tighten the lower bolt to the specified torque.

Chapter 10 Steering and suspension systems

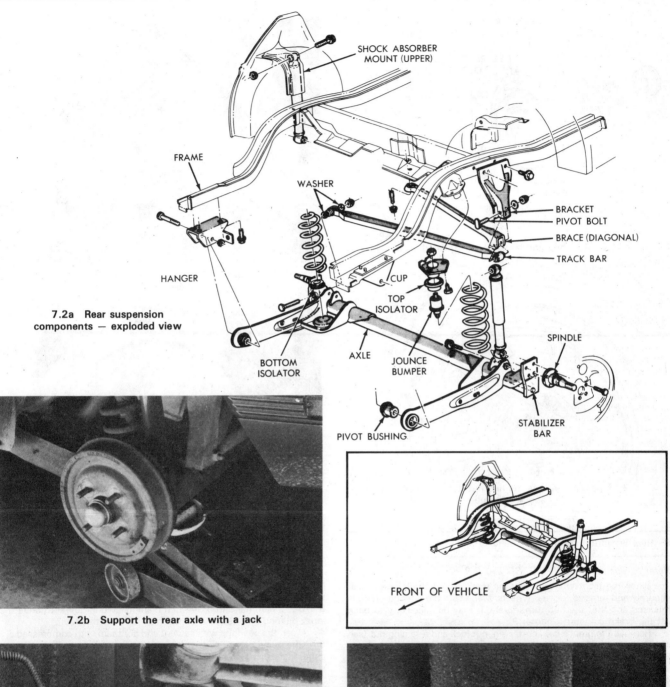

7.2a Rear suspension components — exploded view

7.2b Support the rear axle with a jack

7.3a The rear shock is held in place by one lower . . .

7.3b . . . and one upper bolt

Chapter 10 Steering and suspension systems

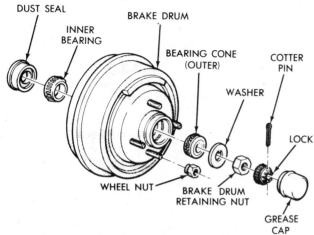

8.2 Rear hub and bearings — exploded view

8.3 Pull out on the brake drum to dislodge the outer bearing while holding it so it doesn't fall out

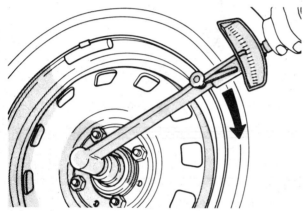

8.7 Rotate the tire as the hub nut is tightened to the specified torque

8.8 Tap the grease cap evenly into place with a hammer

8 Rear hub/drum assembly — removal and installation

Refer to illustrations 8.2, 8.3, 8.7 and 8.8

1 Block the front wheels, raise the rear of the vehicle, support it securely and remove the rear wheels. Make sure the parking brake is released and that the hub turns freely. It may be necessary to back off the brake adjuster (Chapter 9).
2 Remove the grease cap, cotter pin, nut lock and retaining nut (see illustration).
3 Grasp the brake drum and pull it out enough to dislodge the outer wheel bearing and washer (see illustration).
4 Remove the bearing.
5 Withdraw the hub/drum from the axle.
6 Place the hub in position on the axle, install the outer wheel bearing and washer and push the assembly into place.
7 Install the retaining nut and the wheel. Rotate the tire while tightening the nut to the specified torque (see illustration). Stop rotating the tire and back off the nut 1/4-turn, then tighten it finger tight while rotating the tire.
8 Install the nut lock, cotter pin and grease cap (see illustration). Pull out on the drum; the end play in the hub should not exceed 0.001 to 0.003-inch (0.025 to 0.076 mm).
9 Lower the vehicle and tighten the wheel lug nuts.

9 Rear hub and bearings — inspection and lubrication

Refer to illustrations 9.7, 9.9, 9.10a, 9.10b, 9.11a and 9.11b

1 Remove the hub from the vehicle (Section 8).
2 Inspect the bearings for proper lubrication and signs that the grease has been contaminated by dirt (grease will be gritty) or water (grease will have milky-white appearance).
3 Use a hammer and a 3/4-inch diameter wood dowel to drive the inner bearing and seal out of the hub (discard the seal).
4 Clean the bearings with solvent and dry them with compressed air.
5 Check the bearings for wear, pitting and scoring of the roller and cage. Light discoloration of the bearing surfaces is normal, but if the surfaces are badly worn or damaged, replace the bearings with new ones.
6 Clean the hub with solvent and remove the old grease from the hub cavity.
7 Inspect the bearing races for wear, signs of overheating, pitting and corrosion. If the races are worn or damaged, drive them out with a hammer and a drift (see illustration).
8 Drive the new races in with a section of pipe and a hammer, but be very careful not to damage them or get them cocked in the bore.
9 Pack the bearings with high temperature, multi-purpose EP grease prior to installation. Work generous amounts of grease in from the back of the cage so the grease is forced up through the rollers (see illustration).
10 Add a small amount of grease to the hub cavity and to the center of the spindle (see illustrations).
11 Lubricate the outer edge of the new grease seal, insert the bearing and press the seal into position with the lip facing in (see illustration). Make sure the seal is seated completely in the hub by tapping it evenly into place using a hammer and block of wood (see illustration). Apply grease to the seal cavity and lip and the polished sections of the spindle.
12 Install the hub and drum assembly as described in Section 8.

Chapter 10 Steering and suspension systems

9.7 The bearing races can be driven out with a hammer and drift punch (work carefully and do not damage the hub)

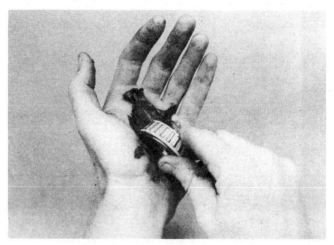

9.9 Work the grease completely into the rollers

9.10a Put a small amount of grease in the hub cavity . . .

9.10b . . . and on the center of the spindle

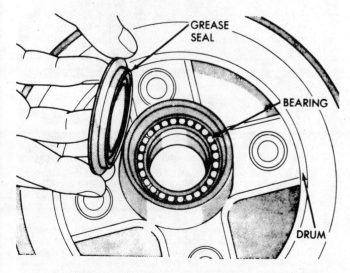

9.11a Make sure the bearing is in place in the hub before installing the grease seal

9.11b Tap the seal into place with a hammer and a wood block

Chapter 10 Steering and suspension systems

10 Rear spindle — inspection, removal and installation

1 Remove the rear hub/drum (Section 8).
2 Clean the axle and inspect the bearing contact surfaces for wear and damage.
3 The axle should be replaced with a new one if it is bent, damaged or worn.
4 Disconnect the parking brake cable.
5 Disconnect and plug the rear brake line at the wheel cylinder (see Chapter 9 if necessary).
6 Remove the four backing plate mounting bolts and lift off the brake assembly and spindle. The bolts may have Torx-type heads. Be sure to mark the location of any spindle shims.
7 To install, place the shim(s) (if equipped), spindle and brake assembly in position, install the bolts and tighten them to the specified torque in a criss-cross pattern.
8 Connect the brake line and parking brake cable.
9 Install the hub/drum (Section 8), bleed the brakes and adjust the parking brake (Chapter 9).

11 Steering system — general information

All models are equipped with rack and pinion steering. The steering gear is bolted to the chassis directly behind the engine and operates the steering arms through tie-rods. The inner ends of the tie-rods are protected by rubber boots, which should be inspected periodically for secure attachment, tears and leaking lubricant.

As an option, some models are equipped with power assisted steering. The power assist system consists of a belt-driven pump and associated lines and hoses. The power steering pump reservoir fluid level should be checked periodically (Chapter 1).

The steering wheel operates the steering shaft, which actuates the steering gear through a universal joint. Looseness in the steering can be caused by wear in the steering shaft universal joint, the steering gear, the tie-rod ends and loose retaining bolts. Inadequate lubrication of the steering shaft seal can cause binding of the steering; the seal should be lubricated periodically (Chapter 1).

12 Steering wheel – removal and installation

Warning: *Later models are equipped with a driver's side air bag installed in the center of the steering wheel. The air bag must be disarmed and removed before the steering wheel is removed (see Chapter 13).*

Refer to illustrations 12.4 and 12.5

1 Disconnect the negative cable at the battery. Place the cable out of the way so it cannot accidentally come in contact with the negative terminal of the battery, as this would once again allow power into the electrical system of the vehicle.
2 Remove the center pad assembly by grasping it securely and pulling out, releasing the clips.
3 Unplug the connector and remove the horn switch.
4 Remove the steering wheel retaining nut **(see illustration)** and mark the relationship of the steering shaft and hub to simplify installation.
5 Use a puller to remove the steering wheel **(see illustration)**. Caution: Do not hammer on the shaft to remove the steering wheel.
6 To install the wheel, align the mark on the steering wheel hub with the mark made on the shaft during removal and slip the wheel onto the shaft. Install the hub nut and tighten it to the specified torque.
7 Install the horn switch and button.
8 Install the center pad assembly.
9 Connect the negative battery cable.

13 Steering column – removal and installation

1 Disconnect the negative cable at the battery. Place the cable out of the way so it cannot accidentally come in contact with the negative terminal of the battery, as this would once again allow power into the electrical system of the vehicle.
2 Unplug the steering column wiring connectors at the column jacket.
3 Remove the steering wheel (Section 12).
4 Remove the instrument panel column cover and the lower reinforcement (Chapter 11).
5 Remove the steering column-to-instrument panel and lower panel bracket retaining nuts. Do not remove the roll pin.
6 Grasp the steering column assembly firmly and pull it toward the rear so the lower stub shaft is disconnected from the steering gear coupling. Reinstall the anti-rattle coupling spring all the way into the lower coupling tube. Be sure it snaps into the slot in the coupling.
7 Remove the assembly carefully from the vehicle. **Note:** *On vehicles equipped with a speed control and manual transaxle, be careful not to damage the clutch pedal speed control switch.*
8 Place the steering column in position with the stub shaft aligned with the lower coupling and insert the shaft into the coupling.
9 Install the retaining nuts (use washers on the breakaway cap-

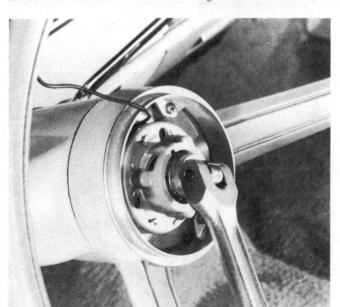

12.4 Removing the steering wheel retaining nut with a breaker bar

12.5 Use a puller to remove the steering wheel — do not hammer on the shaft!

Chapter 10 Steering and suspension systems

sules) and pull the column to the rear while tightening the nuts.
10 Using needle nose pliers, pull the coupling spring up until it touches the spacer on the stub shaft (see illustration 14.2a).
11 Install the steering wheel.
12 Plug in the wiring connectors.
13 Install the panel column cover and lower reinforcement.
14 Connect the negative battery cable and check the operation of the horn and lights.

14 Steering column (fixed) (1984 and 1985 models) — disassembly, inspection and reassembly

Refer to illustrations 14.2a, 14.2b, 14.8, 14.9, 14.15 and 14.23

Note: *The steering column used on some models may differ slightly in appearance and details from the one shown in the illustrations, but the procedure is essentially the same.*

1 Remove the breakaway capsules and clamp the bracket in a vise, using blocks of wood to protect the bracket.
2 Pry the wiring trough retainers loose and remove the trough (see illustrations).
3 Remove the screws, detach the turn signal lever cover and remove the washer/wiper switch assembly.
4 Pull the shroud (if equipped) up the control stalk and remove the two screws that hold the control stalk sleeve to the washer/wiper switch.
5 Rotate the control stalk to the full clockwise position and remove the shaft from the switch by pulling it straight out.
6 Remove the screws and detach the turn signal/flasher switch (lift it up to remove it).
7 Disconnect the horn and key light ground wires and remove the ignition key lamp.
8 Remove the shaft upper bearing retaining screws, then remove the shaft snap-ring (see illustration). Do not allow the steering shaft to slide out of the jacket.
9 Slide the bearing housing, lock plate and spring from the shaft (see illustration). Remove the shaft through the lower end of the column.
10 Remove the screw and lift out the key buzzer assembly.
11 Remove the ignition switch mounting screws, then rotate the switch 90 degrees on the rod and slide it off.
12 Remove the two screws and disengage the dimmer switch from the actuator rod.
13 Remove the mounting screws and slide the bellcrank up in the lock housing until it can be disconnected from the ignition switch actuator rod.
14 Place the cylinder in the Lock position and remove the key. Insert two small diameter screwdrivers into the lock cylinder release holes and push in to release the retainers (the lower release hole is just above the buzzer switch screw hole). At the same time, pull the lock cylinder out of the housing bore. Remove the hex head screws and detach the lock housing and plate.
15 Inspect the steering shaft bearings for wear, looseness and signs of binding. The bearings should be replaced with new ones if there is appreciable wear or rough action. Lubricate the bearing with multi-purpose grease prior to installation (see illustration).

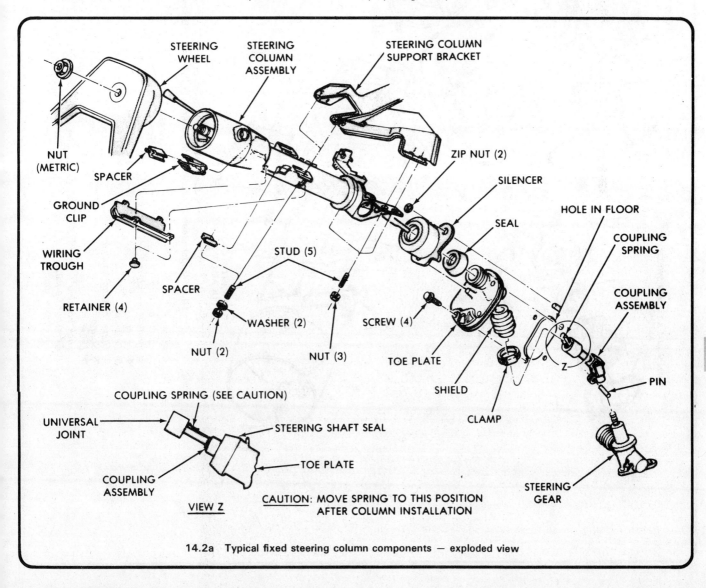

14.2a Typical fixed steering column components — exploded view

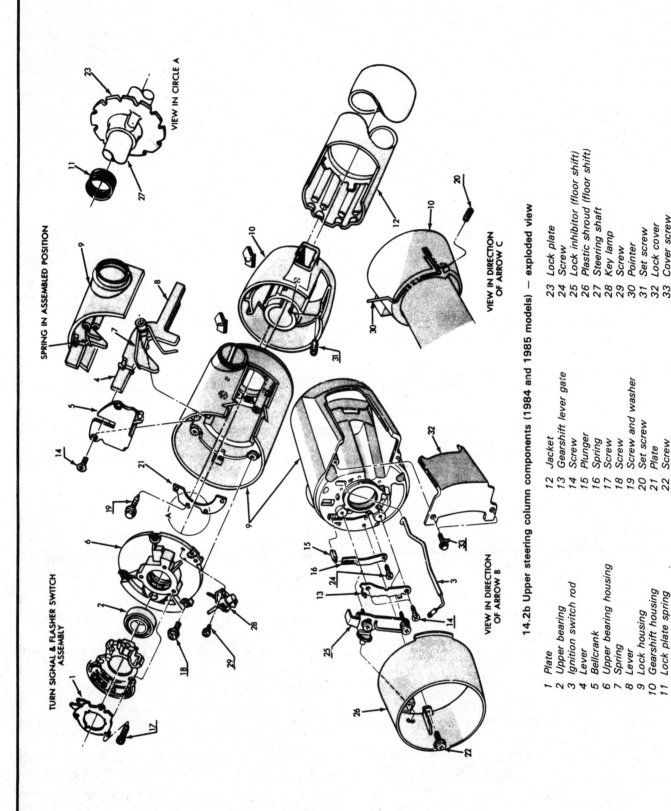

14.2b Upper steering column components (1984 and 1985 models) – exploded view

1 Plate
2 Upper bearing
3 Ignition switch rod
4 Lever
5 Bellcrank
6 Upper bearing housing
7 Spring
8 Lever
9 Lock housing
10 Gearshift housing
11 Lock plate spring
12 Jacket
13 Gearshift lever gate
14 Screw
15 Plunger
16 Spring
17 Screw
18 Screw
19 Screw and washer
20 Set screw
21 Plate
22 Screw
23 Lock plate
24 Screw
25 Lock inhibitor (floor shift)
26 Plastic shroud (floor shift)
27 Steering shaft
28 Key lamp
29 Screw
30 Pointer
31 Set screw
32 Lock cover
33 Cover screw

Chapter 10 Steering and suspension systems

14.8 Removing the upper steering column bearing housing snap-ring

14.9 Lifting out the bearing housing

14.15 Apply grease to the lower bearing outer edge and contact surface

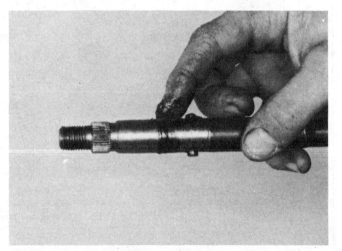

14.23 Cover the O-ring and surrounding shaft area with grease

16 Assemble the key cylinder plunger spring and attach the assembly to the lock housing.
17 Install the dimmer switch (Chapter 12).
18 Lubricate the lock lever assembly and install it in the lock housing. Seat the pin in the bottom of the slots and make sure the lock lever spring leg is in place in the casting notch.
19 Install the ignition switch actuator rod up from the bottom of the lock housing and connect the bellcrank. Position the bellcrank assembly in the lock housing while pulling the switch rod down the column, then attach the bellcrank.
20 Attach the ignition switch to the rod and rotate it 90 degrees so the rod will lock in position.
21 Install the ignition lock, turn the key to the Lock position and remove it, which will cause the buzzer actuating lever to retract. Insert the key again, push in and turn the cylinder until the retainers align and the cylinder snaps into place. Make sure the key cylinder and the ignition switch are in the Lock position and tighten the mounting screws.
22 Push the wires leading from the key buzzer switch down the column through the space between the housing and the jacket. Remove the key and tighten the switch screws.
23 Install the bearing and spring onto the steering shaft. Install the O-ring on the shaft and lubricate it (see illustration).

24 Insert the steering shaft into the column and press the bearing into place in the housing. Push up on the steering shaft so the spring will be compressed and install the snap-ring.
25 Install the lock plate, anti-rattle spring and bearing and housing assembly, retaining it with the snap-ring.
26 Install the bearing housing retaining screws.
27 Install the key lamp assembly, followed by the turn signal switch. Feed the wires down the steering column through the opening between the bearing and lock housing.
28 Install the retainer plate and screws and connect the ground wires.
29 Assemble the wiper switch, shaft, cover, or speed control switch, shroud and knob.
30 Place the washer/wiper switch assembly in the lock housing and feed the wires through the lock and shift housings. Connect the wires to the turn signal switch.
31 Insert the dimmer switch actuating rod up through the housing and connect it to the washer/wiper switch.
32 Adjust the dimmer switch as described in Chapter 12.
33 Install the turn signal lever cover and the breakaway capsules.
34 Install the wiring trough, being careful not to pinch the wires, and install new retainers if needed.

Chapter 10 Steering and suspension systems

15 Steering column (fixed) (1986 models) — disassembly and reassembly

Refer to illustrations 15.1a, 15.1b, 15.2, 15.3, 15.4, 15.5, 15.6, 15.7, 15.8, 15.9, 15.10 and 15.11

1 Clamp the bracket in a vise and remove the upper and lower covers (see illustrations).
2 Remove the retaining screws, disengage the switch rod and remove the dimmer switch (see illustration).
3 Remove the two retaining screws, disengage the rod and lift the ignition switch off (see illustration).
4 Remove the retaining screw and lift off the washer/wiper switch (see illustration).
5 Remove the three screws and remove the turn signal and flasher switch (see illustration).
6 Remove the key-in buzzer switch retaining screws, and lift the switch off (see illustration).
7 Remove the retaining screw and remove the lock cylinder plunger (see illustration).
8 Insert a suitable tool into the release hole and, with the key removed, pull the lock cylinder out of the housing (see illustration).

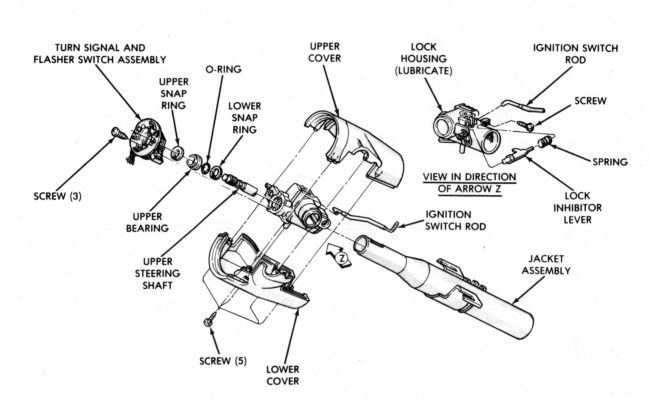

15.1a 1986 model upper steering column — exploded view

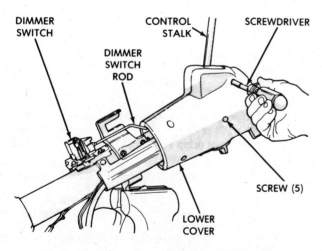

15.1b Removing the column cover (1986 models)

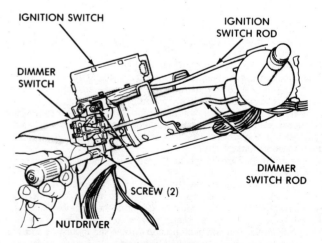

15.2 Dimmer switch removal and installation details (1986 models)

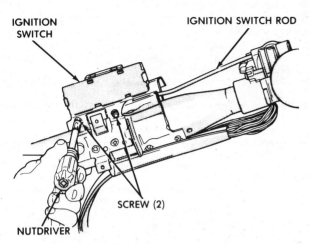

15.3 Ignition switch removal and installation details (1986 models)

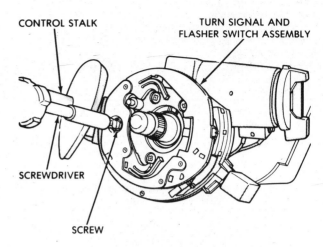

15.4 Washer/wiper removal and installation details (1986 models)

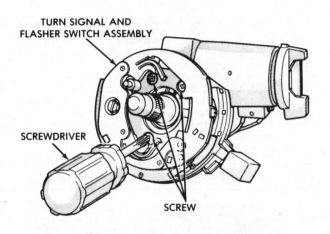

15.5 Turn signal and flasher switch removal and installation details (1986 models)

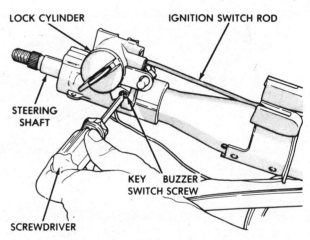

15.6 Key buzzer removal and installation details (1986 models)

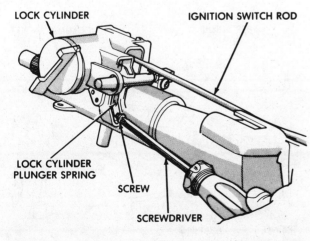

15.7 Lock cylinder plunger removal and installation details (1986 models)

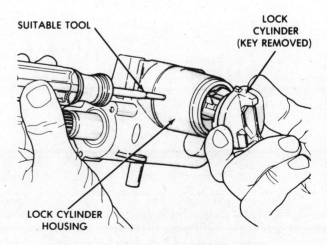

15.8 Lock cylinder removal (1986 models)

198   Chapter 10   Steering and suspension systems

9 Remove the screw, disengage the rod and remove the lock lever inhibitor assembly (see illustration).
10 Use snap-ring pliers to remove the upper bearing snap-ring (see illustration) and slide the bearing from the shaft.
11 Remove the steering shaft through the lower end of the column jacket (see illustration).
12 Reassembly is the reverse of the disassembly procedure.

16 Steering column (tilt wheel) — disassembly, inspection and reassembly

Other than removal and installation of the steering wheel and column assemblies, procedures for the tilt wheel column vary considerably from those for the fixed column. Because of the special tools and expertise required, work on the tilt wheel steering column assembly should be referred to your dealer service department or a repair shop.

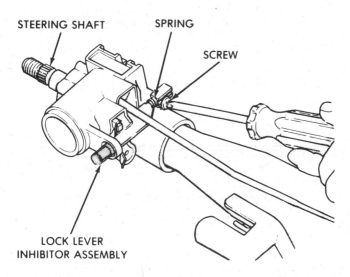

15.9 Lock lever inhibitor assembly details (1986 models)

17 Steering gear — removal and installation

Refer to illustration 17.4

1 Raise the vehicle, support it securely and remove the front wheels. Remove the steering column assembly (Section 13), leaving the lower universal joint with the steering gear.
2 Disconnect the tie-rod ends from the steering knuckles (Section 20).
3 Remove the anti-rotational link or damper from the crossmember.
4 Support the front crossmember with a jack, then remove the four steering gear-to-crossmember bolts (see illustration).
5 Lower the crossmember to gain access to the steering gear.
6 Remove the boot seal and splash shields from the crossmember.
7 On power steering equipped models, disconnect the hoses and drain the fluid into a container.
8 Remove the steering gear mounting bolts and withdraw it from the crossmember to the left side of the vehicle.
9 To install the steering gear, position it in the crossmember, install the bolts and tighten them securely.
10 Attach the tie-rod ends to the steering knuckles.
11 On power steering equipped models, reconnect the hoses (use new O-rings).
12 On manual steering equipped models, check to make sure the master serrations on the steering gear shaft are properly aligned so the steering shaft will be installed in the straight-ahead position.
13 Install the boot seal and splash shields.
14 Raise the crossmember and steering gear into position, install the four crossmember bolts and tighten them securely (the right rear bolt is a pilot bolt and must be tightened first).
15 Install the steering column assembly.
16 Install the front wheels and lower the vehicle.
17 On power steering equipped models, start the engine and bleed the steering system (Section 19). While the engine is running, check for leaks at the hose connections.
18 Have the front end alignment checked by a dealer service department or an alignment shop.

18 Power steering pump — removal and installation

Note: *All fasteners used on the power steering pump mounting brackets are metric.*

1 Open the hood and disconnect the two wires from the air conditioner clutch switch (if equipped).

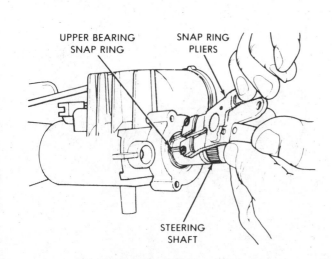

15.10 Removing the upper bearing snap-ring (1986 models)

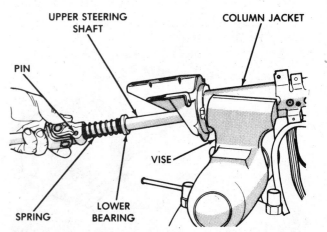

15.11 Slide the steering shaft out of the lower end of the column jacket (1986 models)

Chapter 10 Steering and suspension systems

2 Remove the drivebelt adjustment locking bolt from the front of the pump and (if equipped) the nut from the end hose bracket.
3 Raise the vehicle and support it securely.
4 Remove the pump pressure hose locating bracket at the crossmember, disconnect the hose from the steering gear and drain the fluid through the hose into a container.
5 While the fluid is draining, remove the right side splash shield to expose the drivebelts.
6 Disconnect the hoses from the pump and plug all openings so that dirt cannot enter.
7 Remove the lower stud nut and pivot bolt from the pump, then lower the vehicle.
8 Remove the drivebelt from the pulley, then move the pump to the rear and remove the adjustment bracket.
9 Turn the pump around so the pulley is facing toward the rear of the vehicle and lift it up and out of the engine compartment.
10 Install the adjustment bracket on the pump. Make sure the tab is in the lower left front mounting hole.
11 Lower the pump into position in the engine compartment.
12 Raise the vehicle and support it securely.
13 Install the lower pump bolt and stud nut finger tight.
14 Using new O-rings, attach the hoses to the pump.
15 Place the drivebelt onto the pulley and then lower the vehicle.
16 Install the adjustment bolt and nut, adjust the belt to the proper tension (Chapter 1) and tighten the nut.
17 Raise the vehicle, tighten the lower stud nut and pivot bolt and install the splash shield.
18 Lower the vehicle and connect the air conditioner switch wires.
19 Fill the pump to the top of the filler neck with the specified fluid.
20 Start the engine, bleed the air from the system (Section 19) and check the fluid level.

19 Power steering system — bleeding

1 The power steering system must be bled whenever a line is disconnected.
2 Open the hood and check the fluid level in the reservoir, adding the specified fluid necessary to bring it up to the proper level (see Chapter 1 if necessary).
3 Start the engine and slowly turn the steering wheel several times from left-to-right and back again. Do not turn the wheel fully from lock-to-lock. Check the fluid level, topping it up as necessary until it remains steady and no more bubbles appear in the reservoir.

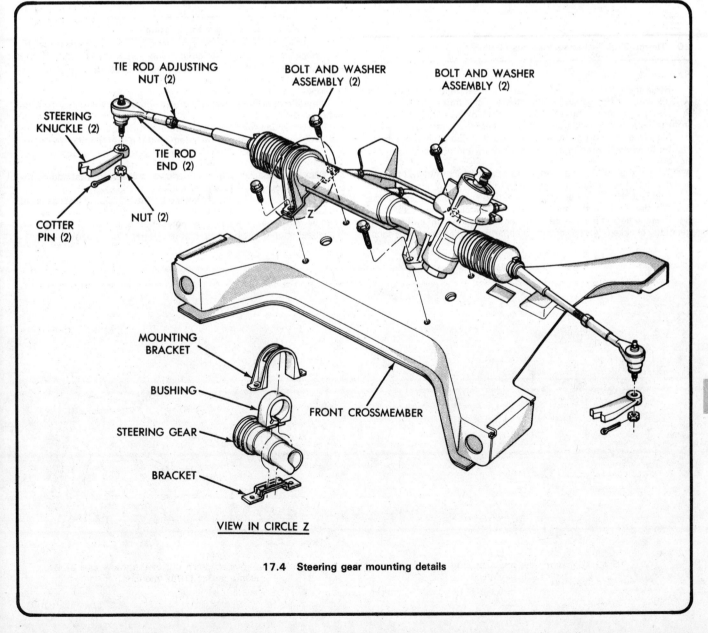

17.4 Steering gear mounting details

20.2 Use a puller to separate the tie-rod ends from the steering knuckle arms

20 Tie-rod ends — removal and installation

Refer to illustration 20.2

1 Raise the front of the vehicle, support it securely, block the rear wheels and set the parking brake. Remove the front wheels.
2 Remove the cotter pin and tie-rod nut, then disconnect the tie-rod from the steering knuckle arm with a puller (see illustration).
3 Mark the location of the jam nut and then loosen the nut sufficiently to allow the rod end to be unscrewed and removed from the tie-rod.
4 Thread the tie-rod end onto the rod until the jam nut will tighten against it to the marked position and tighten the jam nut securely.
5 Connect the tie-rod end to the steering knuckle arm, install the nut and tighten it to the specified torque. Install a new cotter pin.
6 If a new tie-rod end has been installed, have the front end steering geometry checked by a dealer service department or an alignment shop.

21 Steering angles and wheel alignment — general information

1 Proper wheel alignment is essential to proper steering and even tire wear. Symptoms of alignment problems are pulling of the steering to one side or the other and uneven tire wear.
2 If these symptoms are present, check for the following before having the alignment adjusted:
   Loose steering gear mounting bolts
   Damaged or worn steering gear mounts
   Improperly adjusted wheel bearings
   Bent tie-rods
   Worn balljoints
   Insufficient steering gear lubricant
   Improper tire pressure
   Mixing tires of different construction
3 Alignment faults in the rear suspension are manifested in uneven tire wear or uneven tracking of the rear wheels. This can be easily checked by driving the vehicle straight across a puddle of water onto a dry patch of pavement. If the rear wheels do not follow the front wheels exactly, the alignment should be adjusted.
4 Front or rear wheel alignment should be left to a dealer service department or an alignment shop.

22 Wheels and tires — general information

1 Check the tire pressures (cold) weekly.
2 Inspect the sidewalls and treads periodically for damage and signs of abnormal or uneven wear.
3 Make sure the wheel lug nuts are properly tightened.
4 Do not mix tires of dissimilar construction or tread pattern on the same axle.
5 Never include the temporary spare in the tire rotation pattern as it is designed for use only until a damaged tire is repaired or replaced.
6 Periodically inspect the wheels for elongated or damaged lug holes, distortion and nicks in the rim. Replace damaged wheels.
7 Clean the wheel inside and outside and check for rust and corrosion, which could lead to wheel failure.
8 If the wheel and tire are balanced on the vehicle, one wheel stud and lug hole should be marked whenever the wheel is removed so that it can be reinstalled in the original position. If balanced on the vehicle, the wheel should not be moved to a different axle position.

Chapter 11 Body

Contents

Body — maintenance	2
Body repair — major damage	5
Body repair — minor damage	4
Bumpers — removal and installation	19
Console — removal and installation	15
Door glass — removal and installation	10
Door inside handle — removal and installation	13
Door latch assembly — removal and installation	11
Door outside handle and lock cylinder — removal and installation	12
Door trim panel — removal and installation	9
Front fender — removal and installation	21
Front seat recliner — removal and installation	17
General information	1
Glove box — removal and installation	18
Hinges and locks — maintenance	6
Hood — removal and installation	22
Hood latch and cable — removal and installation	23
Liftgate lock cylinder and latch — removal and installation	25
Liftgate — removal and installation	24
Outside rear view mirror — removal and installation	20
Seats — removal and installation	16
Upholstery and carpets — maintenance	3
Weatherstripping — maintenance and replacement	8
Window regulator — removal and installation	14
Windshield and stationary glass — replacement	7

Specifications

Torque specifications	Ft-lbs	Nm
Door glass-to-lift plate bolts	10	13
Door outside handle mounting nuts	65 in-lbs	7
Hood-to-hinge bolts	9.5	12
Hood latch bolts	9.5	12
Energy absorbing unit-to-body bolts		
front	25	34
rear	20	28
Fender mounting bolts	9.5	12
Front seat bolts and nuts	20	28
Liftgate hinge bolts	9.5	12
Liftgate prop bolts	9.5	12
Rear seat back bolts	26	30

1 General information

These models are of unitized construction. The body is designed to provide vehicle rigidity so that a separate frame is not necessary. Front and rear frame side rails integral with the body support the front end sheet metal, front and rear suspension systems and other mechanical components. Due to this type of construction, it is very important that, in the event of collision damage, the underbody be thoroughly checked at a facility with the proper equipment.

Component replacement and repairs possible for the home mechanic are included in this Chapter.

2 Body — maintenance

1 The condition of your vehicle's body is very important, because it is on this that the second hand value will mainly depend. It is much more difficult to repair a neglected or damaged body than it is to repair mechanical components. The hidden areas of the body, such as the fender wells, the frame, and the engine compartment, are equally important, although obviously do not require as frequent attention as the rest of the body.

2 Once a year, or every 12,000 miles, it is a good idea to have the underside of the body steam cleaned. All traces of dirt and oil will be removed and the underside can then be inspected carefully for rust, damaged brake lines, frayed electrical wiring, damaged cables and other problems. The front suspension components should be greased after completion of this job.

3 At the same time, clean the engine and the engine compartment using either a steam cleaner or a water soluble degreaser.

4 The fender wells should be given particular attention, as undercoating can peel away and stones and dirt thrown up by the tires can cause the paint to chip and flake, allowing rust to set in. If rust is found, clean down to the bare metal and apply an anti-rust paint.

5 The body should be washed once a week (or when dirty). Wet the vehicle thoroughly to soften the dirt, then wash it down with a soft sponge and plenty of clean soapy water. If the surplus dirt is not washed off very carefully, it will in time wear down the paint.

6 Spots of tar or asphalt coating thrown up from the road should be removed with a cloth soaked in solvent.

7 Once every six months, give the body and chrome trim a thorough waxing. If a chrome cleaner is used to remove rust from any of the vehicle's plated parts, remember that the cleaner also removes part of the chrome, so use it sparingly.

3 Upholstery and carpets — maintenance

1   Every three months remove the carpets or mats and clean the interior of the vehicle (more frequently if necessary). Vacuum the upholstery and carpets to remove loose dirt and dust.
2   If the upholstery is soiled, apply upholstery cleaner with a damp sponge and wipe it off with a clean, dry cloth.

4 Body repair — minor damage

See photo sequence

Repair of minor scratches

1   If the scratch is superficial and does not penetrate to the metal of the body, repair is very simple. Lightly rub the scratched area with a fine rubbing compound to remove loose paint and built up wax. Rinse the area with clean water.
2   Apply touch-up paint to the scratch, using a small brush. Continue to apply thin layers of paint until the surface of the paint in the scratch is level with the surrounding paint. Allow the new paint at least two weeks to harden, then blend it into the surrounding paint by rubbing with a very fine rubbing compound. Finally, apply a coat of wax to the scratch area.
3   If the scratch has penetrated the paint and exposed the metal of the body, causing the metal to rust, a different repair technique is required. Remove all loose rust from the bottom of the scratch with a pocket knife, then apply rust inhibiting paint to prevent the formation of rust in the future. Using a rubber or nylon applicator, coat the scratched area with glaze-type filler. If required, the filler can be mixed with thinner to provide a very thin paste, which is ideal for filling narrow scratches. Before the glaze filler in the scratch hardens, wrap a piece of smooth cotton cloth around the tip of a finger. Dip the cloth in thinner and then quickly wipe it along the surface of the scratch. This will ensure that the surface of the filler is slightly hollow. The scratch can now be painted over as described earlier in this section.

Repair of dents

4   When repairing dents, the first job is to pull the dent out until the affected area is as close as possible to its original shape. There is no point in trying to restore the original shape completely as the metal in the damaged area will have stretched on impact and cannot be restored to its original contours. It is better to bring the level of the dent up to a point which is about 1/8-inch below the level of the surrounding metal. In cases where the dent is very shallow, it is not worth trying to pull it out at all.
5   If the back side of the dent is accessible, it can be hammered out gently from behind using a soft-face hammer. While doing this, hold a block of wood firmly against the opposite side of the metal to absorb the hammer blows and prevent the metal from being stretched.
6   If the dent is in a section of the body which has double layers, or some other factor makes it inaccessible from behind, a different technique is required. Drill several small holes through the metal inside the damaged area, particularly in the deeper sections. Screw long, self tapping screws into the holes just enough for them to get a good grip in the metal. Now the dent can be pulled out by pulling on the protruding heads of the screws with locking pliers.
7   The next stage of repair is the removal of paint from the damaged area and from an inch or so of the surrounding metal. This is easily done with a wire brush or sanding disk in a drill motor, although it can be done just as effectively by hand with sandpaper. To complete the preparation for filling, score the surface of the bare metal with a screwdriver or the tang of a file or drill small holes in the affected area. This will provide a good grip for the filler material. To complete the repair, see the Section on filling and painting.

Repair of rust holes or gashes

8   Remove all paint from the affected area and from an inch or so of the surrounding metal using a sanding disk or wire brush mounted in a drill motor. If these are not available, a few sheets of sandpaper will do the job just as effectively.
9   With the paint removed, you will be able to determine the severity of the corrosion and decide whether to replace the whole panel, if possible, or repair the affected area. New body panels are not as expensive as most people think and it is often quicker to install a new panel than to repair large areas of rust.
10   Remove all trim pieces from the affected area except those which will act as a guide to the original shape of the damaged body, such as headlight shells, etc. Using metal snips or a hacksaw blade, remove all loose metal and any other metal that is badly affected by rust. Hammer the edges of the hole inward to create a slight depression for the filler material.
11   Wire brush the affected area to remove the powdery rust from the surface of the metal. If the back of the rusted area is accessible, treat it with rust inhibiting paint.
12   Before filling is done, block the hole in some way. This can be done with sheet metal riveted or screwed into place, or by stuffing the hole with wire mesh.
13   Once the hole is blocked off, the affected area can be filled and painted. See the following subsection on filling and painting.

Filling and painting

14   Many types of body fillers are available, but generally speaking, body repair kits which contain filler paste and a tube of resin hardener are best for this type of repair work. A wide, flexible plastic or nylon applicator will be necessary for imparting a smooth and contoured finish to the surface of the filler material. Mix up a small amount of filler on a clean piece of wood or cardboard (use the hardener sparingly). Follow the manufacturer's instructions on the package, otherwise the filler will set incorrectly.
15   Using the applicator, apply the filler paste to the prepared area. Draw the applicator across the surface of the filler to achieve the desired contour and to level the filler surface. As soon as a contour that approximates the original one is achieved, stop working the paste. If you continue, the paste will begin to stick to the applicator. Continue to add thin layers of paste at 20-minute intervals until the level of the filler is just above the surrounding metal.
16   Once the filler has hardened, the excess can be removed with a body file. From then on, progressively finer grades of sandpaper should be used, starting with a 180-grit paper and finishing with 600-grit wet-or-dry paper. Always wrap the sandpaper around a flat rubber or wooden block, otherwise the surface of the filler will not be completely flat. During the sanding of the filler surface, the wet-or-dry paper should be periodically rinsed in water. This will ensure that a very smooth finish is produced in the final stage.
17   At this point, the repair area should be surrounded by a ring of bare metal, which in turn should be encircled by the finely feathered edge of good paint. Rinse the repair area with clean water until all of the dust produced by the sanding operation is gone.
18   Spray the entire area with a light coat of primer. This will reveal any imperfections in the surface of the filler. Repair the imperfections with fresh filler paste or glaze filler and once more smooth the surface with sandpaper. Repeat this spray-and-repair procedure until you are satisfied that the surface of the filler and the feathered edge of the paint are perfect. Rinse the area with clean water and allow it to dry completely.
19   The repair area is now ready for painting. Spray painting must be carried out in a warm, dry, windless and dust free atmosphere. These conditions can be created if you have access to a large indoor work area, but if you are forced to work in the open, you will have to pick the day very carefully. If you are working indoors, dousing the floor in the work area with water will help settle the dust which would otherwise be in the air. If the repair area is confined to one body panel, mask off the surrounding panels. This will help minimize the effects of a slight mismatch in paint color. Trim pieces such as chrome strips, door handles, etc., will also need to be masked off or removed. Use masking tape and several thicknesses of newspaper for the masking operations.
20   Before spraying, shake the paint can thoroughly, then spray a test area until the spray painting technique is mastered. Cover the repair area with a thick coat of primer. The thickness should be built up using several thin layers of primer rather than one thick one. Using 600-grit wet-or-dry sandpaper, rub down the surface of the primer until it is very smooth. While doing this, the work area should be thoroughly rinsed with water and the wet-or-dry sandpaper periodically rinsed as well. Allow the primer to dry before spraying additional coats.
21   Spray on the top coat, again building up the thickness by using several thin layers of paint. Begin spraying in the center of the repair area and then, using a circular motion, work out until the whole repair area and about two inches of the surrounding original paint is covered. Remove all masking material 10 to 15 minutes after spraying on the

Chapter 11 Body

final coat of paint. Allow the new paint at least two weeks to harden, then use a very fine rubbing compound to blend the edges of the new paint into the existing paint. Finally, apply a coat of wax.

5 Body repair — major damage

1 Major damage must be repaired by an auto body shop specifically equipped to perform unibody repairs. These shops have available the specialized equipment required to do the job properly.
2 If the damage is extensive, the underbody must be checked for proper alignment or the vehicle's handling characteristics may be adversely affected and other components may wear at an accelerated rate.
3 Due to the fact that all of the major body components (hood, fenders, etc.) are separate and replaceable units, any seriously damaged components should be replaced rather than repaired. Sometimes these components can be found in a wrecking yard that specializes in used vehicle components, often at considerable savings over the cost of new parts.

6 Hinges and locks — maintenance

Once every 3000 miles, or every three months, the door and hood hinges and locks should be given a few drops of light oil or lock lubricant. The door striker plates can be given a thin coat of grease to reduce wear and ensure free movement.

7 Windshield and stationary glass — replacement

The windshield and stationary window glass on all models is sealed in place with a special butyl compound. Removal of the existing sealant requires the use of an electric knife specially made for the operation and glass replacement is a complex operation.
   In view of this, it is not recommended that stationary glass removal be attempted by the home mechanic. If replacement is necessary due to breakage or leakage, the work should be referred to your dealer or a qualified glass or body shop.

8 Weatherstripping — maintenance and replacement

1 The weatherstripping should be kept clean and free of contaminants such as gasoline or oil. Spray the weatherstripping periodically with silicone lubricant to reduce abrasion, wear and cracking.
2 The weatherstripping is retained to the doors by adhesive above the vehicle beltline and by clips below it.
3 To remove the weatherstripping, release the plastic clip at the bottom of the door. Work your way around the circumference of the door and carefully pull the weatherstripping free.
4 Clean the channel of any residual adhesive or weatherstripping which would interfere with the installation of the new weatherstripping.
5 Apply a thin coat of a suitable adhesive to the upper portion of the door and install the new weatherstripping, making sure to push it fully into the channel and secure the clips.

9 Door trim panel — removal and installation

Refer to illustration 9.1, 9.2a, 9.2b, 9.5, 9.6, 9.8 and 9.9

1 Remove the window crank (see illustration).
2 Pry off the door handle cover (see illustrations).
3 On vehicles with power locks, carefully pry out the power switch bezel.
4 Pry off the outside mirror control (if equipped) and unplug it.

9.1 After removing the retaining screw the window crank can be pulled off

9.2a Use a small screwdriver to pry out the door handle cover, . . .

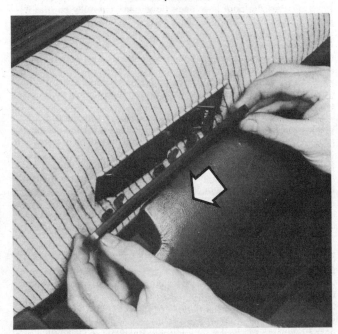

9.2b . . . then pull it out and push it to the rear

204　　　　　　　　　　　　　　　　　　Chapter 11　Body

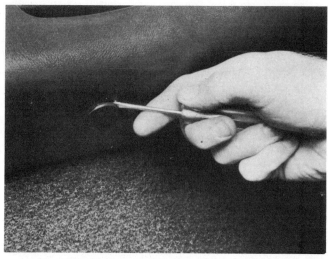

9.5　Pry the screw covers out with a small screwdriver

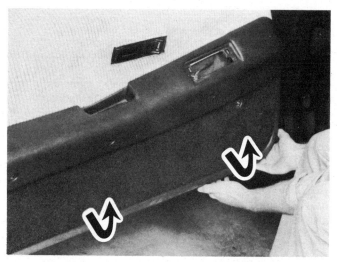

9.6　Pull the door panel out at the bottom and lift up

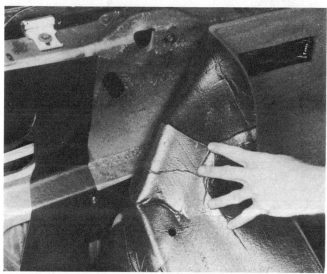

9.8　Peel the door liner back slowly and carefully so that it isn't torn

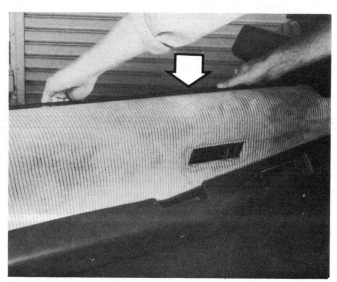

9.9　Start by seating the panel hooks in the top of the door, then push the panel down

5 Pry out the screw covers and remove the six door panel retaining screws (see illustration).
6 Disengage the clips around the edges of the panel using a screwdriver. Rotate the panel up from the bottom and remove it from the door (see illustration).
7 Disconnect the door courtesy light wires (if equipped).
8 If access to the door is required, carefully peel off the plastic and foam liner (see illustration). When reinstalling, it may be necessary to use adhesive to hold the liner to the door.
9 To install, hang the panel in place on the two hooks which go into the door at the top, push down to seat it and then press the lower clips securely into the door (see illustration).
10 The remainder of installation is the reverse of removal.

10　Door glass — removal and installation

Refer to illustrations 10.3 and 10.4

1 Lower the glass as far as possible. On power window models, disconnect the negative battery cable.
2 Refer to Section 9 and remove the trim panel and the liner.
3 Remove the three glass-to-regulator nuts, raising the glass if necessary to gain access to the nuts through the lower access hole in the door. Remove the glass stabilizers (see illustration).

4 Carefully detach the glass from the regulator assembly and remove it from the door through the belt opening (see illustration).
5 Insert the glass into the door and attach the nuts and washers finger tight.
6 Attach the glass stabilizers with the screws finger tight.
7 Loosen the glass track mounting screws.
8 Connect the battery cable (power windows) or install the window crank (manual windows) and raise the glass to the full up position, guiding the glass so it doesn't bind.
9 Beginning with the center nut, tighten the lift plate-to-glass attachment nuts securely.
10 Adjust the inner glass stabilizers so they lightly touch the glass and then tighten them securely.
11 Lower the glass to the full down position and disconnect the battery negative cable or remove the crank.
12 Install the trim panel.

11　Door latch assembly — removal and installation

Refer to illustrations 11.3 and 11.6

1 Lower the window.
2 On electric window models, disconnect the negative battery cable.
3 Refer to Section 9 and remove the door trim panel. Peel the liner

Chapter 11 Body

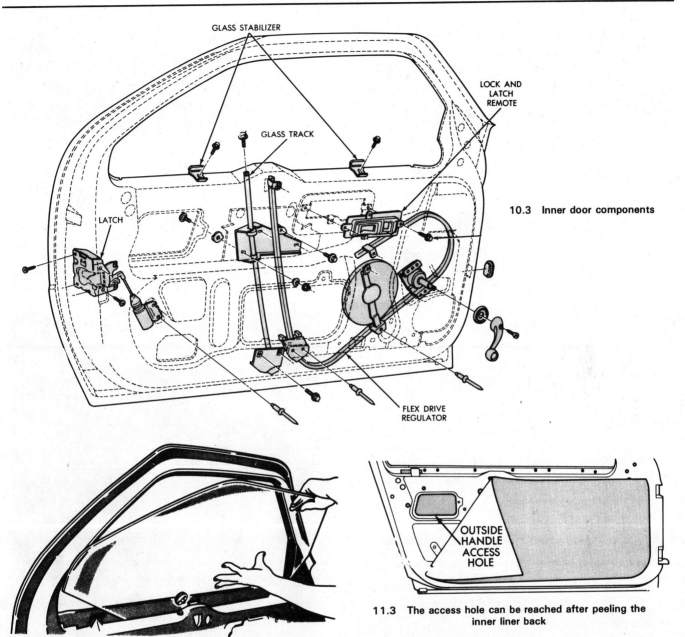

10.3 Inner door components

10.4 Slide the glass up and in through the door opening

11.3 The access hole can be reached after peeling the inner liner back

back partially for access to the outside handle access hole (see illustration).
4 Install the window crank temporarily (manual windows) or connect the battery (electric windows) and raise the window. Disconnect the battery negative cable (electric windows).
5 Disconnect all of the links at the latch (pry the clip off the rod and rotate it to disengage the link).
6 Remove the three Torx screws (see illustration) from the door end panel and detach the latch assembly. On electric window models it will be necessary to drill out the two rivets retaining the lock motor to the door and remove the motor along with the latch.
7 Installation is the reverse of removal. On electric window models, use two short 1/4-20 bolts and nuts to replace the rivets holding the motor to the door.

12 Door outside handle and lock cylinder — removal and installation

Refer to illustration 12.3a, 12.3b and 12.5

1 Lower the window and disconnect the battery.
2 Refer to Section 9 and remove the door trim panel. Peel back the

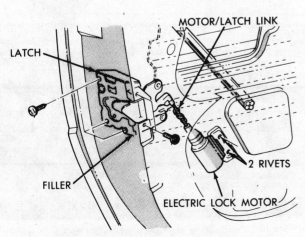

11.6 Door latch connection details

Chapter 11 Body

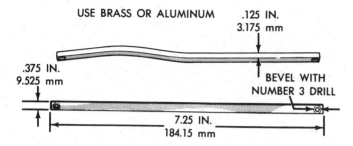

12.3a Illuminated entry switch tool details

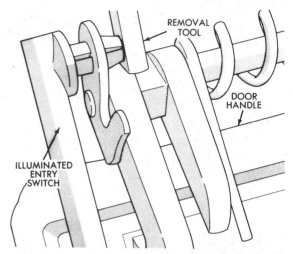

12.3b Removing the illuminated entry switch

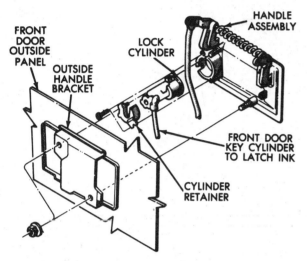

12.5 Outside door handle and lock cylinder details

liner to expose the rear of the major access hole.
3   Remove the illuminated entry switch (if equipped) with a fabricated tool (see illustrations).
4   Disconnect the handle links from the latch and the lock.
5   Remove the two retaining nuts, pull the handle out and remove the lock cylinder from the handle (see illustration).
6   Place the key cylinder in the handle, install the retainer and tighten the screws securely.
7   Install the handle in the door and tighten the nuts securely.
8   Connect the links to the handle and the lock.
9   Install the illuminated entry switch.
10 Install the door trim panel.

13   Door inside handle — removal and installation

Refer to illustrations 13.5a and 13.5b
1   The door inside handle is often called the remote control handle.
2   Raise the glass all the way (fully closed position).
3   Refer to Section 9 and remove the door trim panel and liner.
4   Remove the two retaining screws.
5   Pull the handle assembly out, disconnect the links and rotate the handle away from the door (see illustrations).
6   Installation is the reverse of removal.

14   Window regulator — removal and installation

Refer to illustration 14.5
1   Refer to Section 9 and remove the door trim panel and peel back the liner.
2   Remove the glass from the door (Section 10).
3   Disconnect the electrical regulator wiring harness and remove the clip from inside the panel.
4   Use an electric drill and an appropriate size bit to drill out the

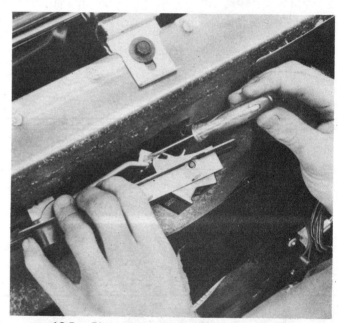

13.5a   Disconnect the front link with a small screwdriver . . .

13.5b   . . . and rotate the handle assembly to disconnect the second link

Chapter 11 Body

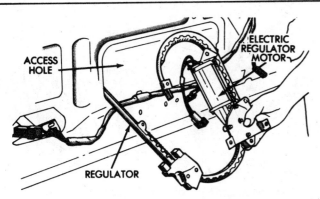

14.5 Lower the regulator assembly through the access hole

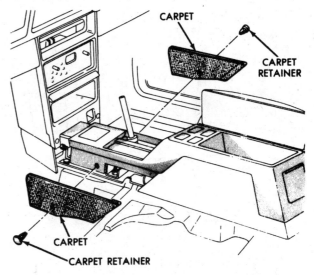

15.1 Console carpet details

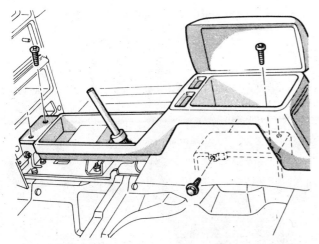

15.2 Console mounting details

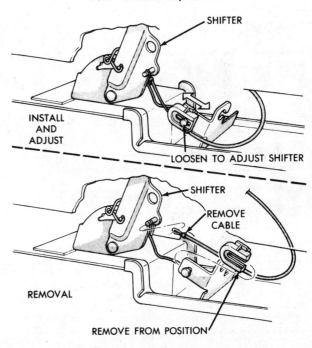

15.5 Automatic transaxle shifter details

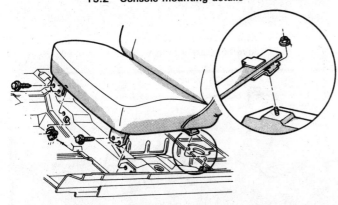

16.2 Front seat mounting bolts

regulator mounting rivet heads (eight for manual regulator and seven for power regulator).
5 Rotate the regulator and withdraw it through the major access hole (see illustration).
6 Installation is the reverse of removal. Make sure the flex drive is clean and well lubricated before installation. Use 1/2-inch long 1/4–20 bolts and nuts in place of the rivets. Tighten them securely.

15 Console — removal and installation

Refer to illustration 15.1, 15.2 and 15.5

1 Pry out the two retainers and remove the carpet from both sides of the console (see illustration)
2 Remove the two bolts and four retaining screws (see illustration).
3 Lift the console sufficiently for access to the wiring and unplug the connectors.
4 On manual transaxle models, unscrew the shift knob.
5 On automatic transaxle models, disconnect the shifter cable (see illustration).
6 Remove the console from the vehicle.
7 Installation is the reverse of removal. Be sure to adjust the automatic transaxle shifter so that the engine starts only in Park or Neutral (Chapter 7).

16 Seats — removal and adjustment

Refer to illustrations 16.2 and 16.5

Front

1 Move the seat to the forward position.
2 Remove the nuts retaining the rear of the seat to the floor, then move the seat to the rear and remove the front seat mounting bolts (see illustration).

208   Chapter 11 Body

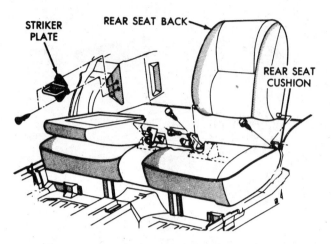

16.5 Rear seat installation details

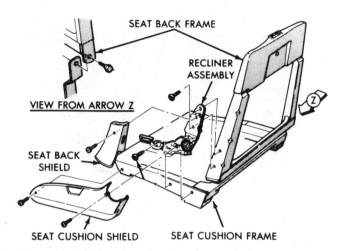

17.1 Front seat recliner details

18.2a The glove box assembly retaining screws are located around the outer edges

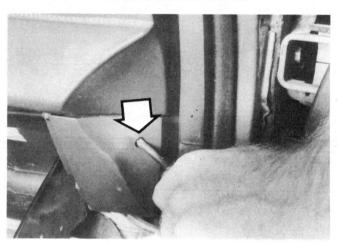

18.2b After prying the plastic cover off, the right side glove box screw (arrow) can be removed

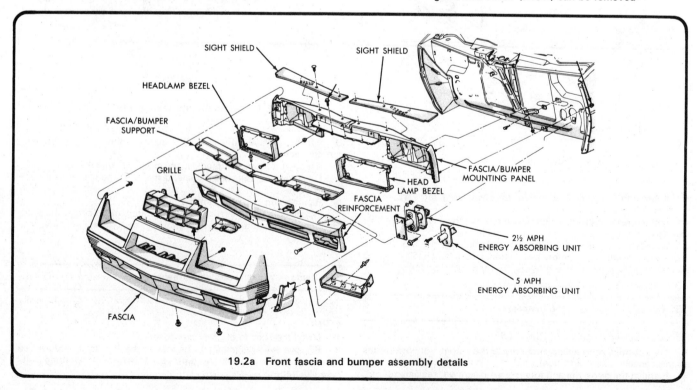

19.2a Front fascia and bumper assembly details

Chapter 11 Body

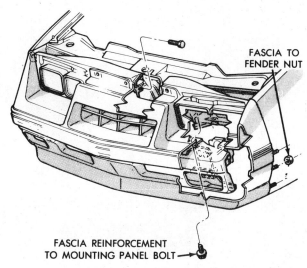

19.2b Fascia fender and reinforcement attachment details

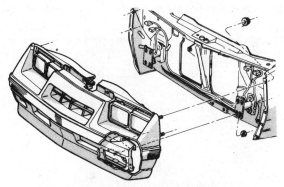

19.5a Standard model fascia-to-fender nuts

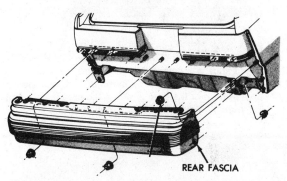

19.12a Rear fascia installation details (standard models)

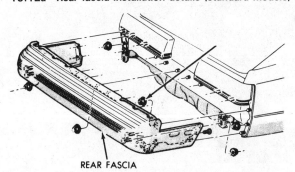

19.12b Turbo Z rear fascia installation details

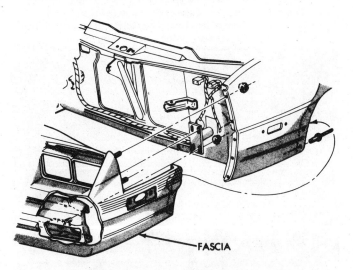

19.5b Turbo Z model fascia details

3 Remove the seat.
4 Installation is the reverse of removal.

Rear

5 Remove the seat cushion-to-floor pan bolts (see illustration).
6 Lift the seat assembly from the vehicle.
7 Installation is the reverse of removal.

17 Front seat recliner — removal and installation

Refer to illustration 17.1

1 Remove the seat cushion shield, seat back shield and recliner mechanism retaining screws (see illustration)
2 Detach the recliner mechanism from the seat.
3 Installation is the reverse of removal.

18 Glove box — removal and installation

Refer to illustration 18.2a and 18.2b

1 Open the glove box.
2 Remove the retaining screws (see illustrations).
3 Detach the electrical connector and lower the glove box from the dash panel.
4 Installation is the reverse of removal.

19 Bumpers — removal and installation

Refer to illustration 19.2a, 19.2b, 19.5a, 19.5b, 19.12a, 19.12b and 19.13

1 Raise the front of the vehicle and support it securely.

Front bumper

2 Remove the upper fascia-to-fascia reinforcement bolt, followed by the two fascia (see illustrations).
3 Remove the side marker lamps.
4 Remove the fascia reinforcement-to-energy absorber nuts.
5 Remove the six lower fascia-to-fender nuts (standard models) or drill out the six facia retaining rivets (Turbo Z models) (see illustrations).
6 Remove the four upper fascia-to-fender nuts.
7 Unplug the headlamp connectors.
8 Remove the fascia assembly.
9 Remove the energy absorbing unit-to-body bolts and withdraw the units from the body.
10 Installation is the reverse of removal, taking care to tighten the energy absorbing unit bolts to the specified torque.

Rear bumper

11 Support the fascia with a jack.
12 Remove the energy absorbing unit-to-fascia nuts and lower the fascia from the vehicle (see illustrations).

These photos illustrate a method of repairing simple dents. They are intended to supplement *Body repair - minor damage* in this Chapter and should not be used as the sole instructions for body repair on these vehicles.

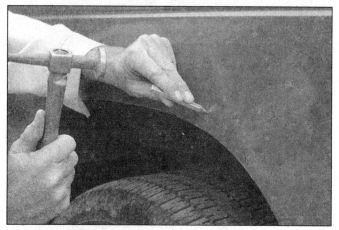

1 If you can't access the backside of the body panel to hammer out the dent, pull it out with a slide-hammer-type dent puller. In the deepest portion of the dent or along the crease line, drill or punch hole(s) at least one inch apart . . .

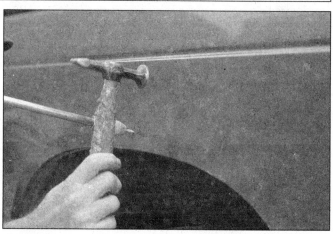

2 . . . then screw the slide-hammer into the hole and operate it. Tap with a hammer near the edge of the dent to help 'pop' the metal back to its original shape. When you're finished, the dent area should be close to its original contour and about 1/8-inch below the surface of the surrounding metal

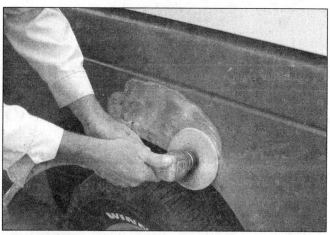

3 Using coarse-grit sandpaper, remove the paint down to the bare metal. Hand sanding works fine, but the disc sander shown here makes the job faster. Use finer (about 320-grit) sandpaper to feather-edge the paint at least one inch around the dent area

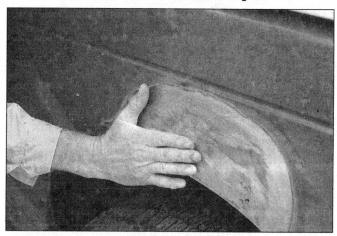

4 When the paint is removed, touch will probably be more helpful than sight for telling if the metal is straight. Hammer down the high spots or raise the low spots as necessary. Clean the repair area with wax/silicone remover

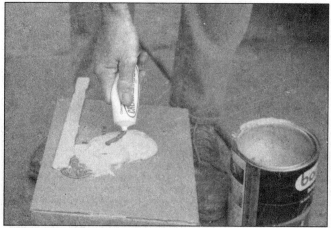

5 Following label instructions, mix up a batch of plastic filler and hardener. The ratio of filler to hardener is critical, and, if you mix it incorrectly, it will either not cure properly or cure too quickly (you won't have time to file and sand it into shape)

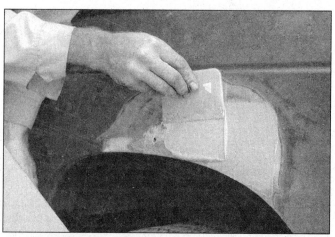

6 Working quickly so the filler doesn't harden, use a plastic applicator to press the body filler firmly into the metal, assuring it bonds completely. Work the filler until it matches the original contour and is slightly above the surrounding metal

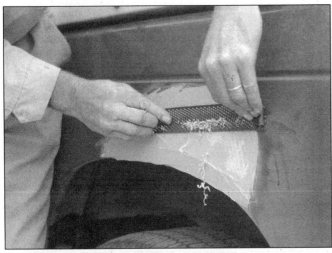

7 Let the filler harden until you can just dent it with your fingernail. Use a body file or Surform tool (shown here) to rough-shape the filler

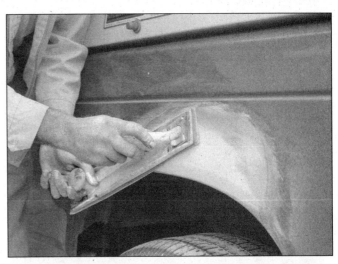

8 Use coarse-grit sandpaper and a sanding board or block to work the filler down until it's smooth and even. Work down to finer grits of sandpaper - always using a board or block - ending up with 360 or 400 grit

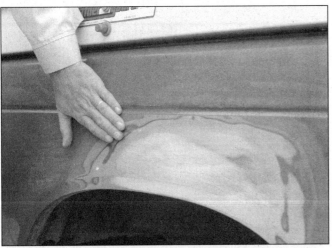

9 You shouldn't be able to feel any ridge at the transition from the filler to the bare metal or from the bare metal to the old paint. As soon as the repair is flat and uniform, remove the dust and mask off the adjacent panels or trim pieces

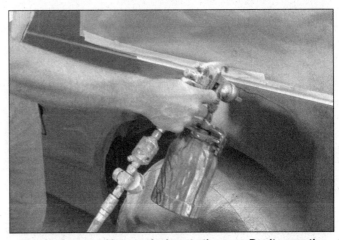

10 Apply several layers of primer to the area. Don't spray the primer on too heavy, so it sags or runs, and make sure each coat is dry before you spray on the next one. A professional-type spray gun is being used here, but aerosol spray primer is available inexpensively from auto parts stores

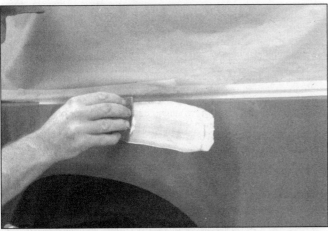

11 The primer will help reveal imperfections or scratches. Fill these with glazing compound. Follow the label instructions and sand it with 360 or 400-grit sandpaper until it's smooth. Repeat the glazing, sanding and respraying until the primer reveals a perfectly smooth surface

12 Finish sand the primer with very fine sandpaper (400 or 600-grit) to remove the primer overspray. Clean the area with water and allow it to dry. Use a tack rag to remove any dust, then apply the finish coat. Don't attempt to rub out or wax the repair area until the paint has dried completely (at least two weeks)

Chapter 11 Body

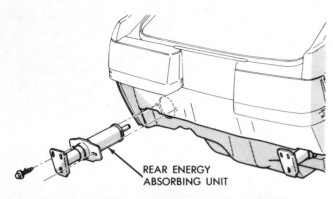

19.13 Rear energy absorbing unit installation details

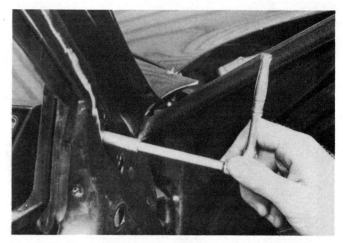

20.3 Use a socket and extension to remove the mirror retaining nuts

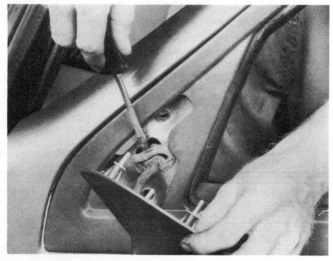

20.4 Pry the wiring grommet out of the door opening with a screwdriver

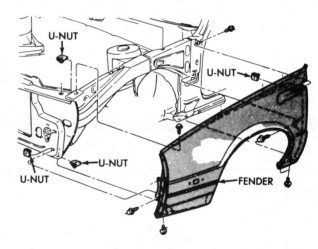

21.6 Fender mounting details

13 Remove the energy absorbing unit-to-body bolts and withdraw the units from the body (see illustration).
14 Installation is the reverse of removal, taking care to tighten the energy absorbing unit bolts to the specified torque.

20 Outside rear view mirror — removal and installation

Refer to illustrations 20.3 and 20.4

1 Remove the door trim panel.
2 Unplug the mirror electrical connector.
3 Pry out the three rubber plugs and remove the mirror retaining nuts (see illustration).
4 Use a screwdriver to pry the rubber grommet out of the door and remove the mirror from the vehicle (see illustration).
5 Installation is the reverse of removal.

21 Front fender — removal and installation

Refer to illustration 21.6

1 Unplug the park, turn signal and headlamp wiring.
2 Cover the forward edge of the door so that it will not be damaged during fender removal.
3 Remove the plastic fasteners retaining the splash shield to the fender.
4 Remove the fascia-to-fender retaining nuts.
5 Raise the hood and secure it in the open position.
6 Remove the fasteners and detach the fender from the vehicle (see illustration).

7 Installation is the reverse of removal. The gaps at the cowl, door front edge and door top edge should be equal.

22 Hood — removal and installation

1 Outline the hinges on the hood to simplify reinstallation and alignment.
2 Place a protective covering (an old blanket should work fine) over the windshield area and make sure it extends down over the ends of both fenders.
3 Place a block of wood between the hood and windshield to prevent sudden rearward movement of the hood.
4 Have an assistant support one side of the hood as the bolts are removed, then remove the remaining bolts and detach the hood. Store it where it will not be damaged.
5 Installation is the reverse of removal. Be sure to align the hinges inside the marks made during removal before tightening the bolts.

23 Hood latch and cable — removal and installation

Refer to illustration 23.1

Latch

1 Disconnect the release cable from the latch, mark the retaining bolt location, remove the bolts and lift the latch away from the radiator brace (see illustration).
2 To install, place the latch in position and install the bolts in the marked locations. Connect the release cable.

Chapter 11 Body

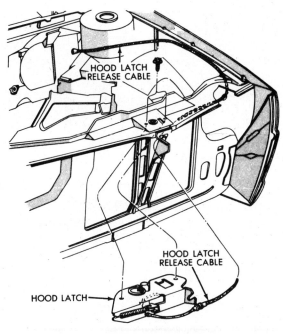

23.1 Hood latch and cable details

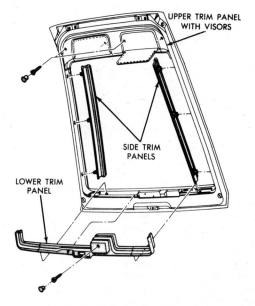

24.4 Liftgate trim panel details

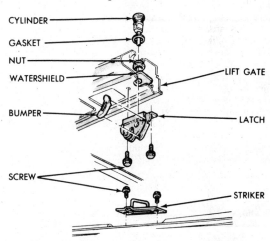

25.1 Liftgate lock cylinder and latch details

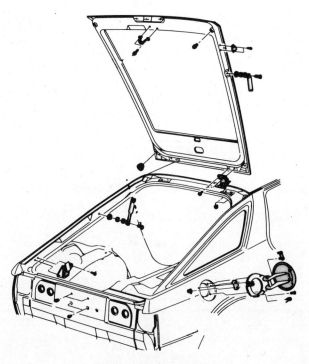

24.6 Liftgate details

24 Liftgate — removal and installation

Refer to illustrations 24.4 and 24.6

1 Support the liftgate in the fully open position.
2 Outline the hinges on the liftgate to simplify reinstallation and alignment.
3 Apply masking tape to the upper edge of the door and the rear edge of the roof to prevent damage to the paint during liftgate removal.
4 Remove the trim panel (see illustration) and disconnect the wires, then thread them out through the liftgate openings.
5 Remove the lift prop fasteners and detach the props. The liftgate must be supported by some other means from this point on.
6 While an assistant is supporting the liftgate, remove the hinge-to-liftgate bolts and detach the liftgate (see illustration).
7 Installation is the reverse of removal. Use rope-type sealer to seal the hinges to the liftgate.

25 Liftgate lock cylinder and latch — removal and installation

Refer to illustration 25.1

1 The lock cylinder and latch are accessible from the inside after removing the trim panel (see illustration).
2 The lock cylinder is held in place with one nut.
3 The latch is held in place with two screws.

Hood latch actuating cable

3 Release the cable from the hood latch by pulling the cable toward the latch and lifting the ball end from the slot.
4 In the passenger compartment, disconnect the cable at the hood latch lever by prying the retaining clip off with a small screwdriver. Connect a piece of thin wire to the cable.
5 Pull the cable through the firewall into the engine compartment.
6 Connect the wire to the new cable and pull it through the firewall into the passenger compartment.
7 Connect the cable at the latch lever and the hood latch.

Chapter 12 Chassis electrical system

Contents

Antenna — removal and installation	12
Dimmer switch — removal, adjustment and installation	15
Electrical troubleshooting — general information	2
Electronic instrument cluster — general information	29
Flasher units — replacement	5
Front exterior bulbs — replacement	9
Fuses and fusible links — replacement	3
General information	1
Headlight — alignment	8
Headlight switch — removal and installation	19
Headlight — removal and installation	6
Headlight adjuster screw — replacement	7
Horn — check and adjustment	4
Ignition lock cylinder (fixed column) — removal and installation	14
Ignition switch — removal and installation	13
Instrument cluster (conventional instrument panel) — removal and installation	24
Instrument cluster bezel (conventional instrument panel) — removal and installation	26
Instrument cluster bulbs (conventional instrument panel) — replacement	28
Instrument cluster mask (conventional instrument panel) — removal and installation	27
Instrument cluster printed circuit board (conventional instrument panel) — removal and installation	25
Key buzzer/chime switch (fixed column) — removal and installation	18
Liftgate wiper motor — removal and installation	31
Oil temperature and pressure gauge (conventional instrument panel) — removal and installation	23
Radio — removal and installation	11
Rear exterior bulbs — replacement	10
Speedometer (conventional instrument panel) — removal and installation	20
Tachometer (conventional instrument panel) — removal and installation	21
Turn signal/hazard warning switch (fixed column) — removal and installation	17
Voltmeter and fuel gauge (conventional instrument panel) — removal and installation	22
Washer/wiper switch (fixed column) — removal and installation	16
Windshield wiper motor — removal and installation	30
Wiring diagrams	See end of Manual

Specifications

Bulb application

Interior	Number
Air conditioner control	
1984 models	74
1985 and 1986 models	161
Ashtray	161
Brake system warning indicator	194
Elapsed timer	161
Instrument cluster	194
Radio	74
Heater	161
Hatch open	74 or 161
Navigator	37
Reading light	912
High beam/turn signal indicators	194
Ignition light	
1984 models	1445
1985 and 1986 models	53
Dome lights	212-2
Door lock illuminated entry	158
Low washer fluid indicator lights	74 or 161
Oil/low fuel/coolant temperature indicator lights	194
Seat belt/door open indicator lights	194
Switch callouts	161
Rear cargo light	212-2
Underhood light	
1984 models	1003
1985 and 1986 models	105
Visor vanity light	194
Voltmeter	194

Chapter 12 Chassis electrical system

Exterior
Headlights
- high beam PH4656
- low beam PH4651

Front parking and turn signal lights 2057
Rear stop and turn signal 2057
Rear license and side marker lights 168
Back-up lights 1156

Fusible link wire color code — Wire gauge
- Black 12
- Red 14
- Dark blue 16
- Gray 18
- Orange 20

1 General information

This Chapter covers repair and service procedures for the various lighting and electrical components not associated with the engine. Information on the battery, alternator, voltage regulator, ignition and starting systems can be found in Chapter 5.

The electrical system is a 12-volt negative ground type. Power for the electrical system and accessories is supplied by a lead/acid type battery which is charged by the alternator. The circuits are protected from overload by a system of fuses and fusible links. **Note:** *Whenever the electrical system is worked on, the negative battery cable should be disconnected to prevent electrical shorts and/or fires.*

2 Electrical troubleshooting — general information

A typical electrical circuit consists of an electrical component, any switches, relays, motors, etc. related to that component and the wiring and connectors that connect the component to both the battery and the chassis. To aid in locating a problem in any electrical circuit, wiring diagrams are included at the end of this book.

Before tackling any troublesome electrical circuit, first study the appropriate diagrams to get a complete understanding of what makes up that individual circuit. Trouble spots, for instance, can often be narrowed down by noting if other components related to that circuit are operating properly or not. If several components or circuits fail at one time, chances are the problem lies in the fuse or ground connection, as several circuits often are routed through the same fuse and ground connections.

Electrical problems often stem from simple causes, such as loose or corroded connections, a blown fuse or melted fusible link. Prior to any electrical troubleshooting, always visually check the condition of the fuse, wires and connections in the problem circuit.

If testing instruments are going to be utilized, use the diagrams to plan ahead of time where you will make the necessary connections in order to accurately pinpoint the trouble spot.

The basic tools needed for electrical troubleshooting include a circuit tester or voltmeter (a 12-volt bulb with a set of test leads can also be used), a continuity tester, which includes a bulb, battery and set of test leads, and a jumper wire, preferably with a circuit breaker incorporated, which can be used to bypass electrical components.

Voltage checks should be performed if a circuit is not functioning properly.

Connect one lead of a circuit tester to either the negative battery terminal or a known good ground. Connect the other lead to a connector in the circuit being tested, preferably nearest to the battery or fuse. If the bulb of the tester goes on, voltage is reaching that point, which means the part of the circuit between that connector and the battery is problem free. Continue checking along the entire circuit in the same fashion. When you reach a point where no voltage is present, the problem lies between there and the last good test point. Most of the time the problem is due to a loose connection. **Note:** *Keep in mind that some circuits receive voltage only when the ignition key is in the Accessory or Run position.*

A method of finding shorts in a circuit is to remove the fuse and connect a test light or voltmeter in its place to the fuse terminals. There should be no load in the circuit. Move the wiring harness from side-to-side while watching the test light. If the bulb goes on, there is a short to ground somewhere in that area, probably where insulation has rubbed off of a wire. The same test can be performed on other components of the circuit, including the switch.

A ground check should be done to see if a component is grounded properly. Disconnect the battery and connect one lead of a self-powered test light, such as a continuity tester, to a known good ground. Connect the other lead to the wire or ground connection being tested. If the bulb goes on, the ground is good. If the bulb does not go on, the ground is not good.

A continuity check is performed to see if a circuit, section of circuit or individual component is passing electricity properly. Disconnect the battery and connect one lead of a self-powered test light, such as a continuity tester, to one end of the circuit. If the bulb goes on, there is continuity, which means the circuit is passing electricity properly. Switches can be checked in the same way.

Remember that all electrical circuits are composed basically of electricity running from the battery, through the wires, switches, relays, etc. to the electrical component (light bulb, motor, etc.). From there it is run to the body (ground), where it is passed back to the battery. Any electrical problem is basically an interruption in the flow of electricity to and from the battery.

3 Fuses and fusible links — replacement

Refer to illustrations 3.2 and 3.6

Caution: *Do not bypass a fuse with metal or aluminum foil as serious damage to the electrical system could result.*

1 The fuse block is located below the dash, to the left of the steering column and directly above the hood release handle, behind a panel. Put your finger in the notch at the bottom of the panel and pull it back sharply to expose the fuse block.

2 With the ignition off, remove each fuse in turn by grasping it and pulling it from the block (see illustration). Replace the blown fuse with a new one of the same value by pushing it into place. Install the cover panel.

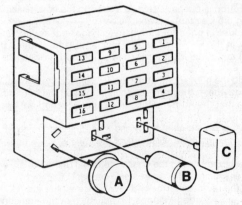

3.2 Fuse block component layout

- A Flasher
- B Ignition time delay relay
- C Horn relay

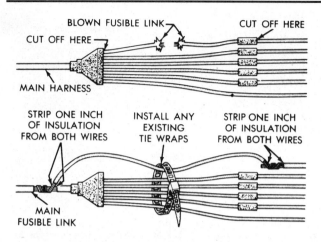

3.6 Multiple fusible link repair details

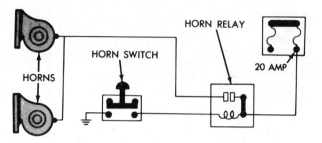

4.2 Conventional horn wiring diagram

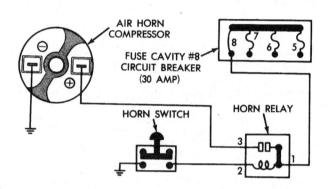

4.12 Air horn wiring diagram

3   If the fuses in the fuse block are not blown and the headlights or other components are inoperable, check the fusible links located in the wiring harness adjacent to the left suspension strut tower. Before replacing a fusible link, determine the reason that it burned out. Replacing the link without finding and correcting the cause for the failure could lead to serious damage to the electrical system.
4   Determine the proper gauge for the replacement link by referring to the Specifications section. Obtain the new link or link wire from your dealer.
5   Disconnect the negative battery cable and cut off all of the remaining burned out fusible link.
6   Strip off one inch of insulation from both ends of the new fusible link and the main harness wire (see illustration).
7   Install the new fusible link by twisting the ends securely around the main link wire. On multiple links, the replacement wire must be connected to the main wires beyond the connector insulators of the old link.
8   Solder the wires together with non-acid core solder. After the connection has cooled, wrap the splice with at least three layers of electrical tape.

4   Horn - check and adjustment

Refer to illustrations 4.2, and 4.12

Conventional horn

1   If the horn will not sound, release the parking brake, place the transaxle lever selector in Park or Neutral and observe the brake light on the dash as you start the engine. If the light does not illuminate, the steering column is not properly grounded to the instrument panel so the horn switch is not grounded.
2   If the brake lamp lights but the horn still does not sound, check for a blown fuse (see illustration). Should the new fuse blow out when the horn button is pushed, there is a short in the horn assembly itself or between the fuse terminal and the horn.
3   If the fuse is good and the horn still does not sound, unplug the connector at the horn and insert a test lamp lead. Ground the other lamp lead and note whether the lamp lights. If it does, the horn is faulty.
4   Should the horn sound continuously, replace the horn relay with a known good one. If the horn still sounds, pull off the horn button and make sure that the horn contact wire is not shorting out against the hub.
5   To adjust the horn loudness and tone, first determine which horn is in need of adjustment. Disconnect the horn which is not being adjusted.
6   Connect the horn to the positive terminal of the battery with a remote starter switch and an ammeter in series.
7   With the remote starter switch depressed, the ammeter should read between 4.5 and 5.5 amps. To adjust, turn the adjusting screw clockwise to decrease or counterclockwise to increase the current. Check the horn for satisfactory tone and current draw after each adjustment.

Air horn

8   If the air horn sounds continuously or intermittently, see conventional horn checking procedures.
9   If the air horn has poor sound or no sound, disconnect the hose to the horn and connect it to an external air supply of 7 to 15 psi. If the horn still has poor sound or no sound, replace the horn.
10   If the horn sounds good, check the compressor for operation when the horn button is activated. It should have an output of 7 to 15 psi. If the compressor is operating but output is low, replace the compressor.
11   If the compressor is inoperative, check for battery voltage at the red wire. Make sure the compressor is grounded properly (black wire). If battery voltage is present at the red wire, the ground is good, and the compressor is still inoperative, replace the compressor.
12   If there is no voltage present at the compressor, check the circuit breaker fuse cavity #8. If the circuit breaker is good, check the relay, horn switch, and related wiring (see illustration).

5   Flasher units - replacement

1   The hazard and turn signal flasher units are located on or adjacent to the fuse block.
2   On 1984 models, the turn signal flasher is located on the lower left corner of the fuse block and can be replaced by removing it and plugging in a new unit (see illustration 3.2).
3   On 1985 and 1986 models the turn signal flasher is clipped to the instrument panel below the fuse block.
4   The hazard flasher on 1984 models is taped to the main wiring harness and on later models it is clipped to the instrument panel below the fuse block.

6   Headlight - removal and installation

Refer to illustrations 6.1 and 6.4

1   Remove the headlight bezel (see illustration).
2   Remove the retaining ring screws, taking care not to disturb the adjustment screws.
3   Withdraw the headlight just enough to gain access to the connector.

Chapter 12 Chassis electrical system

6.1 Headlight bezel mounting screws (arrows)

6.4 Hold on to the headlight securely when detaching the wiring connector from the terminals at the rear

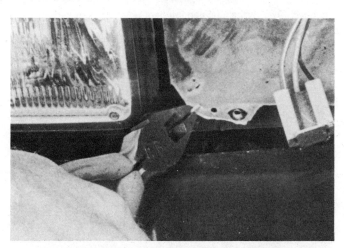

7.2 Pull the headlight bucket spring out to detach it

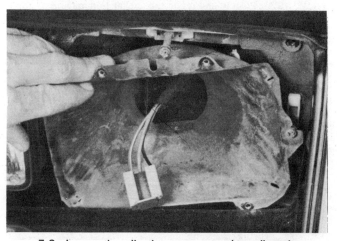

7.3 Loosen the adjusting screw enough to allow the headlight bucket to be removed from the opening

7.4 Use a screwdriver to release the adjusting screw locking tabs

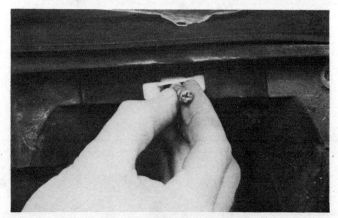

7.5 Press the new adjusting screw in until it locks in place

4 Unplug the connector and remove the headlight (see illustration).
5 To install, plug in the headlight, place it in position and install the retaining screws.
6 Install the headlight bezel.

7 Headlight adjuster screw — replacement

Refer to illustrations 7.2, 7.3, 7.4 and 7.5
1 Remove the headlight (Section 6).
2 Use pliers to remove the retaining spring (see illustration).
3 Remove the headlight bucket (see illustration).
4 Use a small screwdriver to push back the locking tabs and rock the adjusting screw back-and-forth while pulling to remove it (see illustration).
5 To install the new screw, push it securely into place (see illustration).
6 Install the headlight bucket and sealed beam. Adjust the headlights as described in Section 8 or have headlight adjustment checked and adjusted by a properly equipped shop as soon as possible.

Chapter 12 Chassis electrical system

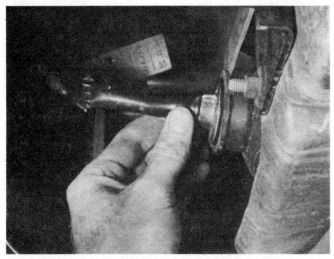

9.1 Push in and rotate the socket to remove it

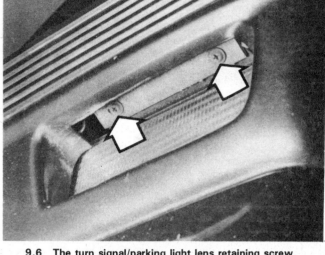

9.6 The turn signal/parking light lens retaining screw locations (arrows)

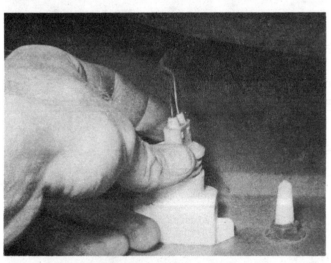

9.7 Turn the front side marker bulb holder counterclockwise to remove it

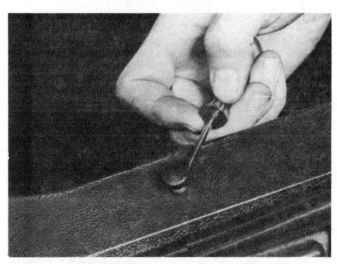

10.2a Pry the upper trim cover clips out . . .

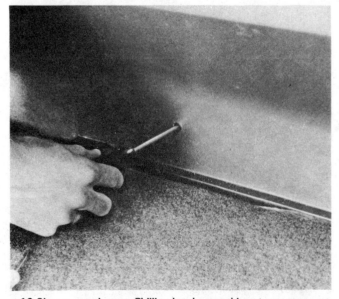

10.2b . . . and use a Phillips head screwdriver to remove the lower cover screws

8 Headlight — alignment

1 It is always best to have the headlights aligned with the proper equipment, but the following procedure may be used.
2 Position the vehicle on level ground 10 feet in front of a dark wall or board. The wall or board must be at right angles to the center line of the vehicle.
3 Draw a vertical line on the wall or board in line with the centerline of the vehicle.
4 Bounce the vehicle on its suspension to ensure that it settles at the proper level and check the tires to make sure that they are at the proper pressure. Measure the height between the ground and the center of the headlights.
5 Draw a horizontal line across the board or wall at this measured height. Mark a cross on the horizontal line on either side of the vertical centerline at the distance between the center of the light and the centerline of the vehicle.
6 Turn the headlights on and switch them to High beam.
7 Use the adjusting screws to align the center of each beam with the crosses which were marked on the horizontal line.
8 Bounce the vehicle on its suspension again to make sure the beams return to the correct position. Check the operation of the dimmer switch.
9 The headlights should be adjusted with the proper equipment at the earliest opportunity.

Chapter 12 Chassis electrical system

10.3 Remove the bulb holder socket by pushing it in and turning it

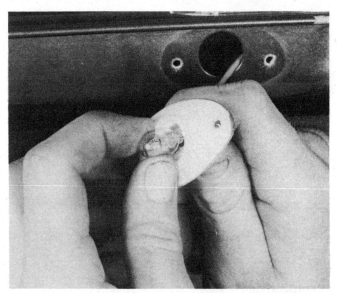

10.7 To remove the license plate light bulb, grasp it securely and withdraw it from the socket

9 Front exterior bulbs — replacement

Refer to illustrations 9.1, 9.6 and 9.7

Turn signal and parking light

1 From under the front of the vehicle, grasp the bulb socket, push in and rotate it in a counterclockwise direction to remove it from the housing (see illustration).
2 Push the bulb in, turn it counterclockwise and withdraw it from the socket.
3 Lubricate the contact area of the new bulb with light grease or petroleum jelly prior to installation.
4 Press the bulb in and rotate it clockwise to install it in the socket.
5 Place the socket in position with the tabs aligned with those in the housing, press in and turn it clockwise.
6 To replace the lens, remove the two screws (see illustration).

Front side marker light

7 From underneath, rotate the bulb holder counterclockwise and withdraw it from the housing (see illustration).
8 Remove the bulb from the holder.

10 Rear exterior bulbs — replacement

Refer to illustrations 10.2a, 10.2b, 10.3 and 10.7

Tail, stop, turn signal and back-up lights

1 Open the liftgate.
2 Remove the trim cover (see illustrations).
3 Remove the appropriate socket by rotating it counterclockwise and withdrawing it from the housing (see illustration). Remove the bulb from the socket.
4 Install the new bulb by pressing in and rotating it clockwise.
5 Install the socket by aligning the tabs, pushing in and turning it clockwise.

License plate light

6 Remove the screws and detach the cover.
7 Grasp the bulb securely and pull it from the socket (see illustration).
8 Insert the new bulb and install the cover and screws.

12.3 Removing the radio antenna mast

11 Radio — removal and installation

1 Disconnect the negative cable at the battery. Place the cable out of the way so it cannot accidentally come in contact with the negative terminal of the battery, as this would once again allow power into the electrical system of the vehicle.
2 Remove the screws from the bottom of the console trim bezel and lift the bezel from the console.
3 Remove the two retaining screws and pull the radio out sufficiently to unplug the connectors and disconnect the ground cable and antenna.
4 Remove the radio from the dash.
5 Installation is the reverse of removal.

12 Antenna — removal and installation

Refer to illustrations 12.3 and 12.4

1 Disconnect the negative cable at the battery. Place the cable out of the way so it cannot accidentally come in contact with the negative terminal of the battery, as this would once again allow power into the electrical system of the vehicle.
2 Remove the radio (Section 11).
3 Use a wrench to unscrew the antenna mast from the cap nut (see illustration).

Chapter 12 Chassis electrical system

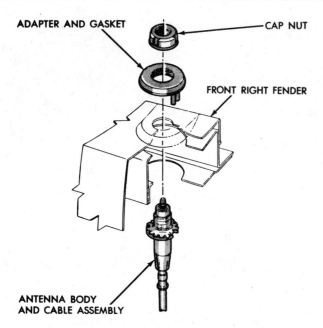

12.4 The antenna lead and body are accessible from inside the fender

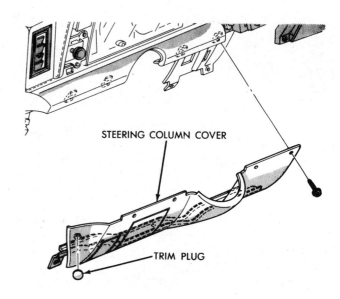

13.4 Steering column cover details

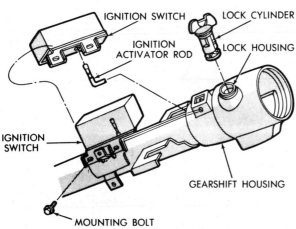

13.6 Ignition switch mounting details

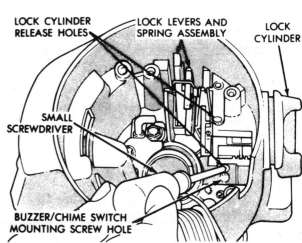

14.9 Ignition lock cylinder removal details (1984 and 1985 models)

4 Use needle nose pliers to unscrew and remove the cap nut and adapter (remove the gasket with the adapter) (see illustration).
5 From under the fender, remove the three inner fender shield-to-fender screws, pull the shield back for access and remove the antenna body and lead assembly.
6 To install the antenna, insert the body and lead assembly into the fender and install the adapter, gasket and cap nut.
7 Install the antenna mast.
8 Install the fender shield screws.
9 Install the radio and connect the negative battery cable. Don't forget to hook the antenna lead to the radio.

13 Ignition switch — removal and installation

Refer to illustrations 13.4 and 13.6

1 Disconnect the negative cable at the battery. Place the cable out of the way so it cannot accidentally come in contact with the negative terminal of the battery, as this would once again allow power into the electrical system of the vehicle.

2 Remove the left lower instrument panel cover.
3 Remove the cruise control clutch speed switch (if equipped).
4 Remove the steering column cover (see illustration) (see Chapter 11 for more information on the steering column, if necessary).
5 Remove the steering column-to-support bracket nuts, then carefully lower the column until the ignition switch is exposed (it is mounted on top of the steering column).
6 Remove the two mounting bolts, rotate the switch 90° and detach the switch from the ignition switch rod (see illustration).
7 Installation is the reverse of removal.

14 Ignition lock cylinder (fixed column) — removal and installation

Refer to illustrations 14.9, 14.19 and 14.20

1 Disconnect the negative cable at the battery. Place the cable out of the way so it cannot accidentally come in contact with the negative terminal of the battery, as this would once again allow power into the electrical system of the vehicle.

Chapter 12 Chassis electrical system

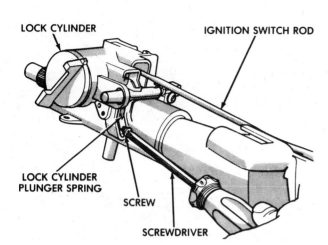

14.19 Lock cylinder plunger details (1986 models)

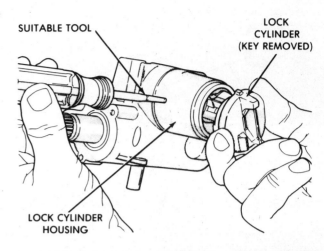

14.20 Lock cylinder removal details (1986 models)

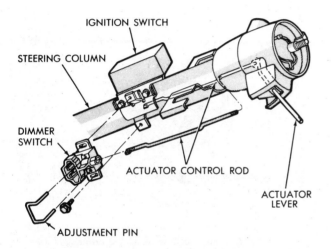

15.4 1984 and 1985 model dimmer switch installation and adjustment details

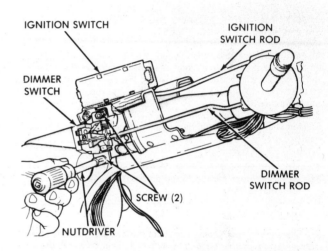

15.5 1986 model dimmer switch installation details

1984 and 1985 models
2 Remove the steering wheel.
3 Remove the turn signal and hazard warning switch.
4 Remove the four retaining screws and the snap-ring and remove the upper shaft bearing and housing.
5 Remove the lock plate and spring.
6 Separate the ignition key buzzer switch wiring connector.
7 Remove the screws from the steering lock bellcrank mechanism.
8 Place the lock cylinder in the Lock position and pull the key out.
9 Insert a small screwdriver into the release holes while pulling out on the lock cylinder to remove it (see illustration).
10 To install, insert the lock cylinder into position and seat the lock lever spring leg securely into the bottom of the notch in the lock casting.
11 Connect the actuator rod to the lock housing and attach the bellcrank. Install the retaining screws.
12 Turn the key to the Lock position and remove it.
13 Push the lock cylinder housing in far enough to contact the switch actuator, insert the key, press it in and rotate the cylinder. When the inner parts of the mechanism are in alignment, the cylinder will move in, the spring loaded retainers will snap into place and the cylinder will be locked into the housing.
14 Plug in the key buzzer connector and install the lock plate and spring.
15 Install the upper bearing housing, screws and snap-ring.
16 Install the turn signal and hazard warning switches.
17 Install the steering wheel.

1986 models
18 Remove the upper and lower steering column covers.
19 Remove the lock cylinder plunger (see illustration).
20 With the key removed, insert a suitable tool into the release hole and withdraw the lock cylinder (see illustration).
21 To install, insert the lock cylinder until it locks into the housing and then install the lock cylinder plunger.

All models
22 Connect the negative battery cable.

15 Dimmer switch — removal, adjustment and installation

Refer to illustrations 15.4 and 15.5
1 Disconnect the negative cable at the battery. Place the cable out of the way so it cannot accidentally come in contact with the negative terminal of the battery, as this would once again allow power into the electrical system of the vehicle.
2 Remove the steering column cover.
3 Unplug the connector, remove the retaining screws and detach the switch from the steering column.
4 On 1984 and 1985 models, install the switch, place it in position, insert the control rod, install the screws finger tight and plug in the connector (see illustration).

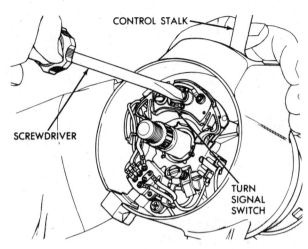

16.3a On 1984 and 1985 models, the turn signal switch screw must be removed to detach the wiper/washer switch

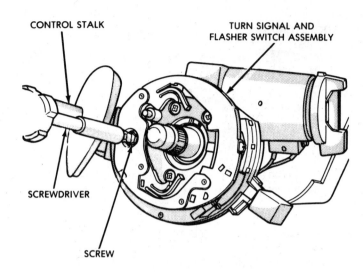

16.3b 1986 model wiper washer switch details

5 On 1986 models, place the switch in position, insert the control rod and install the screws. Tighten the screws securely and plug in the connector (see illustration).
6 On 1984 and 1985 models, adjust the switch by first fabricating an adjustment pin from a piece of wire, then insert the pin ends into the switch (see illustration 15.4). Adjust the switch by pushing it gently to the rear to take up the slack in the control rod and tighten the retaining screws.
7 Remove the adjustment pin.
8 Install the steering column cover and connect the negative battery cable.

16 Washer/wiper switch (fixed column) — removal and installation

Refer to illustrations 16.3a and 16.3b

1 Disconnect the negative cable at the battery. Place the cable out of the way so it cannot accidentally come in contact with the negative terminal of the battery, as this would once again allow power into the electrical system of the vehicle.
2 Remove the steering wheel (Chapter 10) and on 1986 models, the column covers.
3 Remove the screw which retains the turn signal switch to the wiper/washer switch (see illustrations).
4 Pull the shroud up the control stalk. Remove the screws and detach the control stalk.
5 Remove the wiring trough cover and unplug the switch connector.
6 Remove the switch and carefully pull the wiring out of the steering column.
7 To install the switch, insert the wiring harness into the column and thread it down the column and into position. Plug in the connector.
8 Place the wiper/washer switch in position, making sure the dimmer switch actuating rod is securely seated. Install the retaining screw.
9 Install the steering wheel.
10 Attach the stalk and shroud.
11 Install the wiring trough cover and column covers.
12 Connect the negative battery cable.

17 Turn signal/hazard warning switch (fixed column) — removal and installation

Refer to illustrations 17.7 and 17.8

1 Disconnect the negative cable at the battery. Place the cable out of the way so it cannot accidentally come in contact with the negative terminal of the battery, as this would once again allow power into the electrical system of the vehicle.
2 Remove the steering wheel (Chapter 10) and on 1986 models, the column covers.

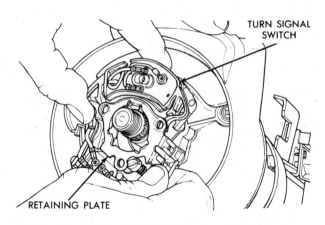

17.7 On 1984 and 1985 models, the turn signal switch can be lifted out after removing the upper bearing retainer screws

3 Remove the wiring trough cover.
4 Remove the lower instrument panel cover from the base of the steering column.
5 Separate the wiring connector.
6 Remove the screw which retains the turn signal switch and washer/wiper switch. Disengage the washer/wiper switch from the column and allow it to hang by the wires.
7 On 1984 and 1985 models, remove the screws retaining the turn signal/hazard warning switch and the upper bearing retainer and lift the assembly from the column (see illustration).
8 On 1986 models, remove the turn signal and hazard flasher warning switch retaining screws (see illustration).
9 Carefully pull the wiring harness up through the column.
10 Prior to installation, lubricate the full circumference of the turn signal switch pivot with light grease.
11 Insert the wiring harness through the hub (1984 and 1985 models) and down the steering column (all models).
12 On 1984 and 1985 models, place the switch assembly and bearing retainer in position and install the screws.
13 On 1986 models, place the switch assembly in position and install the retaining screws.
14 With the washer/wiper switch in position, install the turn signal retaining screw.
15 Plug in the connector and install the wiring trough cover.
16 Install the lower instrument panel cover and steering column panel.
17 Install the steering wheel.
18 Connect the negative battery cable.

Chapter 12 Chassis electrical system

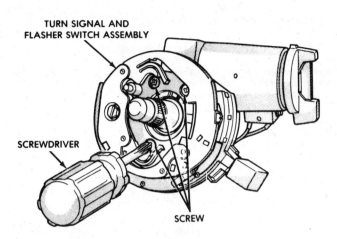

17.8 1986 model turn signal switch screw locations

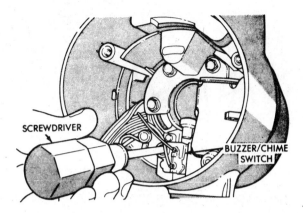

18.5a 1984 and 1985 model buzzer switch removal and installation details

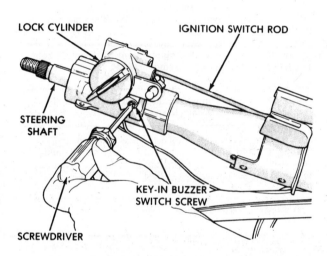

18.5b 1986 model key buzzer installation details

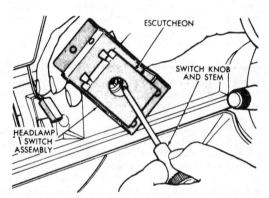

19.5 Depress the headlight switch button to release the knob

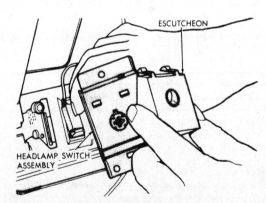

19.6 Grasp the escutcheon securely and unsnap it from the switch

18 Key buzzer/chime switch (fixed column) — removal and installation

Refer to illustrations 18.5a and 18.5b

1 Disconnect the negative cable at the battery. Place the cable out of the way so it cannot accidentally come in contact with the negative terminal of the battery, as this would once again allow power into the electrical system of the vehicle.
2 Remove the steering wheel (Chapter 10) and on 1986 models, the column covers.
3 On 1984 and 1985 models, remove the turn signal/hazard switch, the upper bearing housing and lock plate and spring.
4 Remove the cable trough cover and disconnect the switch connector.
5 Remove the retaining screw and detach the switch and wire from the steering column (see illustrations).
6 To install, insert the wire down through the hub and column, place the switch in position and install the retaining screw.
7 On 1984 and 1985 models, install the lock plate and spring, upper bearing housing and the turn signal/hazard switch.
8 On 1986 models, place the switch in position and install the retaining screw.
9 Plug in the connector and install the wiring trough cover.
10 Install the steering wheel and column covers.
11 Connect the negative battery cable.

19 Headlight switch — removal and installation

Refer to illustrations 19.5, 19.6 and 19.7

1 Disconnect the negative cable at the battery. Place the cable out of the way so it cannot accidentally come in contact with the negative terminal of the battery, as this would once again allow power into the electrical system of the vehicle.
2 Remove the switch assembly trim bezel.
3 Remove the three screws and detach the switch assembly plate from the dash panel.
4 Carefully pull the assembly out and disconnect the wires.
5 Remove the headlight switch knob and stem by depressing the button on the switch (see illustration).
6 Unsnap the headlight switch escutcheon (see illustration).

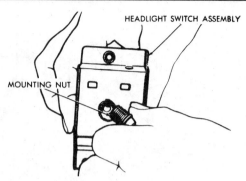

19.7 Unscrew the mounting nut to remove the headlight switch

7 Remove the nut attaching the switch to the mounting plate (see illustration).
8 Installation is the reverse of removal.

20 Speedometer (conventional instrument panel) — removal and installation

1 Disconnect the negative cable at the battery. Place the cable out of the way so it cannot accidentally come in contact with the negative terminal of the battery, as this would once again allow power into the electrical system of the vehicle.
2 Remove the screws and detach the instrument cluster bezel and mask.
3 Remove the speedometer retaining screws.
4 Disconnect the cable in the engine compartment, pull the speedometer out slightly, detach the cable and remove the speedometer.
5 On cruise control equipped vehicles, it will also be necessary to disconnect the speedometer cable from the servo unit located in the engine compartment.
6 To install, place the speedometer in position and press it into the cable assembly until it locks in place. Connect the cable in the engine compartment.
7 Install the retaining screws and the cluster mask and bezel assembly.
8 Connect the cruise control servo to the cable (if equipped).
9 Connect the negative battery cable.

21 Tachometer (conventional instrument panel) — removal and installation

1 Disconnect the negative cable at the battery. Place the cable out of the way so it cannot accidentally come in contact with the negative terminal of the battery, as this would once again allow power into the electrical system of the vehicle.
2 Removes the cluster bezel and mask.
3 Remove the attaching screws, pull the tachometer to the rear and remove it from the cluster.
4 Installation is the reverse of removal.

22 Voltmeter and fuel gauge (conventional instrument panel) — removal and installation

1 Disconnect the negative cable at the battery. Place the cable out of the way so it cannot accidentally come in contact with the negative terminal of the battery, as this would once again allow power into the electrical system of the vehicle.
2 Remove the cluster and mask.
3 Remove the screws retaining the gauge assembly to the cluster.
4 Remove the gauge assembly by pulling it to the rear, out of the cluster.
5 To install, slide the gauge assembly into position and install the retaining screws.
6 Install the cluster bezel and mask.
7 Connect the negative battery cable.

23 Oil temperature and pressure gauge (conventional instrument panel) — removal and installation

1 Disconnect the negative cable at the battery. Place the cable out of the way so it cannot accidentally come in contact with the negative terminal of the battery, as this would once again allow power into the electrical system of the vehicle.
2 Remove the cluster bezel and mask.
3 Remove the speedometer assembly.
4 Remove the retaining screws and pull the gauge assembly toward the rear and out of the cluster.
5 Installation is the reverse of removal.

24 Instrument cluster (conventional instrument panel) — removal and installation

Refer to illustration 24.3

1 Disconnect the negative cable at the battery. Place the cable out of the way so it cannot accidentally come in contact with the negative terminal of the battery, as this would once again allow power into the electrical system of the vehicle.
2 Remove the screws and detach the instrument cluster bezel.
3 Remove the four cluster mounting screws, then carefully pull it out just enough to detach the speedometer cable and wiring connectors from the rear of the cluster (see illustration).
4 Remove the cluster past the right side of the steering column.
5 Installation is the reverse of removal.

25 Instrument cluster printed circuit board (conventional instrument panel) — removal and installation

1 Refer to Section 24 and remove the instrument cluster.
2 Remove the light bulb sockets from the rear of the cluster.
3 Remove the screws and detach the printed circuit board.
4 Installation is the reverse if removal.

26 Instrument cluster bezel (conventional instrument panel) — removal and installation

Refer to illustration 26.1

1 Remove the five screws from the top of the bezel and pull the bezel toward the rear to disengage the three clips along the bottom (see illustration).
2 Place the bezel in position in the cluster and press it in sharply to secure the clips. Install the five screws.

27 Instrument cluster mask (conventional instrument panel) — removal and installation

Refer to illustration 27.3

1 Disconnect the negative cable at the battery. Place the cable out of the way so it cannot accidentally come in contact with the negative terminal of the battery, as this would once again allow power into the electrical system of the vehicle.
2 Remove the cluster bezel (Section 26).
3 Remove the five bayonet clips retaining the mask to the housing (see illustration).
4 Remove the cluster mask by pulling it to the rear.
5 To install, place the mask in position and press in on the clips.
6 Install the cluster bezel.

28 Instrument cluster bulbs (conventional instrument panel) — replacement

Refer to illustration 28.2

1 Remove the instrument cluster (Section 24).
2 Remove the light bulbs from the rear of the cluster by turning the bulbs counterclockwise to release them (see illustration).

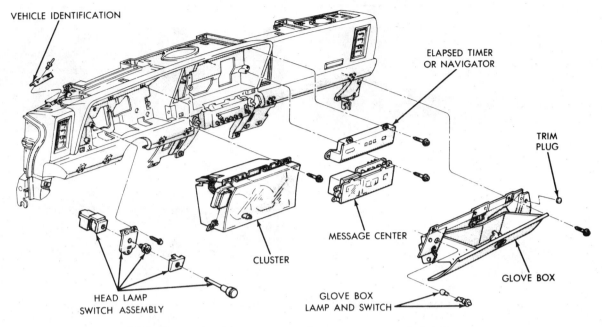

24.3 Conventional instrument panel components — exploded view

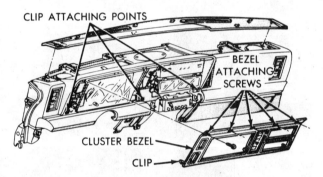

26.1 Instrument cluster bezel components — exploded view

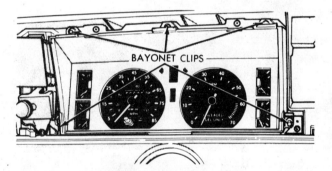

27.3 Instrument cluster mask bayonet clip locations

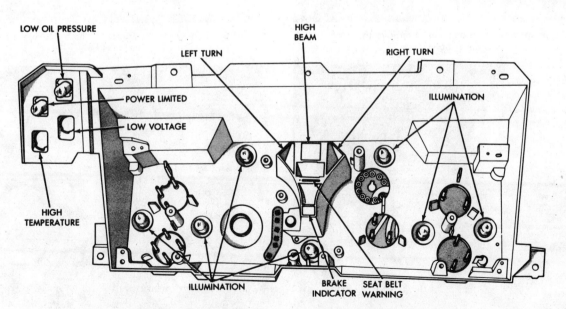

28.2 Conventional instrument panel cluster bulb locations

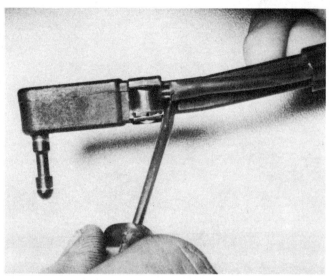

30.3 Push the washer hose off the wiper arm with a small screwdriver

30.4 A forked tool such as the one shown here makes cowl plenum water deflector clip removal easier

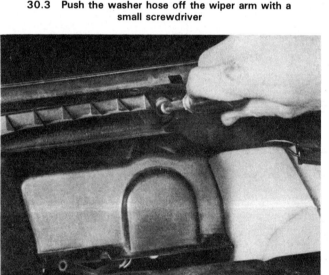

30.5 A nut driver can be used to remove the wiper cover bolts

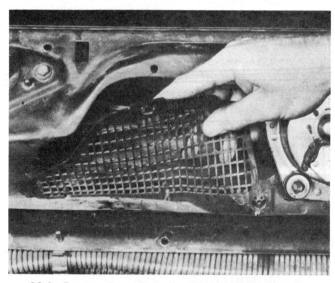

30.6 Remove the perforated grille by pulling it from the access hole

30.7 Remove the wiper pivot bolts

30.8 Wiper motor retaining nuts (arrows)

Chapter 12 Chassis electrical system

29 Electronic instrument cluster — general information

The electronic instrument cluster is removed in the same way as the conventional cluster except that it is not necessary to disconnect the speedometer. All work on the electronic instrument cluster must be left to a dealer or properly equipped shop due to the special tools and techniques required.

30 Windshield wiper motor — removal and installation

Refer to illustrations 30.3, 30.4, 30.5, 30.6, 30.7, 30.8, 30.10a, 30.10b and 30.11

Removal

1 Open the hood.
2 Disconnect the negative cable at the battery. Place the cable out of the way so it cannot accidentally come in contact with the negative terminal of the battery, as this would once again allow power into the electrical system of the vehicle.
3 Remove the wiper arms and disconnect the washer hoses (see illustration).
4 Pry out the clips and remove the cowl plenum water deflector (see illustration).
5 Remove the cover (two bolts) from the wiper motor (see illustration) and unplug the electrical connectors.
6 Remove the perforated grill (see illustration).
7 Remove the wiper pivot mounting bolts (see illustration).
8 Remove the three motor retaining nuts (see illustration).
9 Unplug the motor and relay electrical connectors.
10 Remove the motor, wiper arms and pivot assembly by collapsing the arms and lifting it through the access opening (see illustrations).
11 To remove the wiper arm from the motor, mark the relative position of the arm and motor, support the arm securely in a vise and remove the retaining nut (see illustration).
12 Remove the arm from the motor.

Installation

13 Install the arm on the motor and tighten the nut securely.
14 Insert the wiper assembly into place, extend the arms until the pivots can be pulled into place and install the bolts.
15 Install the motor and relay and install the nuts. Tighten the nuts securely.
16 Plug in the connectors and install the grille and water deflector.
17 Install the wiper arms and washer hoses.
18 Connect the battery.

31 Liftgate wiper motor — removal and installation

Refer to illustrations 31.3, 31.4, 31.5 and 31.6

Removal

1 Disconnect the negative cable at the battery. Place the cable out of the way so it cannot accidentally come in contact with the negative terminal of the battery, as this would once again allow power into the electrical system of the vehicle.
2 Open the liftgate.
3 Remove the two trim cover screws, pry out each plastic pin, insert a screwdriver and pry the cover off (see illustration).

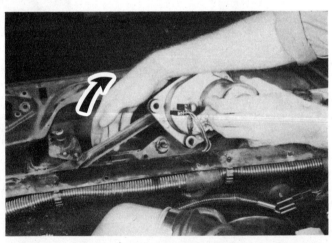

30.10a Collapse the right wiper actuating arm, pull it out of the access hole...

30.10b...and then pull the left arm and the motor assembly out

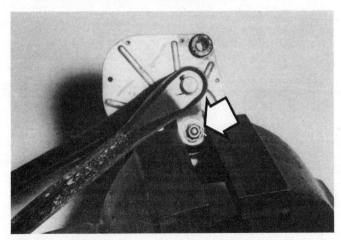

30.11 Make sure the wiper arm is held securely in the vise before removing the nut (arrow)

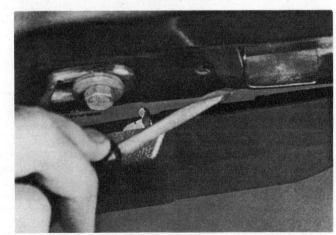

31.3 After removing the screws, pry the trim cover off with a screwdriver

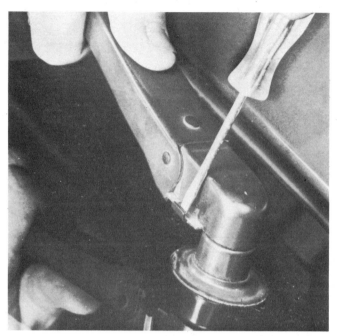

31.4 Use a small screwdriver to pry out the wiper arm release lever

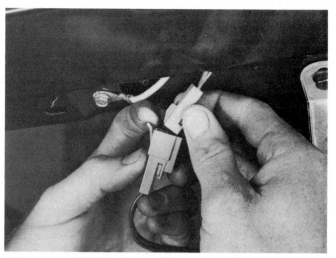

31.5 Pull the retaining clip out and unplug the connector

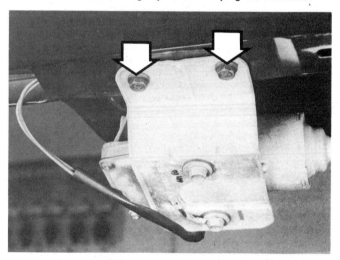

31.6 Rear wiper motor bolt locations (arrows)

4 Remove the wiper arm (see illustration).
5 Unplug the connector (see illustration).
6 Remove the two wiper motor retaining bolts (see illustration).
7 Lower the motor from the liftgate.
8 Unbolt the motor from the mounting plate and transfer the plate to the new motor.

Installation

9 Place the motor in position and install the bolts. Tighten the bolts securely and plug in the connector.
10 After installation, place the trim panel in place and push the retaining pin into the hole at one end. Use the palm of your hand to seat it and work along the entire length to seat them all. Install the plastic clips and screws.

Chapter 13 Supplement: Revisions and information on 1987 and later models

Contents

Introduction 1	**Emissions control systems** 7
Specifications 2	Heated oxygen sensor replacement
Troubleshooting 3	**Manual transaxle** 8
Valve lash adjuster noise diagnosis	General information
Tune-up and routine maintenance 4	Shift cable – adjustment (A-555 heavy duty transaxle)
Cooling system servicing (draining, flushing and refilling)	**Brakes** 9
Turbocharger 5	General information
General information	Rear disc brake pads – replacement
Turbocharger – removal and installation (Turbo II)	**Steering and suspension systems** 10
Engine electrical systems 6	Power steering pump – general information
Single Module Engine Controller (SMEC)	Air bag module – removal and installation
On-board diagnostics	
Fault codes	

1 Introduction

This Supplement contains Specifications and service procedure changes that apply to 1987 and later Dodge Daytona and Chrysler Laser models produced for sale in the US. Also included is material which was not available at the time of the original production of this manual.

Where no differences (or very minor differences) exist between 1987 and later models and previous models, no information is given. In such instances, the original material included in Chapters 1 through 12 should be used. Therefore, owners of vehicles manufactured after 1986 should refer to this Chapter before using the information in the original Chapters of this manual.

2 Specifications

Engine

Crankshaft journal out-of-round
   standard .... 0.0003 in (0.008 mm)
   service limit .... 0.005 in (0.13 mm)
Main bearing oil clearance
   standard .... 0.0004 to 0.0028 in (0.010 to 0.071 mm)
   service limit .... 0.004 in (0.10 mm)

Engine (continued)

Valve spring load
- open .... 195 to 215 lbs. @ 1.22 in (890 to 961 Nm @ 30.99 mm)
- closed .... 108 to 120 lbs. @ 1.65 in (480 to 534 Nm @ 41.91 mm)

Intake valve opens
- 2.2L .... 0-degrees BTDC
- 2.5L .... 4-degrees BTDC

Intake valve closes
- 2.2L .... 56-degrees ABDC
- 2.5L .... 60-degrees ABDC

Exhaust valve opens
- 2.2L .... 44-degrees BBDC
- 2.5L .... 40-degrees BBDC

Exhaust valve closes
- 2.2L .... 8-degrees ATDC
- 2.5L .... 12-degrees ATDC

Valve overlap 8-degrees
Intake valve duration 236-degrees
Exhaust valve duration 232-degrees

A-555 manual transaxle

Lubricant type .... SAE 5W-30 motor oil
Gear ratios:
- 1st 3.00
- 2nd 1.89
- 3rd 1.28
- 4th 0.94
- 5th 0.72
- Reverse 3.14

Final drive ratio 3.85

Torque specifications

	Ft-lbs *(unless otherwise indicated)*	Nm
Water box drain/fill plug	15	20
Turbocharger-to-exhaust manifold nuts	40	54
Turbocharger bracket-to-block bolt	40	54
Turbocharger bracket-to-turbocharger bolt	20	27
Articulated exhaust joint bolts	20	27
Front engine mount through-bolt	40	54
Transaxle cable bolts	70 in-lbs	8
ZF power steering pump discharge fitting	50	68
Rear disc brake caliper mounting bolts	13 to 18	17 to 25

3 Troubleshooting

Valve lash adjuster noise diagnosis

1 Under certain circumstances the valve lash adjusters (hydraulic tappets) can become noisy. Excessive or continued lash adjuster noise can have the following causes:
2 Engine oil level too high, causing the oil to "foam."
3 Insufficient break-in time following cylinder head overhaul. Low speed operation for up to one hour, turning the engine off several times for several minutes during the hour, may be required to properly seat the hydraulic lash adjuster components.
4 Low oil pressure.
5 On 2.5L engines an oil restrictor is pressed into the vertical oil passage to the cylinder head. If this restrictor becomes plugged with sludge or debris it can cause lash adjuster noise.
6 A broken or cracked oil pump pickup can cause aeration (foaming) of the oil.
7 Worn valve guides.
8 Rocker arm ears making contact with the valve spring retainers.
9 Loose rocker arm.
10 Faulty or worn out hydraulic lash adjuster.

4 Tune-up and routine maintenance

Cooling system servicing (draining, flushing and refilling)
Refer to illustration 4.1

1 Later model engines are equipped with a "water box" between the thermostat housing and the engine block, equipped with a drain/fill plug **(see illustration)**.
2 The engine must be cool (no pressure in the cooling system) before starting this procedure.
3 To drain the cooling system, start the engine and turn the heater temperature selector to full on.
4 Shut off the engine before it reaches operating temperature and, *without removing the pressure cap,* open the radiator draincock.
5 Remove the drain and fill plug from the top of the water box.
6 Close the radiator draincock.
7 Add antifreeze/water mix (see Chapter 1 if necessary) through the drain/fill plug hole in the water box until the coolant level reaches the top of the water box.

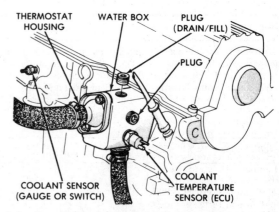

4.1 A water box between the thermostat housing and engine block contains a plug for draining and refilling the cooling system

Chapter 13 Supplement: Revisions and information on 1987 and later models

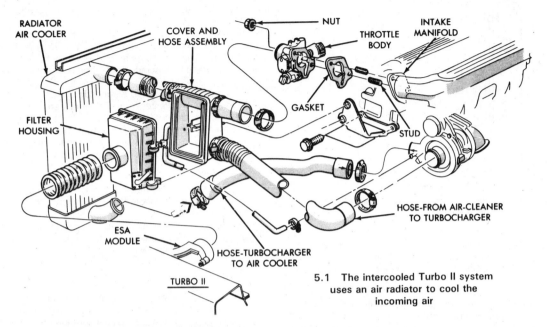

5.1 The intercooled Turbo II system uses an air radiator to cool the incoming air

8 Install the plug and tighten it to the specified torque.
9 Fill the radiator to the bottom of the pressure cap seat and install the pressure cap.
10 Add coolant to the reservoir to the MAX mark.
11 Start the engine, let it reach normal operating temperature and check for coolant leaks.
12 Additional coolant will probably be required for the reservoir during the next few warm-ups as air is expelled from the system.

5 Turbocharger

General information
Refer to illustration 5.1

1 Models equipped with the Turbo II engine option have an intercooled turbocharger, where the incoming air supply for the turbocharger first passes through an air cooling radiator (see illustration).

Turbocharger — removal and installation (Turbo II)

2 Remove the front engine mount through-bolt and pull the top of the engine forward (away from the cowl) until there is sufficient clearance to reach the turbocharger assembly.
3 Remove the air outlet (to air cleaner) hose from the turbocharger.
4 Remove the upper coolant line and fitting from the turbocharger.
5 Disconnect the oil feed line from the turbocharger.
6 Remove the wastegate rod retaining clip.
7 Remove the two upper and lower driver's side turbocharger-to-exhaust manifold mounting nuts. Leave the remaining nut in place.
8 Disconnect the oxygen sensor lead wire.
9 Raise the vehicle and support it securely on jackstands.
10 Remove the right front wheel.
11 Remove the right driveaxle.
12 Remove the turbocharger-to-engine block bracket.
13 Disconnect the oil drain-back hose from the turbocharger.
14 Remove the one remaining turbocharger-to-exhaust manifold nut.
15 Disconnect the articulated exhaust pipe joint from the turbocharger.
16 Remove the lower coolant line.
17 Remove the turbocharger air intake hose.
18 Remove the turbocharger.
19 Position the turbocharger on the exhaust manifold, apply anti-seize compound to the studs and install the lower passenger's side nut. Tighten it to the specified torque.
20 Apply thread sealant to the lower coolant line fitting and attach the coolant line to the turbocharger.
21 Install the oil drain-back hose.
22 Install the turbocharger-to-block bracket, tightening the bolts finger tight.
23 Tighten the block bolt to the specified torque.
24 Tighten the turbocharger housing bolt to the specified torque.
25 Move the exhaust pipe into place and tighten the articulated joint bolts to the specified torque.
26 Install the right driveaxle.
27 Install the right front wheel and lower the vehicle.
28 Install the three remaining turbocharger nuts and tighten them to the specified torque.
29 Reconnect the oxygen sensor.
30 Connect the wastegate rod.
31 Attach the oil feed line to the turbocharger.
32 Install the upper coolant line.
33 Align the front engine mount in the crossmember bracket, install the through-bolt and tighten it to the specified torque.

6 Engine electrical systems

Single Module Engine Controller (SMEC)
Refer to illustrations 6.1, 6.7, 6.9 and 6.134

Description

1 The Single Module Engine Controller (see illustration) is a digital computer which receives input signals from various switches and sensors, which it uses to determine fuel injector pulse width, spark advance, ignition coil dwell, idle speed cooling fan operation and alternator output.

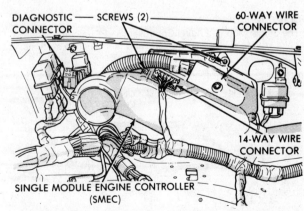

6.1 The Single Module Engine Controller (SMEC) controls the ignition timing, fuel injector operation, cooling fan and alternator output

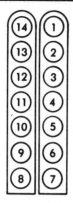

6.7 Single Module Engine Controller 14-pin connector terminal locations

SMEC system coil test

2 Remove the coil wire from the center terminal of the distributor cap. Hold the end of the wire about 1/4-inch from a good engine ground and crank the engine. Check for a consistent spark at the wire as the engine is cranked.

3 If the spark is not constant or there is no spark, connect a voltmeter to the coil positive terminal and crank the engine for five seconds. If the voltage is near zero during this test refer to the on-board diagnostic checks below.

4 If the voltage is at or near battery voltage (approximately 12.4 volts) at the start of the test but drops to zero after one or two seconds of cranking, refer to the on-board diagnostic checks below.

5 If the voltage remains at near battery voltage for the entire five seconds, turn off the ignition key, remove the 14-way connector from the SMEC and check for any spread terminals.

6 Remove the coil positive terminal lead and connect a jumper wire between the battery positive terminal and the coil positive terminal.

7 Using a jumper wire, momentarily ground terminal no. 12 of the 14-way connector (see illustration).

8 If a spark is generated, replace the SMEC.

9 If no spark is generated, connect a special jumper wire (see illustration) to ground the coil negative terminal.

10 If a spark is now produced, check the wiring harness for an open condition.

11 If no spark is produced, replace the coil.

SMEC system coolant temperature sensor test

12 With the ignition off, disconnect the electrical connector from the coolant temperature sensor (see illustration 4.1).

13 Connect an ohmmeter to the coolant temperature sensor (see illustration).

14 With the engine cool (approximately 70~ F) the ohmmeter should read between 7000 and 13,000 ohms.

15 With the engine at normal operating temperature (approximately 200~ F) the ohmmeter should read between 700 and 1000 ohms.

16 For further test procedures refer to the on-board diagnostic checks below.

On-board diagnostics

17 The Single Module Engine Controller (SMEC) monitors many different circuits in the electrical, ignition and fuel injection systems, and if a problem is detected in one of these systems a fault code will be stored in the SMEC.

18 Normally fault codes are displayed by a special diagnostic tool connected to the on-board diagnostic connector in the engine compartment. However, the home mechanic without the special diagnostic tool can use the check engine light in the dashboard to check for fault codes.

19 To activate the on-board diagnostic system, turn the ignition key on-off-on-off within five seconds. The check engine light will come on for two seconds as a bulb check, then will display any fault codes by flashing on and off. There will be a short pause between fault code numbers (all fault codes will have two numbers) and a longer (approximately four second) pause between fault codes.

Fault codes

Code 11	Engine has not been cranked since battery was disconnected
Code 12	Memory standby power lost
Code 13	MAP sensor pneumatic circuit fault
Code 14	MAP sensor electrical circuit fault
Code 15	Vehicle distance sensor fault
Code 16	Loss of battery voltage
Code 17	Engine running too cool
Code 21	Oxygen sensor circuit fault
Code 22	Coolant temperature sensor circuit fault
Code 23	Throttle body temperature sensor circuit fault
Code 24	Throttle position sensor fault
Code 25	ISC motor driver circuit fault
Code 26	Peak injector current has not been reached
Code 27	Fuel injector control problem
Code 31	Canister purge solenoid circuit fault
Code 32	EGR diagnostics fault (California only)
Code 33	Air conditioning cutout relay circuit fault
Code 34	Speed control problem
Code 35	Idle switch circuit fault
Code 36	Air switching solenoid circuit fault
Code 37	Part throttle unlock solenoid driver circuit (automatic only) fault
Code 41	Charging system excess or no field current
Code 42	Auto shutdown relay driver circuit fault
Code 43	Ignition coil control circuit fault
Code 44	Loss of FJ2 to logic board
Code 46	Battery voltage too high
Code 47	Battery voltage too low
Code 51	Lean condition indicated
Code 52	Rich condition indicated
Code 53	Internal module problem
Code 54	Distributor sync pick-up circuit
Code 55	End of fault code readout

7 Emissions control systems

Heated oxygen sensor replacement

Refer to illustration 7.1

1 Late model vehicles are equipped with an electrically heated oxygen sensor (see illustration). A three wire connector (power, ground and sensor) is used in place of the earlier two-wire connector.

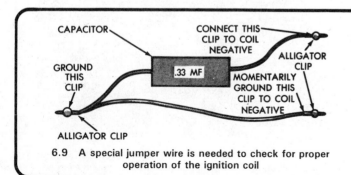

6.9 A special jumper wire is needed to check for proper operation of the ignition coil

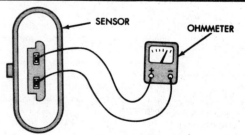

6.13 Testing the coolant temperature sensor with an ohmmeter

Chapter 13 Supplement: Revisions and information on 1987 and later models

2   Removal and installation procedures remain the same as the earlier type sensor.

8   Manual transaxle

General information

1   A new five-speed transaxle, the A-555, with coarse-pitch gears made by Getrag is used in all vehicles equipped with the Turbo II engine package.
2   To reduce internal wear, all manual transaxles in 1987 and later vehicles use SAE 5W-30 motor oil as lubricant.
3   A tag is attached to the top of the transaxle giving the model and assembly number.

Shift cable — adjustment (A-555 heavy duty transaxle)
Refer to illustrations 8.4, 8.10, 8.11a and 8.11b

4   Remove the lock pin from the transaxle selector shaft housing **(see illustration)**.
5   Reverse the lock pin, so the long end is down, and insert it into the same threaded hole in the housing.
6   Push the selector shaft into the selector housing until the hole in the selector shaft aligns with the lock pin, allowing the lock pin to be screwed into the housing.
7   Remove the gearshift knob and pull the boot over the pull-up ring.
8   Remove the console.
9   Loosen the cable adjusting bolts.

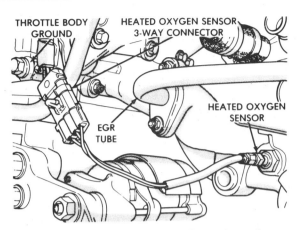

7.1   The heated oxygen sensor has a three-wire electrical connector

10   An adjusting screw tool and spacer block are taped to the shifter support bracket. Remove the tool, which has left-hand threads, and install it in the housing just below the shifter **(see illustration)**. Tighten the adjusting tool to 20 in-lbs (2 Nm).
11   Tighten the selector and crossover cable adjusting bolts to 70 in-lbs (8 Nm) **(see illustrations)**. **Caution:** *It is very important that the cable bolts be tightened to the proper torque.*

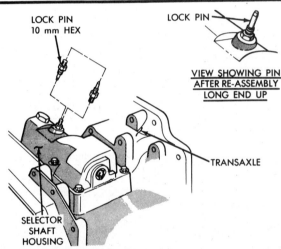

8.4   The lock pin installed in the top of the transaxle housing is reversed to lock the selector shaft before adjusting the shift cables

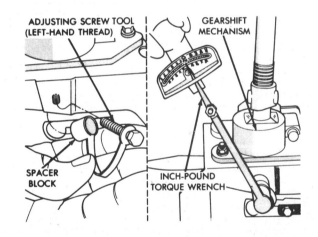

8.10   The adjusting tool and spacer block are taped to the shifter bracket — after installation in the shifter housing, the tool must be tightened to the specified torque

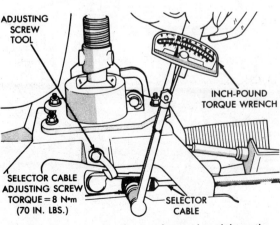

8.11a   A torque wrench must be used to tighten the selector cable bolt . . .

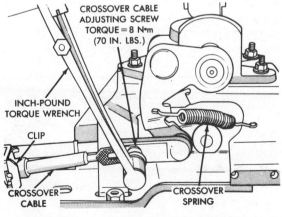

8.11b   . . . and the crossover cable bolt

Chapter 13 Supplement: Revisions and information on 1987 and later models

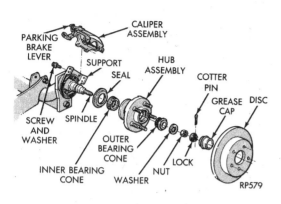

9.1a 1987 and 1988 rear disc brake assembly components - exploded view

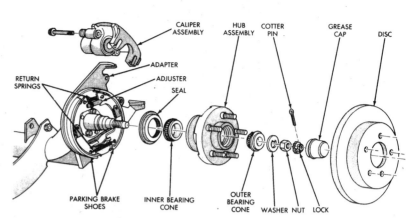

9.1b Rear disc brake assembly components - exploded view

12 Remove the adjusting screw tool and spacer block and tape them to the shifter support bracket for future use.
13 Remove the lock pin from the selector shaft housing and reinstall it with the long end up. Tighten it to 105 in-lb (12 Nm).
14 Check for proper shift operation.
15 Reinstall the console, boot and gearshift knob.

9 Brakes

General information

Refer to illustrations 9.1a and 9.1b

Rear disc brakes are used on 1987 and later models. The 1987 and 1988 model rear disc brake has an automatic adjuster contained in the inside of the piston, which maintains a constant clearance between the brake pads and brake rotor. The parking brake is incorporated into this system and is self adjusting. The 1988 models incorporate a drum-type parking brake (the parking brake shoes expand against the drum cast into the center section of the disc) **(see illustration)**. Periodic adjustment of the parking brake cables is not required because the parking brake handle contains a spring-loaded adjuster mechanism. **Caution:** *Be very careful when working in the vicinity of the parking brake handle assembly because the internal spring is under considerable tension.*

Rear disc brake pads - replacement

Refer to illustrations 9.6 and 9.7

1 Raise the rear of the vehicle, support it securely on jackstands and remove the rear wheels.
2 On 1987 and 1988 models, remove the access plug at the rear of the caliper assembly and insert a 4mm Allen wrench through the access hole.
3 Turn the retraction shaft counterclockwise to make clearance between the rotor and brake pad.
4 Remove the mounting bolts and lift the caliper assembly off the disc.
5 Slide the outer pad off the caliper.
6 Detach the clip and remove the inner pad from the caliper piston.
7 Prior to installation of the new pads, push the piston back into the caliper (to provide room for new, thicker pads).
8 On 1987 and 1988 models, the piston is seated into the caliper by turning the retraction shaft counterclockwise. Install the caliper onto the support.
9 Press the inner pad retaining clip securely into the piston **(see illustration)**.
10 Guide the outer pad into place by sliding the retainer clip over the caliper fingers **(see illustration)**.
11 Place the lower end of the caliper on the mounting adapter, making sure the lower tabs on the pads and caliper projections are under the adapter rail, then rotate the top of the caliper down into position.
12 Install the bolts and tighten them to 13 to 18 ft-lbs (17 to 25 Nm).
13 On 1987 and 1988 models, turn the retraction shaft clockwise until there is a snug fit between the brake pads and the rotor and back off 1/3 of a turn for proper clearance.

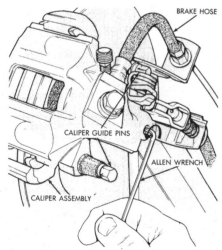

9.2 Insert a 4mm Allen-wrench into the access hole and turn counterclockwise

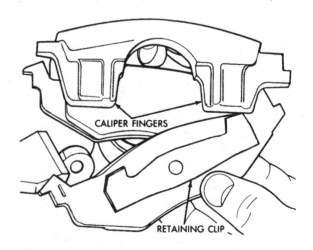

9.10 Slide the outer pad retaining clips over the caliper fingers

Chapter 13 Supplement: Revisions and information on 1987 and later models

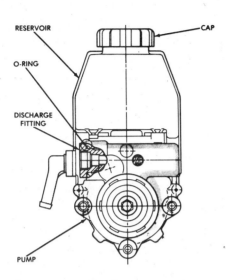

10.3 If the discharge fitting on the ZF power steering pump leaks, the O-ring should be replaced

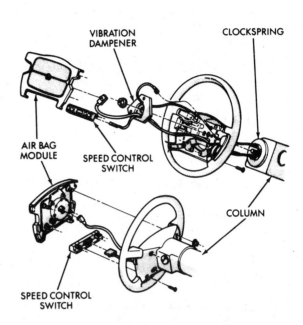

10.7 Air bag module installation details

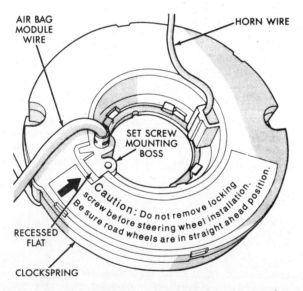

10.9 On models so equipped, install the tethered screw through the mounting boss to hold the clockspring in place if the steering wheel must be removed

10 Steering and suspension systems

Power steering pump – general information
Refer to illustration 10.3

1 Late model vehicles are equipped with a ZF power steering pump.
2 The ZF pump, reservoir and cap are not repairable. Any leakage or improper operation should be corrected by replacing the pump or cap, as needed.
3 If an oil leak develops at the discharge fitting (**see illustration**), tighten the fitting to 37 ft-lbs (50 Nm). If it still leaks, remove the fitting, replace the O-ring, and retighten the fitting.

4 The removal and installation procedure is essentially the same as for the earlier model pump.

Air bag module – removal and installation
Refer to illustrations 10.7 and 10.9
Warning: *Later model vehicles are equipped with a driver's side air bag installed in the center of the steering wheel. The air bag must be disarmed and removed before the steering wheel is removed!*
Caution: *The front wheels must be in the straight-ahead position during this procedure to ensure the clockspring in the steering hub is held in the proper position. The clockspring maintains electrical power to the air bag module during normal use and if not centered, could cause failure of the air bag system.*

5 Removal of the air bag is necessary before the steering wheel can be removed for steering column switch replacement or other service procedures. **Warning:** *Always wear shatter-proof eye protection when servicing the air bag and never probe the air bag electrical connectors.*

Removal
6 The air bag is activated by battery power, so first disconnect the negative cable from the battery.
7 Use a thin-wall 10 mm socket to remove the four air bag module retaining nuts from the back side of the steering wheel (**see illustration**). On some models, Chrysler tool no. 6239 may be required because tamper resistant nuts are used.
8 Detach the clockspring wiring connector and remove the air bag module.
9 If the steering wheel must be removed, make sure the clockspring will remain centered. Some models have autolocking tabs which center the clockspring, while others have a tethered screw which must be installed in the set screw mounting boss at this time (**see illustration**). **Warning:** *Do not remove the set screw until after the steering wheel and nut have been installed.*

Installation
10 Attach the air bag module wiring connector to the clockspring.
11 Install the air bag module in the steering wheel and tighten the four nuts to 80 to 100 in-lbs.

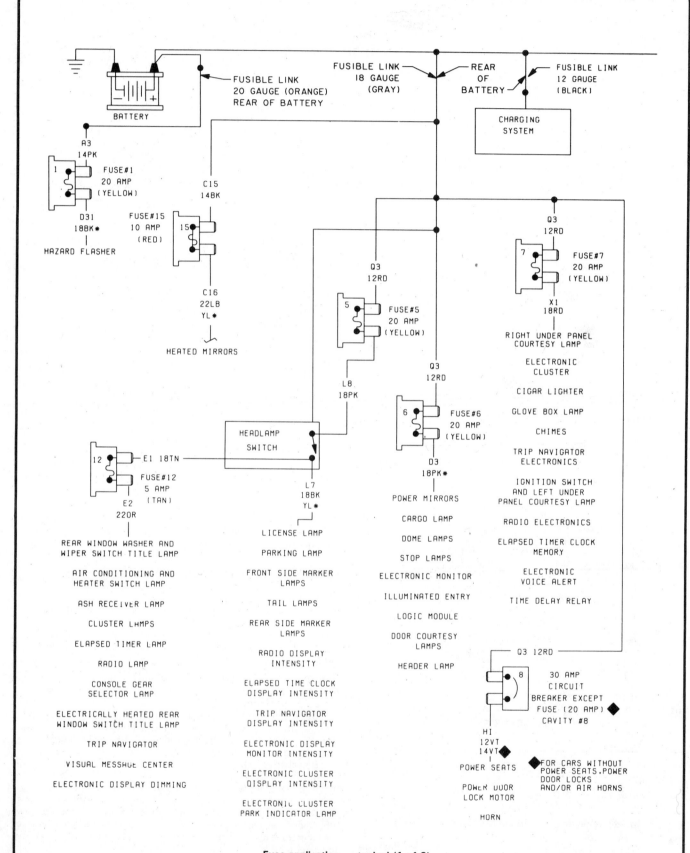

Fuse application — typical (1 of 2)

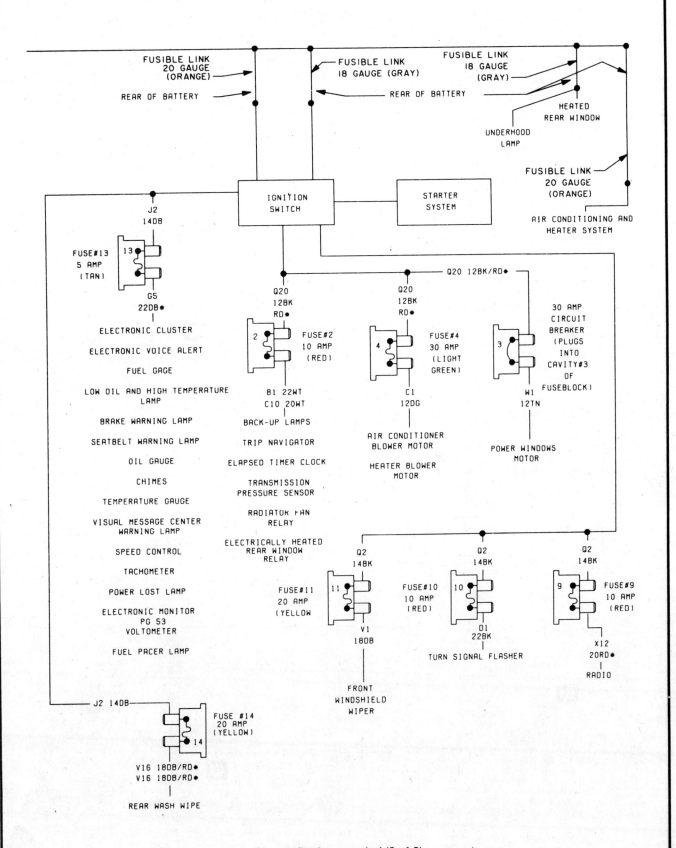

Fuse application — typical (2 of 2)

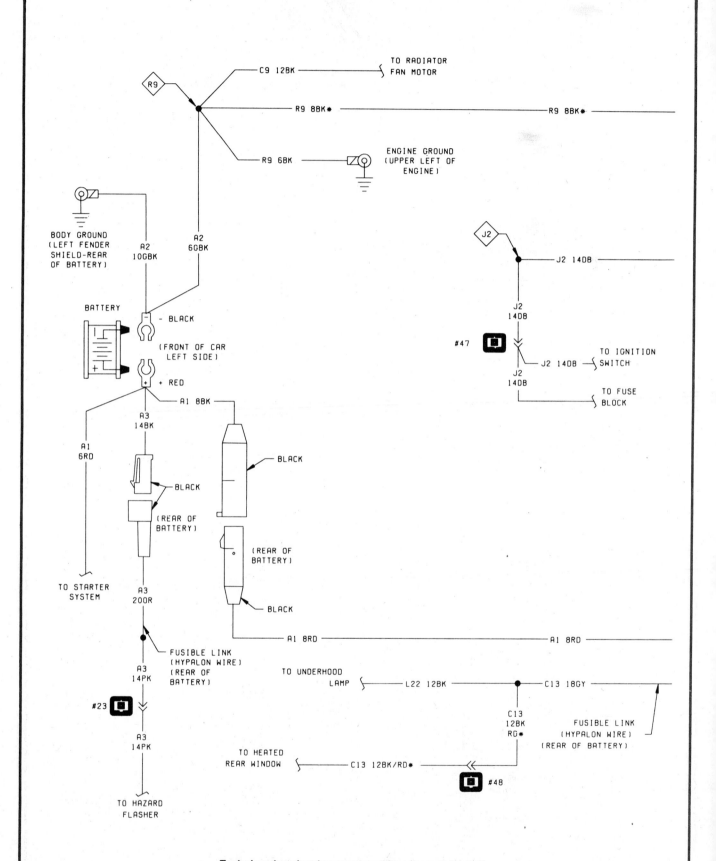

Typical engine charging system wiring diagram (1 of 2)

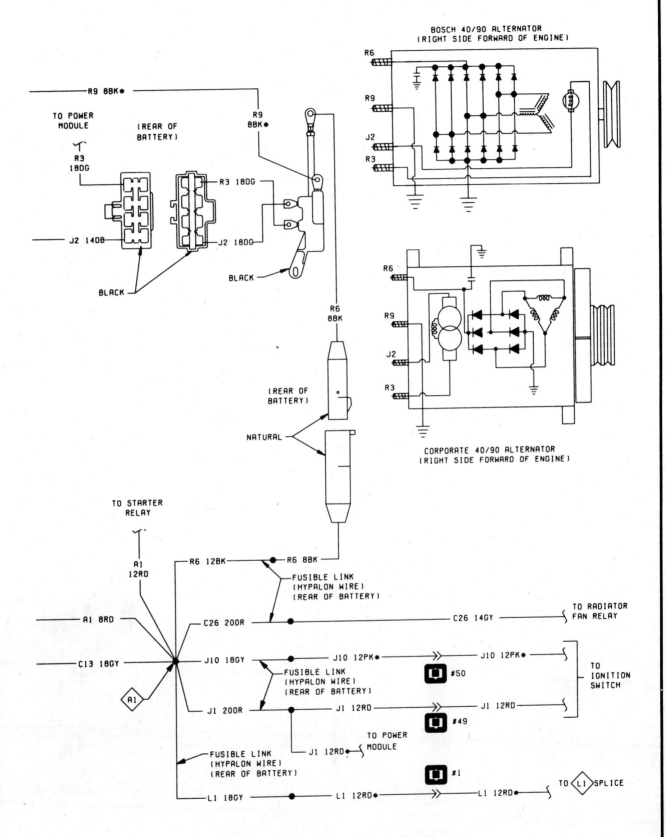

Typical engine charging system wiring diagram (2 of 2)

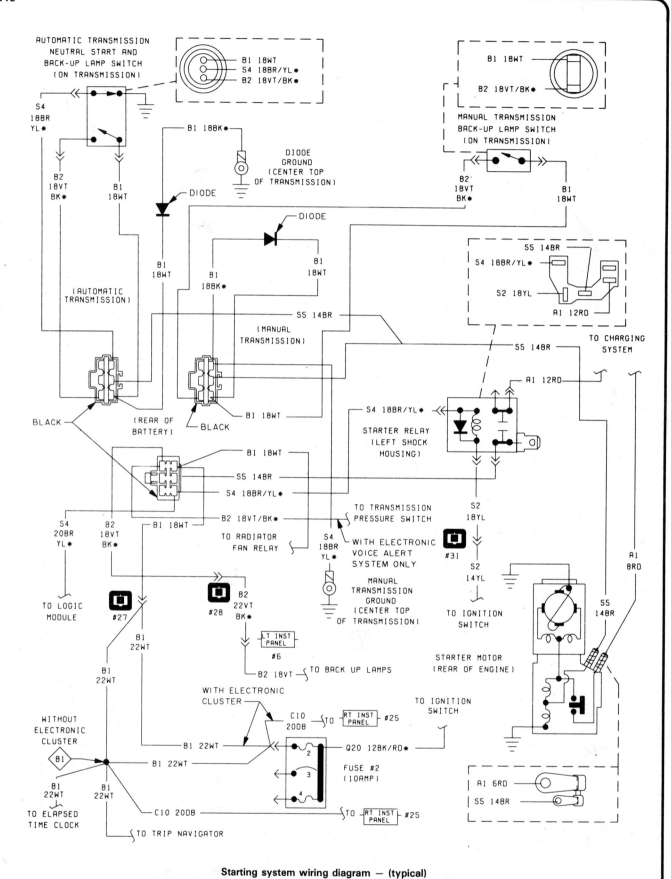

Starting system wiring diagram — (typical)

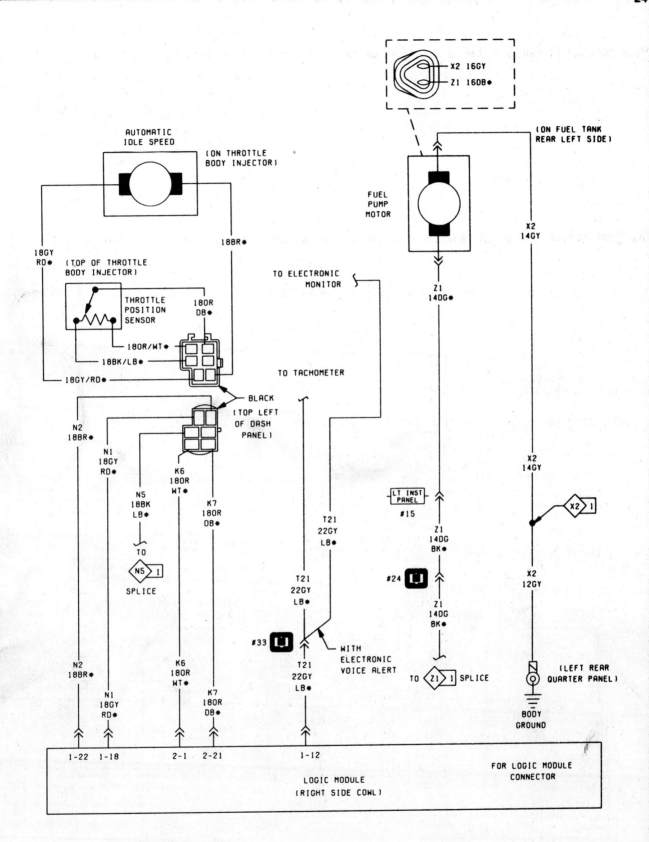

Turbocharged engine EFI ignition system wiring diagram (typical) (1 of 5)

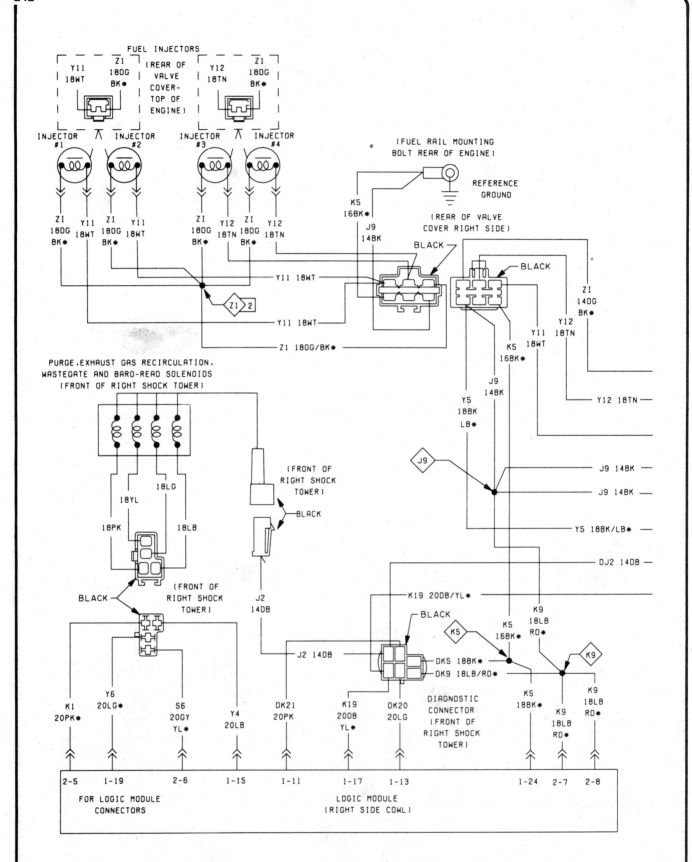

Turbocharged engine EFI ignition system wiring diagram (typical) (2 of 5)

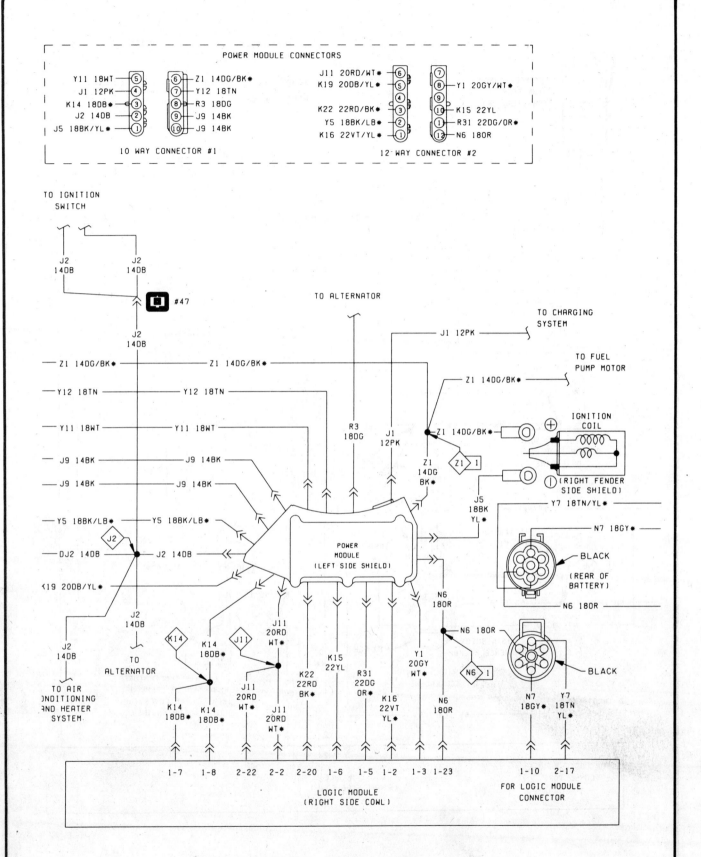

Turbocharged engine EFI ignition system wiring diagram (typical) (3 of 5)

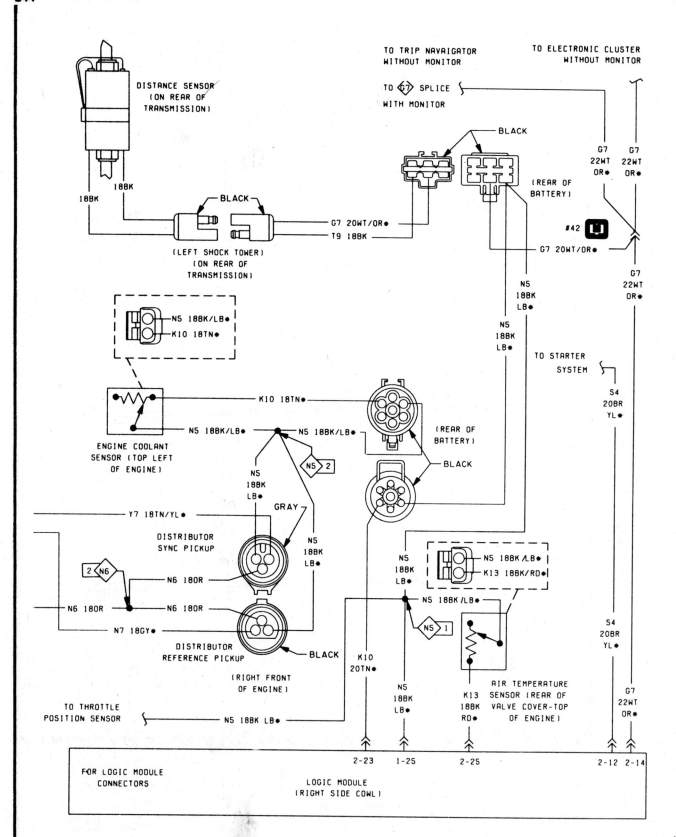

Turbocharged engine EFI ignition system wiring diagram (typical) (4 of 5)

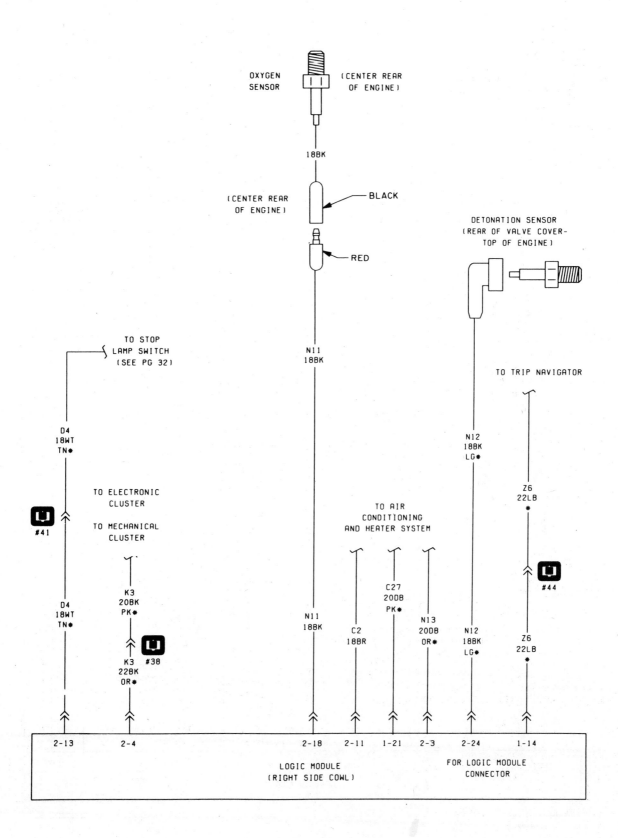

Turbocharged engine EFI ignition system wiring diagram (typical) (5 of 5)

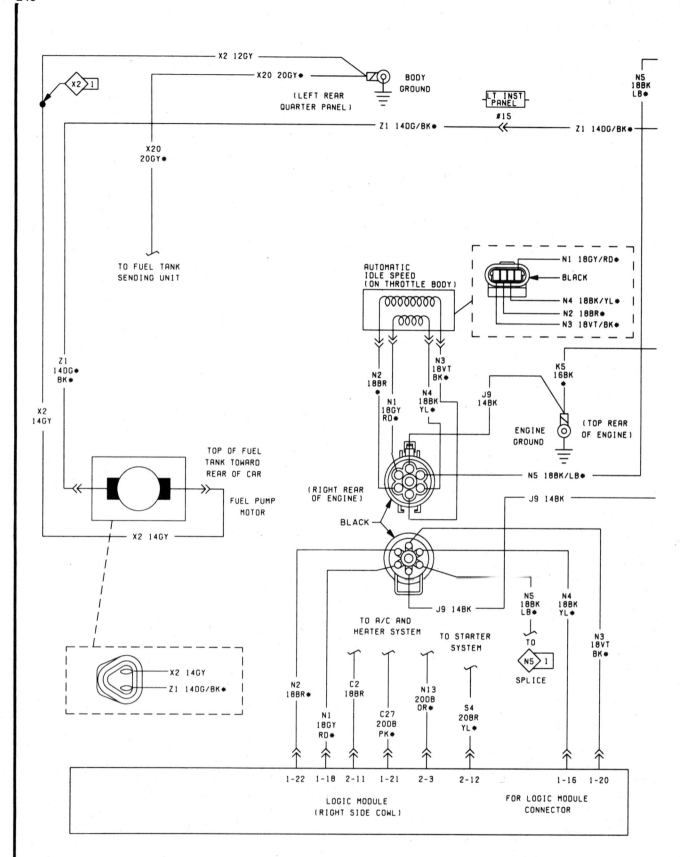

Non-turbocharged engine EFI ignition system wiring diagram (typical) (1 of 4)

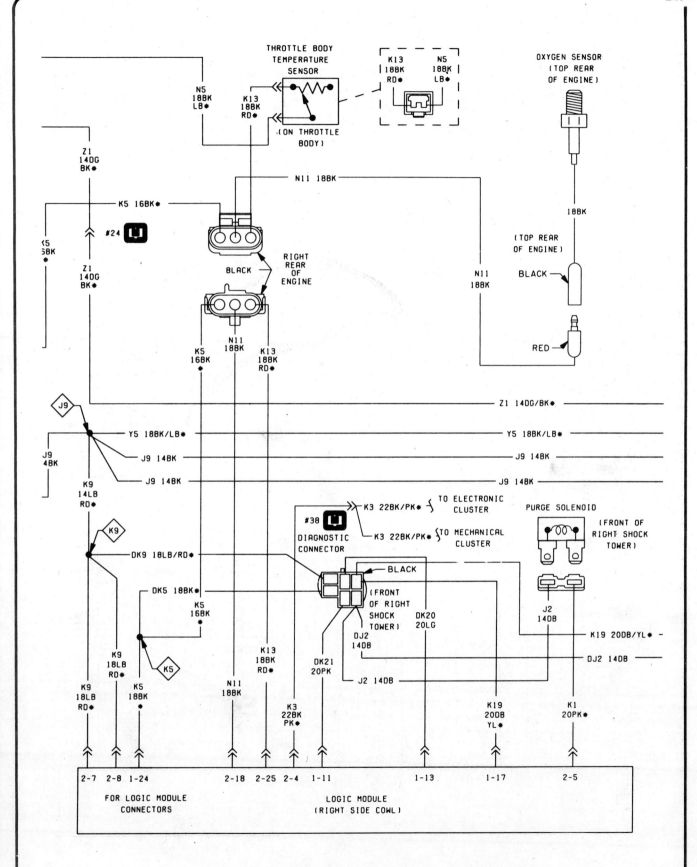

Non-turbocharged engine EFI ignition system wiring diagram (typical) (2 of 4)

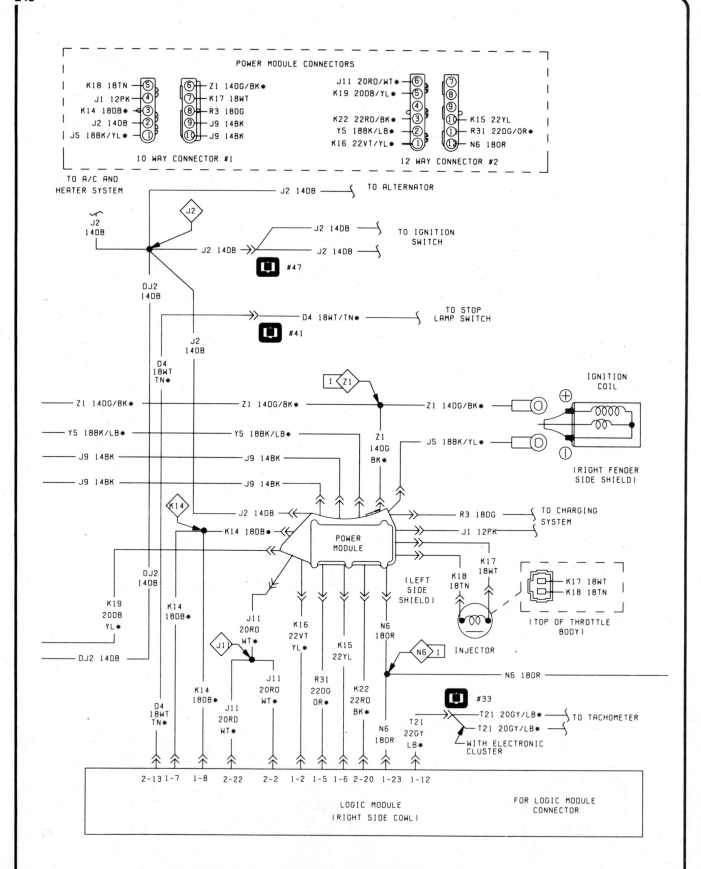

Non-turbocharged engine EFI ignition system wiring diagram (typical) (3 of 4)

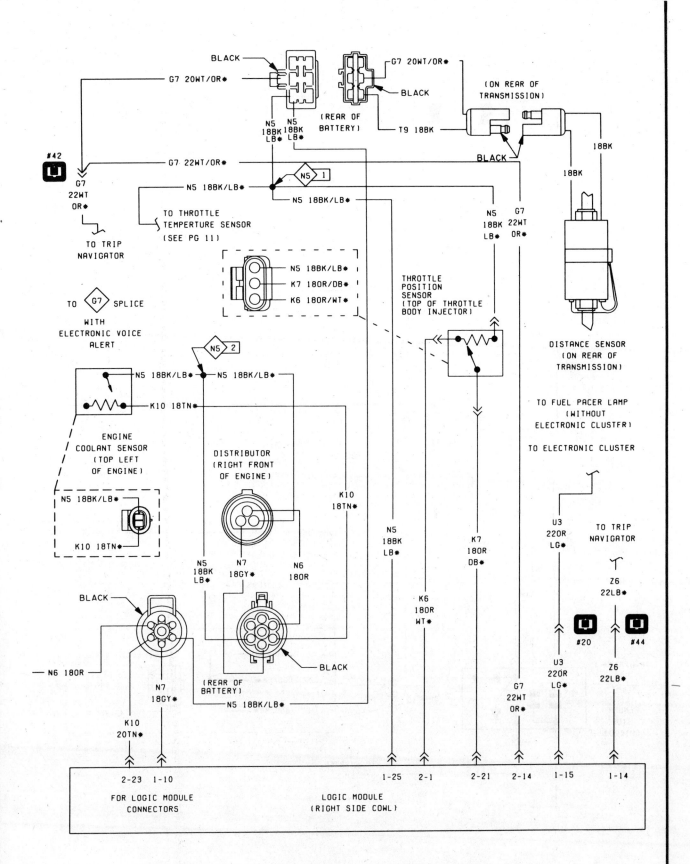

Non-turbocharged engined EFI ignition system wiring diagram (typical) (4 of 4)

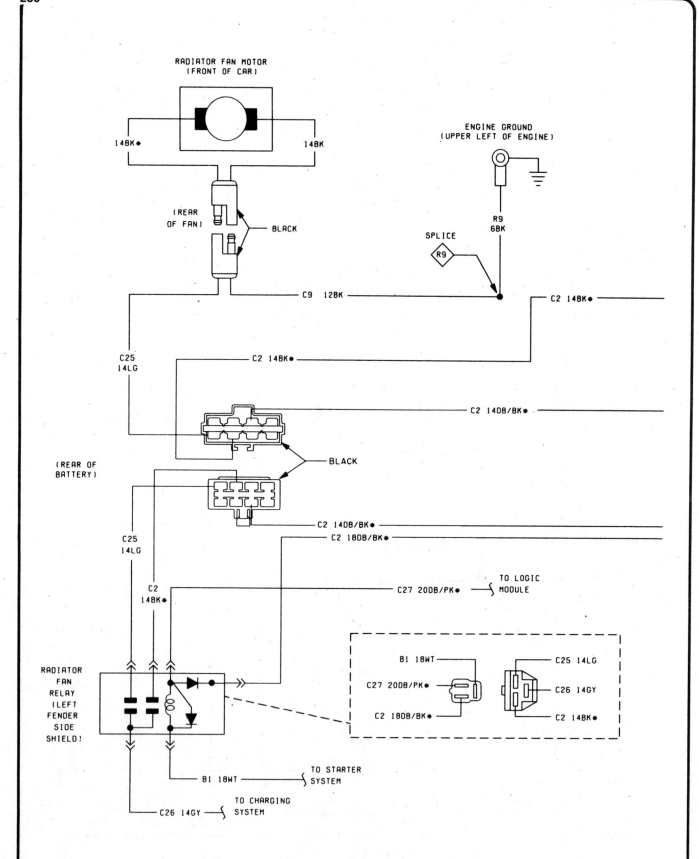

Typical radiator fan motor wiring diagram

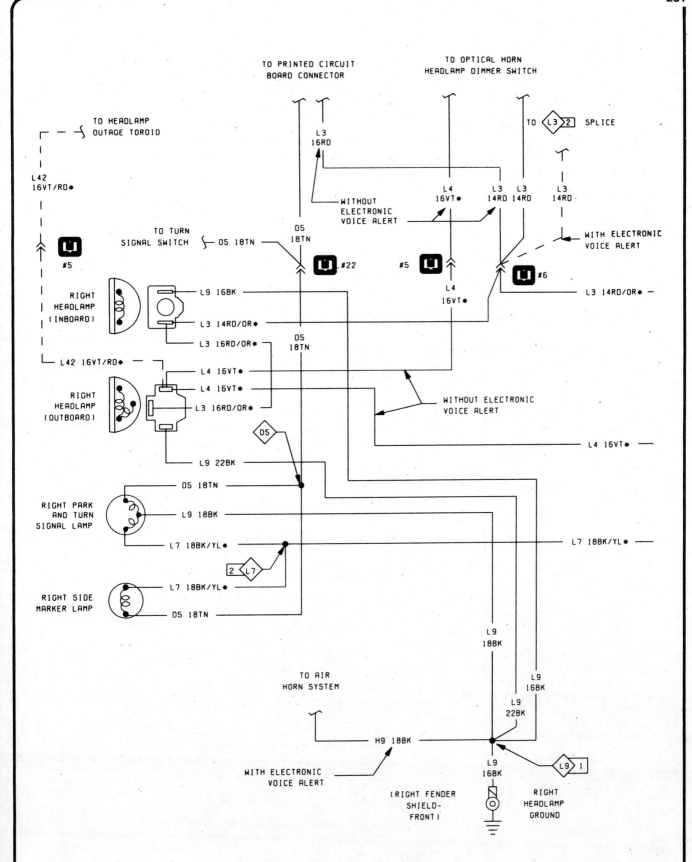

Typical front end lighting system wiring diagram (1 of 2)

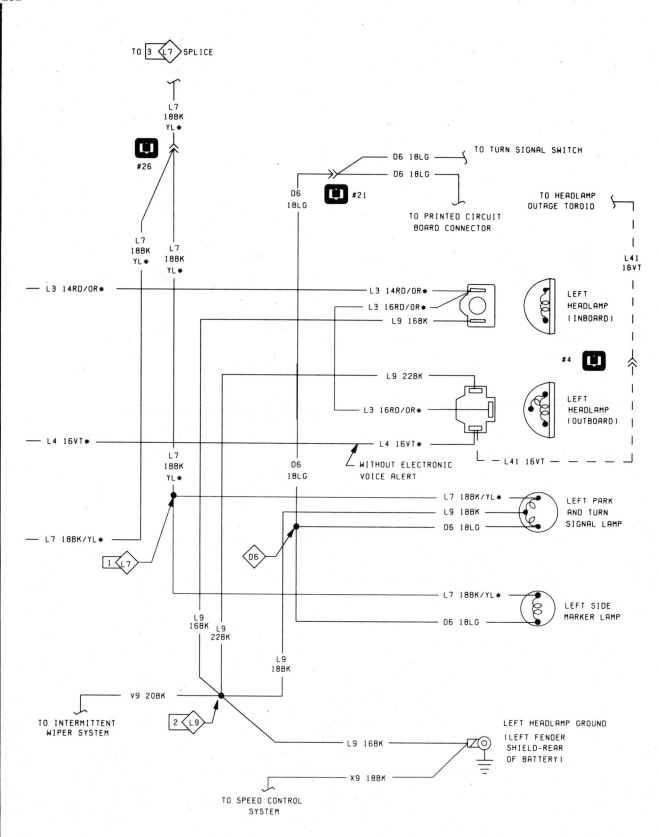

Typical front end lighting system wiring diagram (2 of 2)

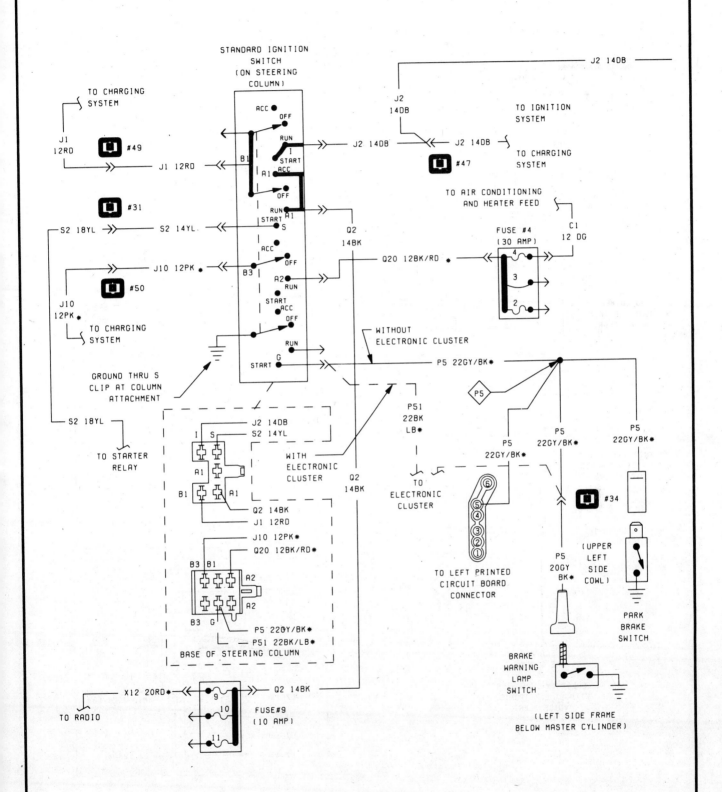

Typical ignition switch wiring diagram

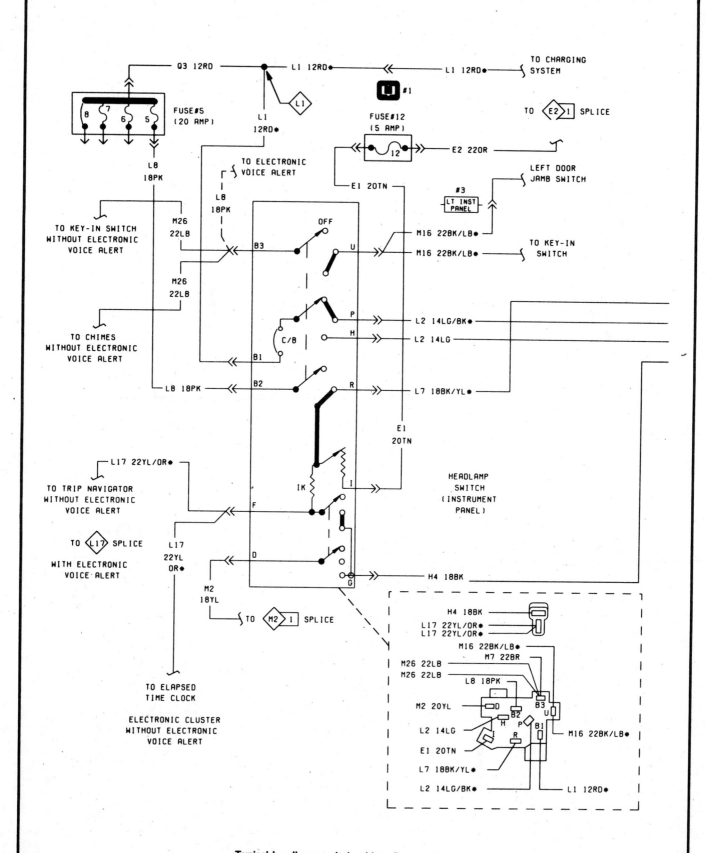

Typical headlamp switch wiring diagram (1 of 2)

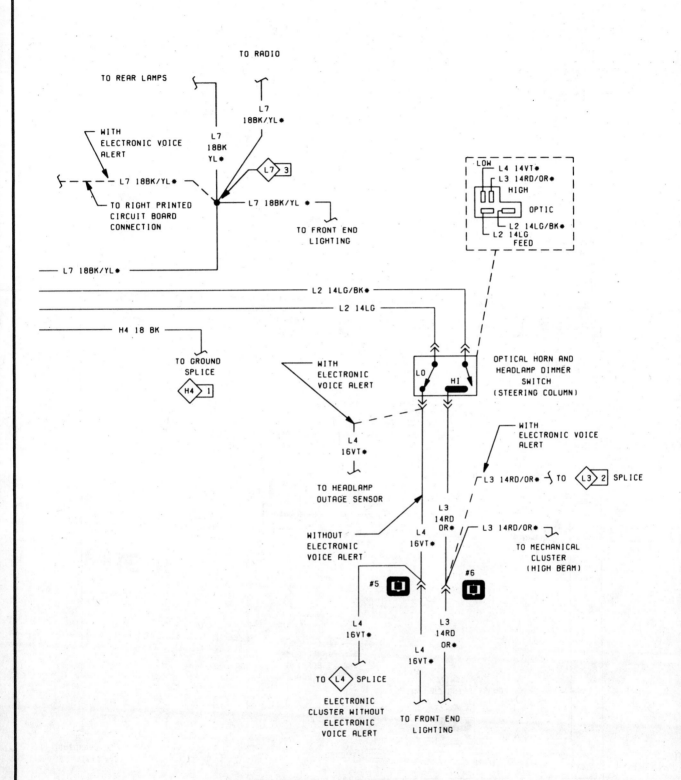

Typical headlamp switch wiring diagram (2 of 2)

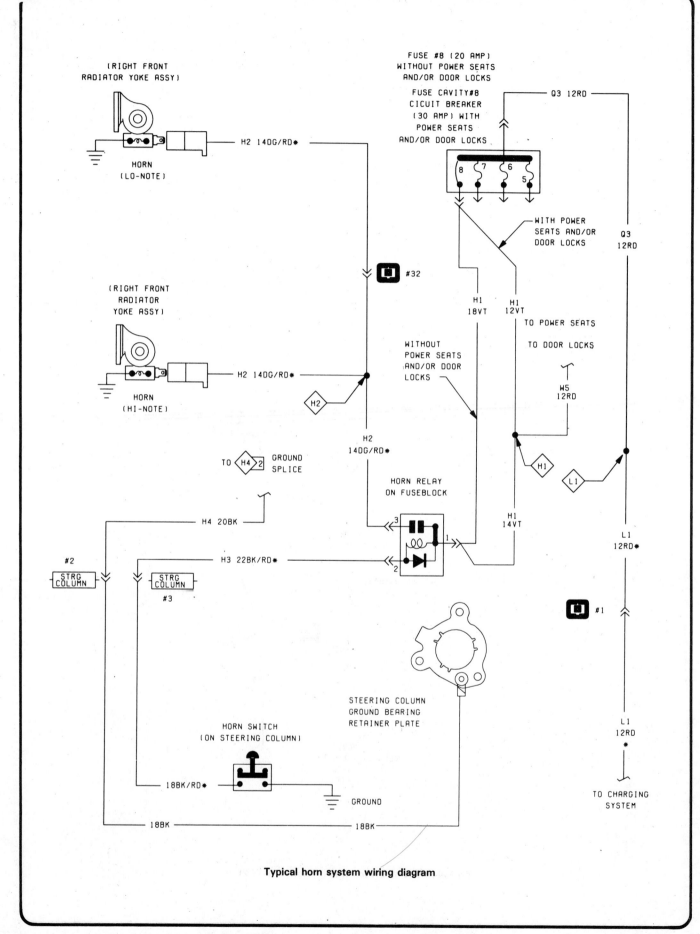
Typical horn system wiring diagram

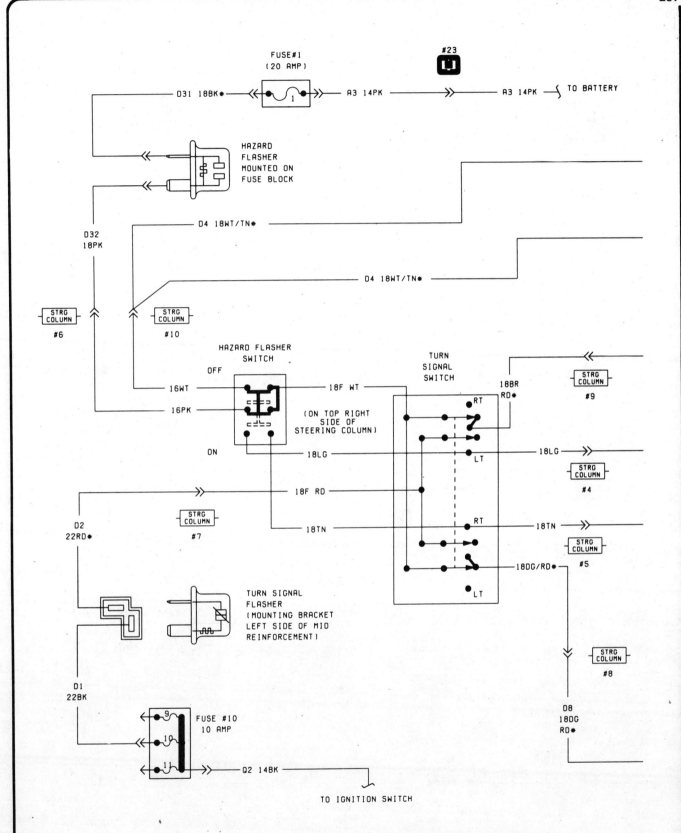

Typical brake light/turn signal/hazard warning flasher wiring diagram (1 of 2)

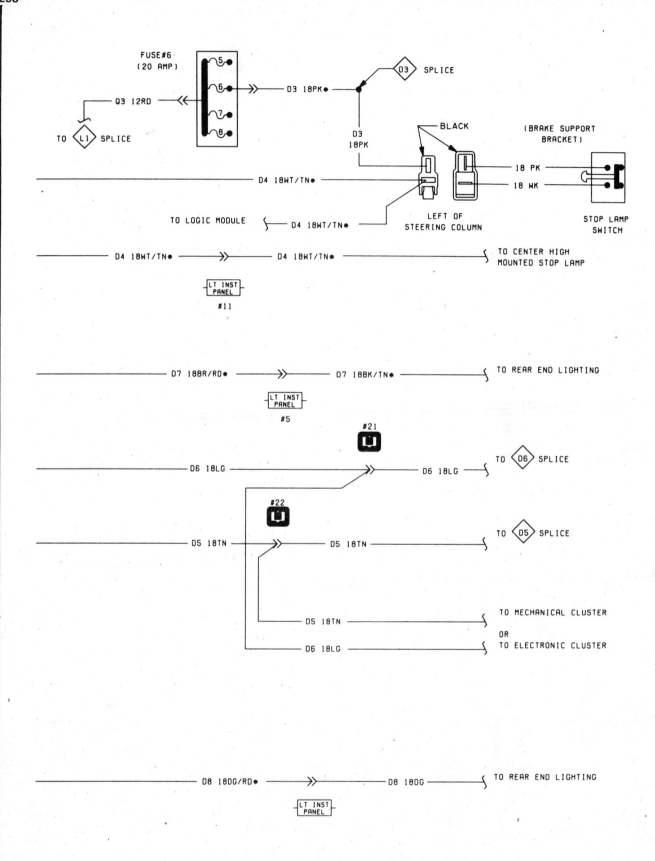

Typical brake light/turn signal/hazard warning flasher wiring diagram (2 of 2)

Index

A

Air bag – 235
Air cleaner
  filter element — 44
  thermo-controlled check — 135
Air conditioning system description — 101
Air filter checking and replacement — 44
Alternator
  belt checking — 44
  brush replacement — 123
  general information and special precautions — 122
  removal and installation — 123
Antenna — 219
Antifreeze
  level checking — 34
  solutions — 92
Automatic transaxle
  checking fluid level — 36
  description — 142
  diagnosis — 21
  driveplate removal and installation — 57
  flexplate removal and installation — 57
  fluid
    capacity — 28
    change — 144
    type — 28
  general information — 142
  removal and installation — 147
  shift linkage — 142
  troubleshooting — 21

B

Backup light switch — 146
Battery
  cables — 39
  charging — 39, 122
  emergency jump starting — 16
  maintenance — 39
  removal and installation — 122
Bearings
  axle — 190
  inspection (engine) — 85, 87
  main (engine) — 85
  rod (engine) — 83
Belt checking and replacement — 44
Body
  bumper — 217
  front fender — 212
  front seat recliner — 209
  general information — 201
  hood — 212
  instrument cluster — 224
  maintenance — 201
  major damage repair — 203
  minor damage repair — 202
  outside mirror — 212
  seats — 207
  weatherstripping — 203
  windshield removal and installation — 203
Body repair photo sequence — 210
Brake system
  bleeding – 177
  checking – 50
  disc brake
    caliper rebuilding – 170
    pad inspection – 50
    pad removal and installation
      front – 168
      rear – 234
    removal and installation – 170
    rotor removal and installation – 169
  drum brake
    checking — 50
    lining — 172
    wheel cylinder — 176
  fluid
    checking level — 35
    type — 28
  general information — 167
  maintenance and inspection — 50
  parking brake cable
    adjustment — 178
    removal and installation — 178
  power booster — 179
  specifications — 167
  stop light switch — 180
  troubleshooting — 21
  wheel cylinder — 168
Bulb replacement — 219

Index

C

Camshaft
  inspection — 78
  removal and installation — 57, 74
Catalytic converter — 132
Charging system — see Electrical system
Chassis lubrication — 42
Chemicals and lubricants — 19
Clutch
  adjustment — 50
  description — 151
  free travel check — 50
  general information — 151
  pressure plate — 153
  removal, servicing and installation — 153
  troubleshooting — 21
Coil — 126
Compression check — 53
Combustion chamber conditioner application — 47
Connecting rod bearings
  inspection — 83
  removal and installation — 79, 88
Cooling system
  air conditioner — 101
  antifreeze solutions — 92
  coolant capacity — 28
  coolant level — 34
  coolant type — 28
  description — 91
  diagnosis — 21
  draining, flushing and refilling — 51, 230
  electric fan — 97
  fan — 97
  general information — 91
  hoses — 46
  radiator — 95
  system check — 47
  thermostat — 92
  troubleshooting — 21
  water pump — 93
Crankshaft
  bearings — 85, 87
  inspection — 85
  removal and installation — 80
CV joint
  boot replacement — 164
  inspection — 153
  removal and installation — 157
Cylinder block
  inspection — 82
  servicing — 81
Cylinder bore
  inspection — 82
  servicing — 81
Cylinder head
  assembly — 79
  cleaning — 77
  disassembly — 77
  inspection — 77
  removal and installation — 61, 68

D

Disc brake — see Brakes
Distributor
  cap — 44
  general description — 126
  removal and installation — 128
Doors — see Body
Driveaxle
  boots — 164
  inspection — 157
  removal and installation — 155

Drivebelt — see Belts
  adjusting — 44
  checking — 44
Driveplate inspection, removal and installation — 57

E

Electrical system
  alternator brush replacement — 123
  backup light switch — 146
  battery
    cables — 39
    charging — 122
    checking — 39
    emergency jump starting — 16
    maintenance — 39
    removal and installation — 122
  bulb replacement — 219
  checking output — 122
  description — 215
  diagnosis — 21
  door locks — 205
  fan — 97
  front lights
    headlights — 216
    parking lights — 219
    turn signal lights — 219
  fuses — 215
  fusible links — 215
  general information — 122, 215
  headlight
    adjuster screw — 217
    adjustment — 218
    dimmer switch — 221
    replacement — 216
    switch — 223
  horn — 216
  ignition switch — 220
  instrument cluster — 224
  radio antenna — 219
  rear lights — 219
  removal and installation — 123
  side marker lights — 219
  special precautions — 122
  starter motor — 124
  stop light switch — 180
  troubleshooting — 21
  turn signal and hazard flashers — 216
  window regulator — 206
  windshield washer — 35
  windshield wiper
    arm — 228
    blade — 39
  wiring diagrams — 229 thru 252
Electronic fuel injection
  description — 104
  fuel injector — 108, 112
  fuel lines — 105
  logic module — 126, 231
  manifold absolute pressure sensor — 127
  mounting torque — 46
  power module — 126, 231
  removal and installation — 105, 108 thru 112
Emission control systems
  air aspirator — 136
  air cleaner element — 44
  catalytic converter — 132
  charcoal canister — 132
  description — 130
  evaporation control system — 132
  Exhaust Gas Recirculation (EGR) system — 133
  fault codes — 126, 232
  heated air intake system — 135
  information label — 131
  oxygen sensor — 134, 232

Index

Positive Crankcase Ventilation (PCV) system — 131
  troubleshooting — 21
  vacuum hose — 46
Engine
  assembly — 76
  bearing inspection — 85
  block
    cleaning — 81
    inspection — 82
  camshaft — 61, 78
  connecting rod
    bearing replacement — 88
    inspection — 83
    removal and installation — 79, 78
  crankshaft
    bearing replacement — 87
    inspection — 85
    removal and installation — 80, 87
  cylinder
    boring — 82
    honing — 82
    inspection — 82
    ridge removal — 79
  cylinder compression — 53
  cylinder head
    cleaning — 79
    disassembly — 77
    inspection — 77
    reassembly — 79
    removal and installation — 61, 68
  description — 55
  diagnosis — 27
  disassembly — 77
  firing order — 28
  flywheel/driveplate — 57
  general information — 55
  installation — 71
  main bearing inspection — 85
  oil
    change — 40
    filter — 40
    level check — 34
  oil pan removal and installation — 61
  oil pump — 66
  overhaul — 68
  piston and connecting rod
    inspection — 83
    removal and installation — 79, 88
  piston ring removal and installation — 79, 85
  reassembly — 88
  rebuilding alternatives — 75
  removal and installation — 56, 71
  start-up after major rebuild or overhaul — 89
  timing belt removal, inspection and installation — 58
  troubleshooting — 21
  valve inspection and servicing — 79
Exhaust system
  catalytic converter — 32
  checking — 48
  description — 102
  Exhaust Gas Recirculation (EGR) system — 133
  general information — 102
  inspection — 48
  muffler — 115
  removal and installation — 115

F

Fan removal and installation — 98
Filters
  air — 44
  fuel — 40
  oil — 40
  PCV — 44

Firing order — 28
Fluid level checks — 33
Fluids — 33
Flywheel inspection — 153
Fuel filter replacement — 41
Fuel pump removal and installation — 112
Fuel tank — 112
Fuses — 215
Fusible links — 215

G

Gauges — 224

H

Headlight
  adjustment — 218
  removal and installation — 216
  switch — 223
Heated air intake system — 135
Heater
  blower motor — 100
  core — 100
  hoses — 46
  removal and installation — 100
Hood
  latch and cable — 216
  removal and installation — 212
Horn checking — 216
Hoses — 46

I

Identification numbers — 7
Intermediate shaft — 153
Ignition switch removal and installation — 220
Ignition system — 126
Instrument cluster removal and installation — 224

J

Jacking — 16
Jump starting — 16

L

Liftgate — 213
Lights (bulb replacement) — 219
Lubricants — 28, 230
Lubrication — 42

M

Maintenance, routine — 29, 230
Manual transaxle
  checking lubricant level — 33
  fluid change — 53
  general information — 137, 233
  linkage — 137, 233
  lubricant type — 33, 230
  overhaul — 141
  removal and installation — 139
  troubleshooting — 21
Mufflers — 115

Index

O

Oil
 changing — 40
 checking — 33
 filter replacement — 40
 pan — 61
 pump — 66
 type — 27
Oil pan removal and installation — 61
Oil pump — 66

P

Parking brake — 178
PCV (Positive Crankcase Ventilation) system — 131
Power brake booster removal and installation — 179
Power steering
 belt — 44
 fluid level — 37
 fluid type — 28
 pump removal and installation — 198, 234
Pump
 fuel — 104
 water — 93

R

Radiator
 coolant level check — 33
 filling — 51, 230
 hoses — 46
 inspection — 97
 removal, servicing and installation — 95
 thermostat — 92
Radio
 antenna — 219
 removal and installation — 219
Rotor — 169
Routine maintenance — 28, 230

S

Safety first! — 18
Shock absorber
 inspection — 50
 removal and installation — 188
Side marker bulbs — 219
Spark plug
 gapping — 43
 general information — 43
 replacement — 43
 wires — 44
Speedometer
 cable replacement — 224
 removal and installation — 224
Starter motor
 checking — 129
 general information — 124
 removal and installation — 124

Steering
 checking — 50
 description — 192
 linkage and balljoints — 186
 wheel — 192
Suspension
 checking — 186
 description — 182
 shock absorber — 188
 sway bar — 182

T

Thermostat
 checking — 92
 removal and installation — 92
Timing
 belt — 58
 ignition — 52
Tire
 changing — 200
 checking — 200
 pressure — 38
 rotation — 38
Tools — 9
Towing — 16
Transmission (automatic) — see Automatic transaxle
Transmission (manual) — see Manual transaxle
Troubleshooting — 21, 230
Tune-up and routine maintenance — 28, 230
Turbocharger — 117, 231
Turn signal and hazard flasher — 216

U

Upholstery maintenance — 202

V

Valves — 79, 230
Vehicle identification — 7

W

Water pump
 inspection — 93
 removal and installation — 93
Weatherstripping — 203
Wheel alignment — 200
Wheel bearing
 checking — 190
 packing — 190
Window glass — 203
Windshield removal and installation — 203
Windshield washer fluid — 35
Windshield wiper — 39
Wiring diagrams — 236 thru 258

Haynes Automotive Manuals

NOTE: New manuals are added to this list on a periodic basis. If you do not see a listing for your vehicle, consult your local Haynes dealer for the latest product information.

ACURA
- *1776   Integra '86 thru '89 & Legend '86 thru '90

AMC
-         Jeep CJ - see JEEP (412)
- 694   Mid-size models, Concord, Hornet, Gremlin & Spirit '70 thru '83
- 934   (Renault) Alliance & Encore '83 thru '87

AUDI
- 615   4000 all models '80 thru '87
- 428   5000 all models '77 thru '83
- 1117   5000 all models '84 thru '88

AUSTIN-HEALEY
-         Sprite - see MG Midget (265)

BMW
- *2020   3/5 Series not including diesel or all-wheel drive models '82 thru '92
- 276   320i all 4 cyl models '75 thru '83
- 632   528i & 530i all models '75 thru '80
- 240   1500 thru 2002 except Turbo '59 thru '77

BUICK
-         Century (front wheel drive) - see GM (829)
- *1627   Buick, Oldsmobile & Pontiac Full-size (Front wheel drive) all models '85 thru '95
Buick Electra, LeSabre and Park Avenue;
Oldsmobile Delta 88 Royale, Ninety Eight and Regency; Pontiac Bonneville
- 1551   Buick Oldsmobile & Pontiac Full-size (Rear wheel drive)
Buick Estate '70 thru '90, Electra '70 thru '84, LeSabre '70 thru '85, Limited '74 thru '79
Oldsmobile Custom Cruiser '70 thru '90, Delta 88 '70 thru '85, Ninety-eight '70 thru '84
Pontiac Bonneville '70 thru '81, Catalina '70 thru '81, Grandville '70 thru '75, Parisienne '83 thru '86
- 627   Mid-size Regal & Century all rear-drive models with V6, V8 and Turbo '74 thru '87
Regal - see GENERAL MOTORS (1671)
Riviera - see GENERAL MOTORS (38030)
Skyhawk - see GENERAL MOTORS (766)
Skylark '80 thru '85 - see GM (38020)
Skylark '86 on - see GM (1420)
Somerset - see GENERAL MOTORS (1420)

CADILLAC
- *751   Cadillac Rear Wheel Drive all gasoline models '70 thru '93
Cimarron - see GENERAL MOTORS (766)
Eldorado - see GENERAL MOTORS (38030)
Seville '80 thru '85 - see GM (38030)

CHEVROLET
- *1477   Astro & GMC Safari Mini-vans '85 thru '93
- 554   Camaro V8 all models '70 thru '81
- 866   Camaro all models '82 thru '92
Cavalier - see GENERAL MOTORS (766)
Celebrity - see GENERAL MOTORS (829)
- 24017   Camaro & Firebird '93 thru '96
- 625   Chevelle, Malibu & El Camino all V6 & V8 models '69 thru '87
- 449   Chevette & Pontiac T1000 '76 thru '87
- 550   Citation all models '80 thru '85
- *1628   Corsica/Beretta all models '87 thru '96
- 274   Corvette all V8 models '68 thru '82
- *1336   Corvette all models '84 thru '91
- 1762   Chevrolet Engine Overhaul Manual
- 704   Full-size Sedans Caprice, Impala, Biscayne, Bel Air & Wagons '69 thru '90
Lumina - see GENERAL MOTORS (1671)
Lumina APV - see GENERAL MOTORS (2035)
- 319   Luv Pick-up all 2WD & 4WD '72 thru '82
- 626   Monte Carlo all models '70 thru '88

- 241   Nova all V8 models '69 thru '79
- *1642   Nova and Geo Prizm all front wheel drive models, '85 thru '92
- 420   Pick-ups '67 thru '87 - Chevrolet & GMC, all V8 & in-line 6 cyl, 2WD & 4WD '67 thru '87; Suburbans, Blazers & Jimmys '67 thru '91
- *1664   Pick-ups '88 thru '95 - Chevrolet & GMC, all full-size pick-ups, '88 thru '95; Blazer & Jimmy '92 thru '94; Suburban '92 thru '95; Tahoe & Yukon '95
- 831   S-10 & GMC S-15 Pick-ups '82 thru '93
- *24071   S-10 & GMC S-15 Pick-ups '94 thru '96
- *1727   Sprint & Geo Metro '85 thru '94
- *345   Vans - Chevrolet & GMC, V8 & in-line 6 cylinder models '68 thru '96

CHRYSLER
- 25025   Chrysler Concorde, New Yorker & LHS, Dodge Intrepid, Eagle Vision, '93 thru '96
- 2114   Chrysler Engine Overhaul Manual
- *2058   Full-size Front-Wheel Drive '88 thru '93
K-Cars - see DODGE Aries (723)
Laser - see DODGE Daytona (1140)
- *1337   Chrysler & Plymouth Mid-size front wheel drive '82 thru '95
Rear-wheel Drive - see Dodge (2098)

DATSUN
- 647   200SX all models '80 thru '83
- 228   B - 210 all models '73 thru '78
- 525   210 all models '79 thru '82
- 206   240Z, 260Z & 280Z Coupe '70 thru '78
- 563   280ZX Coupe & 2+2 '79 thru '83
        300ZX - see NISSAN (1137)
- 679   310 all models '78 thru '82
- 123   510 & PL521 Pick-up '68 thru '73
- 430   510 all models '78 thru '81
- 372   610 all models '72 thru '76
- 277   620 Series Pick-up all models '73 thru '79
        720 Series Pick-up - see NISSAN (771)
- 376   810/Maxima all gasoline models, '77 thru '84
        Pulsar - see NISSAN (876)
        Sentra - see NISSAN (982)
        Stanza - see NISSAN (981)

DODGE
-         400 & 600 - see CHRYSLER Mid-size (1337)
- *723   Aries & Plymouth Reliant '81 thru '89
- 1231   Caravan & Plymouth Voyager Mini-Vans all models '84 thru '95
- 699   Challenger/Plymouth Saporro '78 thru '83
        Challenger '67-'76 - see DODGE Dart (234)
- 610   Colt & Plymouth Champ (front wheel drive) all models '78 thru '87
- *1668   Dakota Pick-ups all models '87 thru '96
- 234   Dart, Challenger/Plymouth Barracuda & Valiant 6 cyl models '67 thru '76
- *1140   Daytona & Chrysler Laser '84 thru '89
        Intrepid - see CHRYSLER (25025)
- *30034   Neon all models '94 thru '97
- *545   Omni & Plymouth Horizon '78 thru '90
- *912   Pick-ups all full-size models '74 thru '93
- *30041   Pick-ups all full-size models '94 thru '96
- *556   Ram 50/D50 Pick-ups & Raider and Plymouth Arrow Pick-ups '79 thru '93
- 2098   Dodge/Plymouth/Chrysler rear wheel drive '71 thru '89
- *1726   Shadow & Plymouth Sundance '87 thru '94
- *1779   Spirit & Plymouth Acclaim '89 thru '95
- *349   Vans - Dodge & Plymouth V8 & 6 cyl models '71 thru '96

EAGLE
-         Talon - see Mitsubishi Eclipse (2097)
-         Vision - see CHRYSLER (25025)

FIAT
- 094   124 Sport Coupe & Spider '68 thru '78
- 273   X1/9 all models '74 thru '80

FORD
- 10355   Ford Automatic Trans. Overhaul
- *1476   Aerostar Mini-vans all models '86 thru '96

- 268   Courier Pick-up all models '72 thru '82
- 2105   Crown Victoria & Mercury Grand Marquis '88 thru '96
- 1763   Ford Engine Overhaul Manual
- 789   Escort/Mercury Lynx all models '81 thru '90
- *2046   Escort/Mercury Tracer '91 thru '96
- *2021   Explorer & Mazda Navajo '91 thru '95
- 560   Fairmont & Mercury Zephyr '78 thru '83
- 334   Fiesta all models '77 thru '80
- 754   Ford & Mercury Full-size, Ford LTD & Mercury Marquis ('75 thru '82); Ford Custom 500, Country Squire, Crown Victoria & Mercury Colony Park ('75 thru '87); Ford LTD Crown Victoria & Mercury Gran Marquis ('83 thru '87)
- 359   Granada & Mercury Monarch all in-line, 6 cyl & V8 models '75 thru '80
- 773   Ford & Mercury Mid-size, Ford Thunderbird & Mercury Cougar ('75 thru '82); Ford LTD & Mercury Marquis ('83 thru '86); Ford Torino, Gran Torino, Elite, Ranchero pick-up, LTD II, Mercury Montego, Comet, XR-7 & Lincoln Versailles ('75 thru '86)
- 231   Mustang 4 cyl, V6 & V8 models '74 thru '78
- 357   Mustang V8 all models '64-1/2 thru '73
- *654   Mustang & Mercury Capri all models Mustang, '79 thru '93; Capri, '79 thru '86
- *36051   Mustang all models '94 thru '97
- 788   Pick-ups & Bronco '73 thru '79
- *880   Pick-ups & Bronco '80 thru '96
- 649   Pinto & Mercury Bobcat '75 thru '80
- 1670   Probe all models '89 thru '92
- *1026   Ranger/Bronco II gasoline models '83 thru '92
- *36071   Ranger '93 thru '96 & Mazda Pick-ups '94 thru '96
- *1421   Taurus & Mercury Sable '86 thru '95
- *1418   Tempo & Mercury Topaz all gasoline models '84 thru '94
- 1338   Thunderbird/Mercury Cougar '83 thru '88
- *1725   Thunderbird/Mercury Cougar '89 and '96
- 344   Vans all V8 Econoline models '69 thru '91
- *2119   Vans full size '92-'95

GENERAL MOTORS
- *10360   GM Automatic Transmission Overhaul
- *829   Buick Century, Chevrolet Celebrity, Oldsmobile Cutlass Ciera & Pontiac 6000 all models '82 thru '96
- *1671   Buick Regal, Chevrolet Lumina, Oldsmobile Cutlass Supreme & Pontiac Grand Prix front wheel drive models '88 thru '95
- *766   Buick Skyhawk, Cadillac Cimarron, Chevrolet Cavalier, Oldsmobile Firenza & Pontiac J-2000 & Sunbird '82 thru '94
- 38020   Buick Skylark, Chevrolet Citation, Olds Omega, Pontiac Phoenix '80 thru '85
- 1420   Buick Skylark & Somerset, Oldsmobile Achieva & Calais and Pontiac Grand Am all models '85 thru '95
- 38030   Cadillac Eldorado '71 thru '85, Seville '80 thru '85, Oldsmobile Toronado '71 thru '85 & Buick Riviera '79 thru '85
- *2035   Chevrolet Lumina APV, Olds Silhouette & Pontiac Trans Sport all models '90 thru '95
        General Motors Full-size Rear-wheel Drive - see BUICK (1551)

GEO
-         Metro - see CHEVROLET Sprint (1727)
-         Prizm - '85 thru '92 see CHEVY Nova (1642), '93 thru '96 see TOYOTA Corolla (1642)
- *2039   Storm all models '90 thru '93
-         Tracker - see SUZUKI Samurai (1626)

GMC
-         Safari - see CHEVROLET ASTRO (1477)
-         Vans & Pick-ups - see CHEVROLET (420, 831, 345, 1664 & 24071)

(Continued on other side)

Listings shown with an asterisk (*) indicate model coverage as of this printing. These titles will be periodically updated to include later model years - consult your Haynes dealer for more information.

Haynes North America, Inc., 861 Lawrence Drive, Newbury Park, CA 91320 • (805) 498-6703

Haynes Automotive Manuals (continued)

NOTE: New manuals are added to this list on a periodic basis. If you do not see a listing for your vehicle, consult your local Haynes dealer for the latest product information.

HONDA
- 351 — Accord CVCC all models '76 thru '83
- 1221 — Accord all models '84 thru '89
- 2067 — Accord all models '90 thru '93
- 42013 — Accord all models '94 thru '95
- 160 — Civic 1200 all models '73 thru '79
- 633 — Civic 1300 & 1500 CVCC '80 thru '83
- 297 — Civic 1500 CVCC all models '75 thru '79
- 1227 — Civic all models '84 thru '91
- *2118 — Civic & del Sol '92 thru '95
- *601 — Prelude CVCC all models '79 thru '89

HYUNDAI
- *1552 — Excel all models '86 thru '94

ISUZU
- *1641 — Trooper & Pick-up, all gasoline models Pick-up, '81 thru '93; Trooper, '84 thru '91
- Hombre - see CHEVROLET S-10 (24071)

JAGUAR
- *242 — XJ6 all 6 cyl models '68 thru '86
- *49011 — XJ6 all models '88 thru '94
- *478 — XJ12 & XJS all 12 cyl models '72 thru '85

JEEP
- *1553 — Cherokee, Comanche & Wagoneer Limited all models '84 thru '96
- 412 — CJ all models '49 thru '86
- 50025 — Grand Cherokee all models '93 thru '95
- 50029 — Grand Wagoneer & Pick-up '72 thru '91 Grand Wagoneer '84 thru '91, Cherokee & Wagoneer '72 thru '83, Pick-up '72 thru '88
- *1777 — Wrangler all models '87 thru '95

LINCOLN
- 2117 — Rear Wheel Drive all models '70 thru '96

MAZDA
- 648 — 626 (rear wheel drive) all models '79 thru '82
- *1082 — 626/MX-6 (front wheel drive) '83 thru '91
- 370 — GLC Hatchback (rear wheel drive) '77 thru '83
- 757 — GLC (front wheel drive) '81 thru '85
- *2047 — MPV all models '89 thru '94
- Navajo - see Ford Explorer (2021)
- 267 — Pick-ups '72 thru '93
- Pick-ups '94 thru '96 - see Ford Ranger (36071)
- 460 — RX-7 all models '79 thru '85
- *1419 — RX-7 all models '86 thru '91

MERCEDES-BENZ
- *1643 — 190 Series four-cyl gas models, '84 thru '88
- 346 — 230/250/280 6 cyl sohc models '68 thru '72
- 983 — 280 123 Series gasoline models '77 thru '81
- 698 — 350 & 450 all models '71 thru '80
- 697 — Diesel 123 Series '76 thru '85

MERCURY
See FORD Listing

MG
- 111 — MGB Roadster & GT Coupe '62 thru '80
- 265 — MG Midget, Austin Healey Sprite '58 thru '80

MITSUBISHI
- *1669 — Cordia, Tredia, Galant, Precis & Mirage '83 thru '93
- *2097 — Eclipse, Eagle Talon & Plymouth Laser '90 thru '94
- *2022 — Pick-up '83 thru '96 & Montero '83 thru '93

NISSAN
- 1137 — 300ZX all models including Turbo '84 thru '89
- *72015 — Altima all models '93 thru '97
- *1341 — Maxima all models '85 thru '91
- *771 — Pick-ups '80 thru '96 Pathfinder '87 thru '95
- 876 — Pulsar all models '83 thru '86
- *982 — Sentra all models '82 thru '94
- *981 — Stanza all models '82 thru '90

OLDSMOBILE
- Achieva - see GENERAL MOTORS (1420)
- Bravada - see CHEVROLET S-10 (831)
- Calais - see GENERAL MOTORS (1420)
- Custom Cruiser - see BUICK RWD (1551)
- *658 — Cutlass V6 & V8 gas models '74 thru '88
- Cutlass Ciera - see GENERAL MOTORS (829)
- Cutlass Supreme - see GM (1671)
- Delta 88 - see BUICK Full-size RWD (1551)
- Delta 88 Brougham - see BUICK Full-size FWD (1551), RWD (1627)
- Delta 88 Royale - see BUICK RWD (1551)
- Firenza - see GENERAL MOTORS (766)
- Ninety-eight Regency - see BUICK Full-size RWD (1551), FWD (1627)
- Ninety-eight Regency Brougham - see BUICK Full-size RWD (1551)
- Omega - see GENERAL MOTORS (38020)
- Silhouette - see GENERAL MOTORS (2035)
- Toronado - see GENERAL MOTORS (38030)

PEUGEOT
- 663 — 504 all diesel models '74 thru '83

PLYMOUTH
- Laser - see MITSUBISHI Eclipse (2097)
- *For other PLYMOUTH titles, see DODGE.*

PONTIAC
- T1000 - see CHEVROLET Chevette (449)
- J-2000 - see GENERAL MOTORS (766)
- 6000 - see GENERAL MOTORS (829)
- Bonneville - see Buick FWD (1627), RWD (1551)
- Bonneville Brougham - see Buick (1551)
- Catalina - see Buick Full-size (1551)
- 1232 — Fiero all models '84 thru '88
- 555 — Firebird V8 models except Turbo '70 thru '81
- 867 — Firebird all models '82 thru '92
- Firebird '93 thru '96 - see CHEVY Camaro (24017)
- Full-size Front Wheel Drive - see BUICK, Oldsmobile, Pontiac Full-size FWD (1627)
- Full-size Rear Wheel Drive - see BUICK Oldsmobile, Pontiac Full-size RWD (1551)
- Grand Am - see GENERAL MOTORS (1420)
- Grand Prix - see GENERAL MOTORS (1671)
- Grandville - see BUICK Full-size (1551)
- Parisienne - see BUICK Full-size (1551)
- Phoenix - see GENERAL MOTORS (38020)
- Sunbird - see GENERAL MOTORS (766)
- Trans Sport - see GENERAL MOTORS (2035)

PORSCHE
- *264 — 911 except Turbo & Carrera 4 '65 thru '89
- 239 — 914 all 4 cyl models '69 thru '76
- 397 — 924 all models including Turbo '76 thru '82
- *1027 — 944 all models including Turbo '83 thru '89

RENAULT
- 141 — 5 Le Car all models '76 thru '83
- Alliance & Encore - see AMC (934)

SAAB
- 247 — 99 all models including Turbo '69 thru '80
- *980 — 900 all models including Turbo '79 thru '88

SATURN
- 2083 — Saturn all models '91 thru '96

SUBARU
- 237 — 1100, 1300, 1400 & 1600 '71 thru '79
- *681 — 1600 & 1800 2WD & 4WD '80 thru '89

SUZUKI
- *1626 — Samurai/Sidekick & Geo Tracker '86 thru '96

TOYOTA
- 1023 — Camry all models '83 thru '91
- 92006 — Camry all models '92 thru '95
- 935 — Celica Rear Wheel Drive '71 thru '85
- *2038 — Celica Front Wheel Drive '86 thru '93
- 1139 — Celica Supra all models '79 thru '92
- 361 — Corolla all models '75 thru '79
- 961 — Corolla all rear wheel drive models '80 thru '87
- 1025 — Corolla all front wheel drive models '84 thru '92
- *92036 — Corolla & Geo Prizm '93 thru '96
- 636 — Corolla Tercel all models '80 thru '82
- 360 — Corona all models '74 thru '82
- 532 — Cressida all models '78 thru '82
- 313 — Land Cruiser all models '68 thru '82
- *1339 — MR2 all models '85 thru '87
- 304 — Pick-up all models '69 thru '78
- *656 — Pick-up all models '79 thru '95
- *2048 — Previa all models '91 thru '95
- 2106 — Tercel all models '87 thru '94

TRIUMPH
- 113 — Spitfire all models '62 thru '81
- 322 — TR7 all models '75 thru '81

VW
- 159 — Beetle & Karmann Ghia '54 thru '79
- 238 — Dasher all gasoline models '74 thru '81
- 96017 — Golf & Jetta all models '93 thru '97
- *884 — Rabbit, Jetta, Scirocco, & Pick-up gas models '74 thru '91 & Convertible '80 thru '92
- 451 — Rabbit, Jetta & Pick-up diesel '77 thru '84
- 082 — Transporter 1600 all models '68 thru '79
- 226 — Transporter 1700, 1800 & 2000 '72 thru '79
- 084 — Type 3 1500 & 1600 all models '63 thru '73
- 1029 — Vanagon all air-cooled models '80 thru '83

VOLVO
- 203 — 120, 130 Series & 1800 Sports '61 thru '73
- 129 — 140 Series all models '66 thru '74
- *270 — 240 Series all models '76 thru '93
- 400 — 260 Series all models '75 thru '82
- *1550 — 740 & 760 Series all models '82 thru '88

TECHBOOK MANUALS
- 2108 — Automotive Computer Codes
- 1667 — Automotive Emissions Control Manual
- 482 — Fuel Injection Manual, 1978 thru 1985
- 2111 — Fuel Injection Manual, 1986 thru 1996
- 2069 — Holley Carburetor Manual
- 2068 — Rochester Carburetor Manual
- 10240 — Weber/Zenith/Stromberg/SU Carburetors
- 1762 — Chevrolet Engine Overhaul Manual
- 2114 — Chrysler Engine Overhaul Manual
- 1763 — Ford Engine Overhaul Manual
- 1736 — GM and Ford Diesel Engine Repair Manual
- 1666 — Small Engine Repair Manual
- 10355 — Ford Automatic Transmission Overhaul
- 10360 — GM Automatic Transmission Overhaul
- 1479 — Automotive Body Repair & Painting
- 2112 — Automotive Brake Manual
- 2113 — Automotive Detailing Manual
- 1654 — Automotive Eelectrical Manual
- 1480 — Automotive Heating & Air Conditioning
- 2109 — Automotive Reference Manual & Dictionary
- 2107 — Automotive Tools Manual
- 10440 — Used Car Buying Guide
- 2110 — Welding Manual
- 10450 — ATV Basics

SPANISH MANUALS
- 98903 — Reparación de Carrocería & Pintura
- 98905 — Códigos Automotrices de la Computadora
- 98910 — Frenos Automotriz
- 98915 — Inyección de Combustible 1986 al 1994
- 99040 — Chevrolet & GMC Camionetas '67 al '87 Incluye Suburban, Blazer & Jimmy '67 al '91
- 99041 — Chevrolet & GMC Camionetas '88 al '95 Incluye Suburban '92 al '95, Blazer & Jimmy '92 al '94, Tahoe y Yukon '95
- 99042 — Chevrolet & GMC Camionetas Cerradas '68 al '95
- 99055 — Dodge Caravan & Plymouth Voyager '84 al '95
- 99075 — Ford Camionetas y Bronco '80 al '94
- 99077 — Ford Camionetas Cerradas '69 al '91
- 99083 — Ford Modelos de Tamaño Grande '75 al '87
- 99088 — Ford Modelos de Tamaño Mediano '75 al '86
- 99095 — GM Modelos de Tamaño Grande '70 al '90
- 99118 — Nissan Sentra '82 al '94
- 99125 — Toyota Camionetas y 4-Runner '79 al '95

* *Listings shown with an asterisk (*) indicate model coverage as of this printing. These titles will be periodically updated to include later model years - consult your Haynes dealer for more information.*

Over 100 Haynes motorcycle manuals also available

5-97

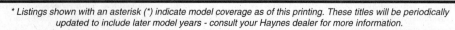

Haynes North America, Inc., 861 Lawrence Drive, Newbury Park, CA 91320 • (805) 498-6703